TRANSVERSITY 2005

TRANSVERSITY 2005

Como, Italy 7 – 10 September 2005

editors

Vincenzo Barone

Università del Piemonte Orientale & INFN, Italy

Philip G Ratcliffe

Università dell'Insubria & INFN, Italy

World Scientific

NEW JERSEY • LONDON • SINGAPORE • BEIJING • SHANGHAI • HONG KONG • TAIPEI • CHENNAI

Published by

World Scientific Publishing Co. Pte. Ltd.
5 Toh Tuck Link, Singapore 596224
USA office: 27 Warren Street, Suite 401-402, Hackensack, NJ 07601
UK office: 57 Shelton Street, Covent Garden, London WC2H 9HE

British Library Cataloguing-in-Publication Data
A catalogue record for this book is available from the British Library.

TRANSVERSITY 2005

ISBN 981-256-846-8

Printed in Singapore by World Scientific Printers (S) Pte Ltd

Preface

The notion of transversity in hadron physics has been with us now for over a quarter of a century. However, for much of that time, intriguing though it might have been, transversity has remained an intangible and remote object, of interest principally to only a few theoreticians. In recent years transversity and transverse-spin effects in general have grown as both theoretical and experimental areas of active research. This increasing attention, while initially riding to some extent in the wake of the famous proton-spin anomaly, has now matured into a thriving field with a driving force of its own. The ever-growing bulk of data on asymmetries in collisions involving transversely polarised hadrons now demands a more solid and coherent theoretical basis for its description. Indeed, it now appears rather clear that transversity and other closely related properties play a significant role in generating such phenomena.

It was within such a context that a two-year University, Education and Research Ministry-funded inter-university Research Project of National Interest (PRIN) in Italy, aimed very specifically at the study of this aspect of hadronic physics, was approved and financed for the two-year period 2003–5. As part of the project, the present workshop was planned to gather together experimentalists and theoreticians engaged in investigating the nature of transverse spin in hadronic physics, with the intent of favouring the exchange of up-to-date theoretical and experimental ideas and news on the subject and also the strengthening of ties, already fostered by the PRIN, between existing experimental and theory groups.

The workshop site was Villa Olmo on the idyllic shores of Lake Como (birthplace and home to Alessandro Volta) in Northern Italy. This beautiful 18th. century mansion, set in a magnificent Italian garden, with its wonderful view across the lake and superintended by the Centro di Cultura Scientifica "Alessandro Volta", is well equipped as a conference centre, having a permanent, experienced, technical and secretarial staff.

Over 70 physicists took part, of which 39 made presentations. Very nearly all the major experiments involved in transverse-spin studies were officially represented, as too were the main theory groups working in the field. New results and new analyses sparked many interesting and lively discussions, the two round-table sessions being particularly fruitful. Indeed, one of the contributions in this volume is the result of a combined comparative analysis stimulated by the interactions that took place precisely during the workshop.*

The all-important sources of finance for the workshop, besides the Ministry funded PRIN, were the Universities of Ferrara, Insubria, Piemonte Orientale, Torino and Trieste; the INFN Frascati National Laboratories; the INFN sections of Ferrara and Milano; the 6th. European Framework Programme Project Hadron Physics I3; and finally the International Committee for Spin Physics Symposia.

The project proposers and workshop organisers, spurred on by the success of this edition, fully intend to repeat the exercise two years on, *i.e.* in the year 2007. We all look forward to seeing the participants here in Como once again, along with a few new faces too.

Vincenzo Barone — Università del Piemonte Orientale and INFN
Philip G. Ratcliffe — Università dell'Insubria and INFN

*Further information on the workshop may found at the following permanent website: `http://scienze-como.uninsubria.it/phil/WS2005/`

Committees

Local Organising Committee

Philip G. Ratcliffe (Chair)	Università dell'Insubria
Vincenzo Barone	Università del Piemonte Orientale
Maria Pia Bussa	Università di Torino
Delia Hasch	INFN, Laboratori Nazionali di Frascati
Paolo Lenisa	Università di Ferrara
Anna Martin	Università di Trieste

International Advisory Committee

Franco Bradamante (Chair)	Università di Trieste
Mauro Anselmino	Università di Torino
Stanley J. Brodsky	SLAC
John C. Collins	Penn State University
Anatoli Efremov	JINR, Dubna
Robert L. Jaffe	MIT
Elliot Leader	London University
Piet Mulders	Vrije Universiteit, Amsterdam
Frank Rathmann	Forschungszentrum Jülich
Klaus Rith	Universität Erlangen-Nürnberg
Naohito Saito	Kyoto University
Jacques Soffer	CNRS, Marseille

International Committee for Spin Physics Symposia

Thomas Roser (Chair)	Brookhaven
Alan D. Krisch (Past-Chair)	Michigan
Ken'ichi Imai (Chair-Elect)	Kyoto
Franco Bradamante	Trieste
Owen Chamberlain*	Berkeley
Ernest D. Courant*	Brookhaven
Donald G. Crabb	Virginia
Anatoli V. Efremov	JINR
Giuseppe Fidecaro*	CERN
Gerhard Mallot	CERN
Willy Haeberli*	Wisconsin
Kichiji Hatanaka	RCNP
Richard G. Milner	MIT
Yoshiharu Mori	KEK
Charles Y. Prescott	SLAC
Frank Rathmann	COSY
Yury M. Shatunov	Novosibirsk
Volker Soergel*	Heidelberg
Erhard Steffens	Erlangen
Edward J. Stephenson	Indiana
Nikolai E. Tyurin	IHEP
Willem T.H. van Oers*	Manitoba

* Honorary Members

CONTENTS

Preface v

Committees vii

Contents ix

Introduction

Purpose and status of the Italian Transversity Project 3
F. Bradamante

Opening Lecture

Transversity 9
M. Anselmino

Experimental Lectures

Azimuthal single-spin asymmetries from polarized and unpolarized hydrogen targets at HERMES 23
G. Schnell (for the HERMES Collaboration)

Collins and Sivers asymmetries on the deuteron from COMPASS data 34
I. Horn (for the COMPASS Collaboration)

First measurement of interference fragmentation on a transversely polarized hydrogen target 42
P.B. van der Nat (for the HERMES Collaboration)

Two-hadron asymmetries at the COMPASS experiment 49
A. Mielech (for the COMPASS Collaboration)

Measurements of chiral-odd fragmentation functions at Belle 55
R. Seidl, M. Grosse Perdekamp, D. Gabbert, A. Ogawa & K. Hasuko

Lambda asymmetries 61
A. Ferrero (for the COMPASS Collaboration)

Transverse spin at PHENIX: Results and prospects 68
C. Aidala (for the PHENIX Collaboration)

Transverse spin and RHIC 77
L. Bland

Studies of transverse spin effects at JLab 88
H. Avakian, P. Bosted, V. Burkert & L. Elouadrhiri (for the CLAS Collaboration)

Neutron transversity at Jefferson Lab 96
J.P. Chen, X. Jiang, J.-C. Peng & L. Zhu (for the Jefferson Lab Hall A Collaboration)

$\mathcal{PAX}$: **p**olarized **a**ntiproton e**x**periments 104
M. Contalbrigo

Single and double spin N–$\bar{N}$ interactions at GSI 112
M. Maggiora (for the ASSIA Collaboration)

Spin filtering in storage rings 120
N.N. Nikolaev & F.F. Pavlov

Theory Lectures

Single-spin asymmetries and transversity in QCD 139
S.J. Brodsky

The relativistic hydrogen atom: a theoretical laboratory for structure functions 154
X. Artru & K. Benhizia

GPD's and SSA's 162
M. Burkardt

Time reversal odd distribution functions in chiral models 170
A. Drago

Soffer bound and transverse spin densities from lattice QCD 173
M. Diehl, M. Göckeler, Ph. Hägler, R. Horsley, D. Pleiter, P.E.L. Rakow, A. Schäfer, G. Schierholz & J.M. Zanotti

Single-spin asymmetries and Qiu–Sterman effect(s) 181
A. Bacchetta

Sivers function: SIDIS data, fits and predictions 188
M. Anselmino, M. Boglione, U. D'Alesio, A. Kotzinian, F. Murgia & A. Prokudin

Twist-3 effects in semi-inclusive deep inelastic scattering 196
M. Schlegel, K. Goeke & A. Metz

Quark and gluon Sivers functions 204
I. Schmidt

Sivers effect in semi-inclusive deeply inelastic scattering and Drell–Yan 212
J.C. Collins, A.V. Efremov, K. Goeke, M. Grosse Perdekamp, S. Menzel, B. Meredith, A. Metz & P. Schweitzer

Helicity formalism and spin asymmetries in hadronic processes 220
M. Anselmino, M. Boglione, U. D'Alesio, E. Leader, S. Melis & F. Murgia

Including Cahn and Sivers effects into event generators 228
A. Kotzinian

Comparing extractions of Sivers functions 236
M. Anselmino, M. Boglione, J.C. Collins, U. D'Alesio, A.V. Efremov, K. Goeke, A. Kotzinian, S. Menzel, A. Metz, F. Murgia, A. Prokudin, P. Schweitzer, W. Vogelsang & F. Yuan

Anomalous Drell–Yan asymmetry from hadronic or QCD vacuum effects 244
D. Boer

"T-odd" effects in transverse spin and azimuthal asymmetries in SIDIS 252
L.P. Gamberg & G.R. Goldstein

T-odd effects in unpolarized Drell–Yan scattering 260
G.R. Goldstein & L.P. Gamberg

Alternative approaches to transversity: How convenient and feasible are they? 268
M. Radici

Relations between single and double transverse asymmetries 276
O.V. Teryaev

Cross sections, error bars and event distributions in simulated Drell–Yan azimuthal asymmetry measurements 284
A. Bianconi

Next-to-leading order QCD corrections for transversely polarized pp and $\bar{p}p$ collisions 292
A. Mukherjee, M. Stratmann & W. Vogelsang

Double transverse-spin asymmetries in Drell–Yan and J/ψ production from proton–antiproton collisions 298
M. Guzzi, V. Barone, A. Cafarella, C. Corianò & P.G. Ratcliffe

The quark–quark correlator: Theory and phenomenology 306
E. Di Salvo

Chiral quark model spin filtering mechanism and hyperon polarization 314
S.M. Troshin & N.E. Tyurin

Closing Lecture

where we've been ....
.... and where we're going 325
G. Bunce

Author Index 337

INTRODUCTION

PURPOSE AND STATUS OF THE ITALIAN TRANSVERSITY PROJECT

F. Bradamante

Dipartimento di Fisica dell'Università degli Studi di Trieste and INFN, Sezione di Trieste, Via A. Valerio 2, I-34127 Trieste, Italy
E-mail: Franco.Bradamante@trieste.infn.it

The Italian Ministry for Education, University and Research (MIUR, Ministero dell'Istruzione, dell'Università e della Ricerca) each year supports two-year Research Projects of National Interest ("Progetti di Ricerca di Interesse Nazionale", PRIN). In 2003 a number of theoretical physicists and experimental physicists, all involved in the physics of transversity, proposed a research project entitled "Measurement of the Nucleon Transversity". The proponents belonged to six Italian Institutes, located in Alessandria, Como, Ferrara, Frascati National Laboratory of the INFN, Torino, and Trieste, and totalled 51 physicists. The project witnessed the participation of several Italian theoreticians working in the field, and many experimentalists active either in the COMPASS or in HERMES experiments. RHIC experiments were not represented since no Italian group is involved in RHIC.

The aim of the project was to set up a solid and efficient collaboration between the Italian theoretical and experimental physicists deeply involved in transversity at DESY and COMPASS at CERN.

In spite of the strong competition (in the area of Physics, only 42 projects out of a total of 147 proposals were approved and funded), our project was approved. Actually, an excellent rating placed it in fifth position among the 42 approved projects!

The primary goal of our project was to identify unambiguous transverse-spin phenomena of the nucleon through the study of experimental asymmetries in SIDIS of high-energy leptons on transversely polarized proton and deuteron targets. To achieve this goal, certain specific objectives are to be realized:

- at the experimental level: a more precise quantitative determination of the transverse-spin observables and the study of new channels in SIDIS and of new reactions;
- at the theoretical level: a better understanding of the observed effects and the identification of further candidate processes for measuring transversity.

Like other groups working in the field, we had an even more ambitious goal, i.e. a global analysis of transversity and of related observables. In short, this project aims to create a general and homogeneous framework, capable of including all the different processes that measure, in a direct or indirect manner, different combinations of the transverse-spin distribution and of the fragmentation functions.

Another important goal of our project was to promote the measurement of transversity at the new antiproton facility at FAIR (GSI) in Darmstadt via the measurement of single- and double-spin asymmetries of Drell–Yan pairs in the process $\bar{p}p \to l^+l^-X$. Indeed, when we wrote our proposal, the physics case for that measurement was just being developed, and most of us were involved in either the ASSIA or the PAX Letter of Intent. The plans at GSI have progressed very quickly, and the prospects of a polarized antiproton beam are fascinating. To fully exploit the potentialities of the Drell–Yan measurement at GSI, we have investigated the possibilities of converting HESR, the antiproton storage ring presently foreseen at GSI, into a proton–antiproton collider. More specifically, in collaboration with Novosibirsk, we have initiated a feasibility study for a polarized proton–polarized antiproton collider at GSI, both in a "green-field" symmetric option (15 GeV + 15 GeV) with one or two rings, and in the asymmetric option (15 GeV + 3.5 GeV). In the mean time, the PAX component of our project is finalizing the experimental set-up.

The organization of this workshop is part of the project. The Local Organizing Committee comprises one member from each participating Institute, and the financial support is granted by all of them. It is a great pleasure and very rewarding for us to see that the community has responded to our call and that most of the actors in the field of transversity have come to Como. Although, as we all know very well, we still have a long way to go before we can claim we have measured the nucleon transversity, in the past two years there have been very important advances in our field, and during the workshop we will surely have many occasions to check how far we can go with our ambitious programme.

In conclusion,

- we are grateful to MIUR for the support;
- let's look forward to a most interesting and useful workshop.

OPENING LECTURE

TRANSVERSITY

M. Anselmino*

Dipartimento di Fisica Teorica, Università di Torino and INFN, Sezione di Torino, Via P. Giuria 1, I-10125 Torino, Italy

Some introductory remarks on parton transverse spin distributions and transverse motion are presented. The issue of accessing experimental information on the transversity distributions h_1^q is discussed. In particular, direct information could be obtained from Drell–Yan processes with polarized protons and antiprotons (planned at GSI), while indirect information, coupling h_1^q to the Collins fragmentation function, is being gathered by HERMES and COMPASS collaborations. The related issue of transverse Single Spin Asymmetries (SSA) measured in several processes and the role of the Sivers distribution function is also discussed.

1. Towards transversity distributions: theory and experiments

The partonic structure of protons and neutrons is well known concerning the longitudinal degrees of freedom; these refer to the longitudinal momentum and spin carried by the partons inside unpolarized or longitudinally polarized fast-moving nucleons. Instead, much less is known about the transverse – with respect to the direction of motion – degrees of freedom: the intrinsic motion of partons ($\boldsymbol{k}_\perp$) and the transverse spin distributions of quarks inside a transversely polarized nucleon (transversity). Their knowledge – including possible spin–$\boldsymbol{k}_\perp$ correlations – is crucial if we aim at having a full understanding of the nucleon structure in terms of spin and orbital motion of quarks and gluons.

Transversity is the last leading-twist missing information on the quark spin structure of the nucleon;[1] whereas the unpolarized quark distributions, $q(x, Q^2)$ or $f_{q/p}(x, Q^2)$, are very well known, and good information

*Work partially supported by the European Community – Research Infrastructure Activity under the FP6 "Structuring the European Research Area" programme (HadronPhysics, contract number RII3-CT-2004-506078)

is now available on the quark helicity distributions $\Delta q(x, Q^2)$, nothing is experimentally known on the nucleon transversity distribution $h_1^q(x, Q^2)$ [also denoted by $\Delta_T q(x, Q^2)$ or $\delta q(x, Q^2)$]. The reason h_1^q, despite its fundamental importance, has never been measured is that it is a chiral-odd function, and consequently it decouples from inclusive Deep Inelastic Scattering (DIS), which is our usual main source of information on the nucleon partonic structure. Since electroweak and strong interactions conserve chirality, h_1^q cannot occur alone, but has to be coupled to a second chiral-odd quantity.

This is possible, for example, in polarized Drell–Yan processes,[2] where one measures the product of two transversity distributions, and in semi-inclusive DIS, where one couples h_1^q to a new unknown, chiral-odd, fragmentation function, the so-called Collins function.[3]

1.1. h_1 *in Drell–Yan processes*

Measurement of transversity is planned at RHIC, in Drell–Yan processes with transversely polarized protons, $p^\uparrow p^\uparrow \to \ell^- \ell^+ X$, via the measurement of the double-spin asymmetry:

$$A_{TT} \equiv \frac{d\sigma^{\uparrow\uparrow} - d\sigma^{\uparrow\downarrow}}{d\sigma^{\uparrow\uparrow} + d\sigma^{\uparrow\downarrow}}, \tag{1}$$

which reads at leading order in the parton model

$$A_{TT}^{pp} = \hat{a}_{TT} \, \frac{\sum_q e_q^2 \left[h_1^q(x_1, M^2)\, h_1^{\bar{q}}(x_2, M^2) + h_1^{\bar{q}}(x_1, M^2)\, h_1^q(x_2, M^2) \right]}{\sum_q e_q^2 \left[q(x_1, M^2)\, \bar{q}(x_2, M^2) + \bar{q}(x_1, M^2)\, q(x_2, M^2) \right]}, \tag{2}$$

where $q = u, d, s$; M is the invariant mass of the lepton pair and $\hat{a}_{TT}$ is the double-spin asymmetry of the elementary QED process $q\,\bar{q} \to \ell^- \ell^+$, see Eq. (5) below. In this case one measures the product of two transversity distributions, one for a quark and one for an anti-quark. The latter (in a proton) is expected to be small; moreover, the QCD evolution of transversity is such that, in the kinematical regions of RHIC data, $h_1^q(x, Q^2)$ is much smaller than the corresponding values of $\Delta q(x, Q^2)$ and $q(x, Q^2)$. All this makes the Drell–Yan double-spin asymmetry A_{TT}^{pp} expected at RHIC very small, no more than a few percent.[4,5]

Definite and direct information on transversity could be best obtained by measuring the double transverse spin asymmetry A_{TT} in the collision of polarized protons and antiprotons. The kinematical regions should be such that $h_1^q(x, Q^2)$ is expected to be sizeable, and the cross section allows to reach a good statistics. Such an ideal situation can be obtained by having polarized anti-protons colliding on polarized protons in the High Energy Storage Ring at GSI, as proposed by the PAX collaboration.[6] The expected ranges of energy and M^2 are:

$$(30 \lesssim s \lesssim 210)\ \mathrm{GeV}^2; \quad M \gtrsim 2\ \mathrm{GeV}/c^2; \quad \tau = x_1 x_2 = \frac{M^2}{s} \gtrsim 0.02\,. \tag{3}$$

One has

$$\begin{aligned} A_{TT}^{p\bar{p}} &= \hat{a}_{TT}\, \frac{\sum_q e_q^2 \left[h_1^q(x_1)\, h_1^q(x_2) + h_1^{\bar{q}}(x_1)\, h_1^{\bar{q}}(x_2)\right]}{\sum_q e_q^2 \left[q(x_1)\, q(x_2) + \bar{q}(x_1)\, \bar{q}(x_2)\right]} \\ &\simeq \hat{a}_{TT}\, \frac{h_1^u(x_1)\, h_1^u(x_2)}{u(x_1)\, u(x_2)} \quad (\text{at large } x)\,, \end{aligned} \tag{4}$$

where all quark distributions refer to protons and $\hat{a}_{TT}$ is the elementary double-spin asymmetry for the $q\,\bar{q} \to \ell^- \ell^+$ process (see, *e.g.*, Ref. 1 for the definition of the scattering angles):

$$\hat{a}_{TT}(\theta, \varphi) = \frac{\sin^2\theta}{1 + \cos^2\theta}\, \cos(2\varphi)\,. \tag{5}$$

Predictions,[7,8] based on assuming at some initial scale $h_1^q = \Delta q$ or on a chiral quark–soliton model, indicate large expected values of $A_{TT}^{p\bar{p}}$. Also the QCD corrections have been computed;[9] although they are huge for the cross sections, they affect very little the asymmetry, as shown in Fig. 1.

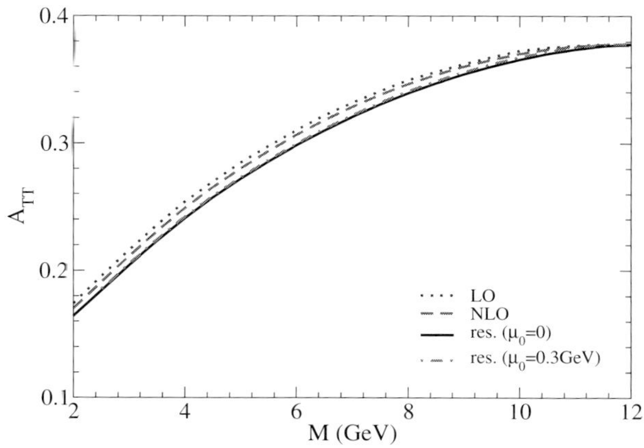

Fig. 1. Expected values of $A_{TT}^{p\bar{p}}$ for the proposed PAX–GSI experiments with polarized proton and antiprotons. The figure is taken from Ref. 9: $s = 210\,(\mathrm{GeV})^2$, the transversity distributions are given by saturation of the Soffer bound, $2|h_1^q(x)| = q(x) + \Delta q(x)$, and higher order perturbative QCD corrections have been computed, including all-order soft-gluon resummations.

1.2. h_1 *from SIDIS processes and extraction of the Collins function*

The Collins mechanism describes the fragmentation function of a transversely polarized quark (with spin vector $\boldsymbol{S}_q$) into an unpolarized or spinless hadron h (typically a pion):

$$\hat{D}_{h/q^\uparrow}(z, \boldsymbol{p}_\perp) = \hat{D}_{h/q}(z, p_\perp) + \tfrac{1}{2}\,\Delta^N \hat{D}_{h/q^\uparrow}(z, p_\perp)\, \boldsymbol{S}_q \cdot (\hat{\boldsymbol{p}}_q \times \hat{\boldsymbol{p}}_\perp)\,, \tag{6}$$

where $\boldsymbol{p}_\perp$ is the hadron transverse momentum, with respect to the fragmenting quark direction $\boldsymbol{p}_q$. $\hat{\boldsymbol{p}}_q$ and $\hat{\boldsymbol{p}}_\perp$ denote unit vectors.

Eq. (6) shows that there might be a spin–$\boldsymbol{p}_\perp$ correlation in the fragmentation function, leading to an azimuthal asymmetry (for a unit transverse polarization vector, $S_q = 1$),

$$\begin{aligned} D_{h/q^\uparrow}(z, \boldsymbol{p}_\perp) - D_{h/q^\downarrow}(z, \boldsymbol{p}_\perp) &= \Delta^N \hat{D}_{h/q^\uparrow}(z, p_\perp)\, \boldsymbol{S}_q \cdot (\hat{\boldsymbol{p}}_q \times \hat{\boldsymbol{p}}_\perp) \\ &= \Delta^N \hat{D}_{h/q^\uparrow}(z, p_\perp)\, \sin(\Phi_S - \Phi_h)\,, \end{aligned} \tag{7}$$

where Φ_S and Φ_h are respectively the azimuthal angles, around the fragmenting quark direction, of the quark spin and the hadron transverse motion. $\Delta^N \hat{D}_{h/q^\uparrow}(z, p_\perp)$ is referred to as the Collins function; another common notation in the literature due to the Amsterdam group [10] is

$$\Delta^N \hat{D}_{h/q^\uparrow} = 2\, \frac{p_\perp}{z M_h}\, H_1^{\perp q}\,. \tag{8}$$

Such a basic quark asymmetry can originate an observable azimuthal asymmetry in polarized SIDIS processes, $\ell\, N^\uparrow \to \ell\, h\, X$, see Fig. 2 for kinematical configurations and notations. The virtual photon scatters off a transversely polarized quark inside the transversely polarized nucleon (in amounts described by h_1^q) and the quark polarization gets modified (it is reduced in size and tilted symmetrically with respect to the normal to the lepton plane, according to QED interactions). The final polarized quark then fragments according to Eq. (6). All this translates into a single-spin asymmetry $(A_N^h)_{\rm Coll}$:

$$(A_N^h)_{\rm Coll} = \frac{\sum_q e_q^2 \int h_1^q(x)\, (1-y/(xy^2)\, \Delta^N \hat{D}_{h/q^\uparrow}(z, p_\perp)}{2 \sum_q e_q^2 \int f_{q/p}(x)\, [1+(1-y)^2]/(xy^2)\, \hat{D}_{q/p}(z, p_\perp)}\, \sin(\Phi_h + \Phi_S) \tag{9}$$

which combines the transversity distribution and the Collins function. The integrals are usually performed in order to study the dependence of the asymmetry on one kinematical variable at a time, taking into account the experimental setups and the kinematical cuts. In the above expression we have neglected the intrinsic motion in the distribution functions; this also might originate azimuthal asymmetries, for example with a $\sin(\Phi_h - \Phi_S)$ dependence, as we shall see in Subsection 2.1. Indeed, the numerator of the observed asymmetry,

$$A_N^h = \frac{d\sigma^\uparrow - d\sigma^\downarrow}{d\sigma^\uparrow + d\sigma^\downarrow}\,, \tag{10}$$

contains several contributions. However, each of them has a different azimuthal dependence, so that one can extract the desired contribution by appropriately averaging A_N^h. The experimental data sensitive to the Collins mechanism, according to the so-called Trento conventions,[11] are presented as:

$$\begin{aligned} A_{UT}^{\sin(\Phi_h+\Phi_S)} &\equiv 2\,\frac{\int d\Phi_h\, d\Phi_S\,[d\sigma^\uparrow - d\sigma^\downarrow]\,\sin(\Phi_h+\Phi_S)}{\int d\Phi_h\, d\Phi_S\,[d\sigma^\uparrow + d\sigma^\downarrow]} \\ &= \frac{\sum_q e_q^2 \int d\Phi_h\, d\Phi_S\, h_1^q(x)\,(1-y/(xy^2)\,\Delta^N\hat{D}_{h/q^\uparrow}(z,p_\perp)\,\sin^2(\Phi_h+\Phi_S)}{\sum_q e_q^2 \int d\Phi_h\, d\Phi_S\, f_{q/p}(x)\,[1+(1-y)^2]/(xy^2)\,\hat{D}_{q/p}(z,p_\perp)}\,. \end{aligned} \tag{11}$$

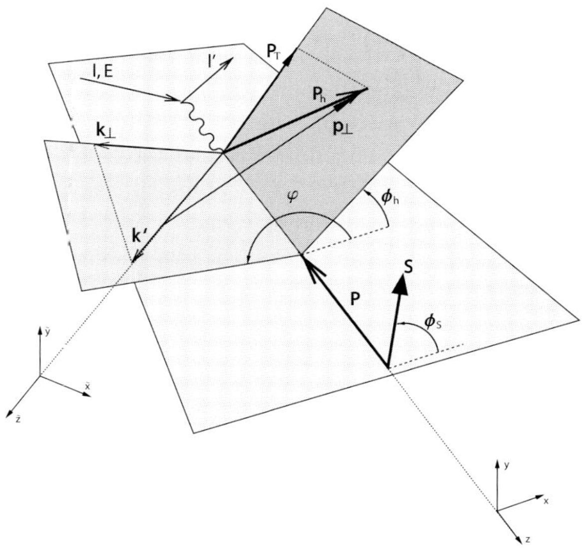

Fig. 2. Three dimensional kinematics of the SIDIS process.

Data on $A_{UT}^{\sin(\Phi_h+\Phi_S)}$ have been obtained by HERMES[12] and COMPASS[13] collaborations. The HERMES data and Eq. (11) have been used in Ref. 14, assuming a Soffer saturated value of $2|h_1^q(x)| = q(x) + \Delta q(x)$, to extract the Collins favoured, $\Delta^N \hat{D}_{\pi^+/u^\uparrow}$ and unfavoured, $\Delta^N \hat{D}_{\pi^+/d^\uparrow}$, fragmentation functions. It is the first attempted extraction of the Collins functions, which still have, given the quality and amount of the experimental information available, large uncertainties.

Further information on the Collins function, which would greatly help in accessing the transversity distributions, is coming from BELLE collaboration. They can measure the combined effects of two Collins mechanisms by looking, event by event, at azimuthal correlations between hadrons in opposite jets in $e^+e^- \to h_1\, h_2\, X$ processes; first results [15] clearly indicate non zero Collins functions, in qualitative agreement with the Collins functions extracted from SIDIS data.

1.3. *Alternative accesses to transversity*

We have seen how the chiral-odd transversity distribution h_1^q must couple to another chiral-odd function in order to form a measurable quantity and to be experimentally accessible. The most promising approach seems to go via the double transverse spin asymmetry $A_{TT}^{p\bar{p}}$ in Drell–Yan processes with polarized protons *and antiprotons*; however, such an experiment is only at the proposal stage,[6] with many difficulties yet to overcome.[16] At present, the most doable approach seems to be the azimuthal transverse SSA in SIDIS, in conjunction with data from BELLE. $A_{UT}^{\sin(\Phi_h+\Phi_S)}$ depends on the convolution of transversity and Collins functions; they are both unknown, but they depend on different variables; moreover, BELLE data should supply information on the product of two Collins functions alone.

Let us consider briefly other possible ways of coupling h_1^q to a chiral-odd function.

- *Inclusive Λ production and measurement of its polarization*
  One should consider the electromagnetic SIDIS process $\ell\, p^\uparrow \to \gamma^* \to \ell\, \Lambda\, X$; the measurement of the Λ polarization $\boldsymbol{P}_\Lambda$, allows to study the spin transfer from the polarized quark to the polarized Λ. For transverse spins one has, schematically [17]

$$P_\Lambda \sim \sum_q h_1^q(x) \otimes \Delta_T D_{\Lambda/q}(z)\,, \tag{12}$$

  which couples h_1^q to its (unknown) analogous fragmentation function, $\Delta_T D_{\Lambda/q} = D_{\Lambda^\uparrow/q^\uparrow} - D_{\Lambda^\downarrow/q^\uparrow}$. Notice that such an observable, a double-spin asymmetry, does not vanish in the absence of intrinsic motions.

- *Two pion production and interference fragmentation functions*
It was suggested,[18,19] as an alternative to the Collins fragmentation mechanism, to consider the process $\ell\, p^\uparrow \to \ell\, \pi\, \pi\, X$, selecting two pions (or two hadrons) in the final state. In such a case, there might easily be a non zero chiral-odd, unknown, interference fragmentation function, δq_I. One should, like for the single pion Collins mechanism, look at a dependence on $\sin(\Phi_R + \Phi_S)$, where Φ_R is the azimuthal angle of the relative momentum $\boldsymbol{R}$ of the two pions:
$$A_{UT} \sim \sum_q \sin(\Phi_R + \Phi_S)\, h_1^q \otimes \delta q_I\,. \tag{13}$$
This asymmetry has been observed by HERMES [20] and the interference fragmentation functions might be independently measured at BELLE.
- *Vector meson production in SIDIS*
The off-diagonal element $\rho_{1,0}$ of the helicity density matrix of vector mesons produced in SIDIS processes is measurable and is related to h_1^q and some new (unknown) product of fragmentation amplitudes.[17]
- *Inclusive hadronic production, $p^\uparrow\, p \to \pi\, X$*
SSA in inclusive hadronic processes, $p^\uparrow\, p \to \pi\, X$, might depend on the convolution of transversity distributions and the Collins function. However, it has been shown recently that a careful treatment of the non collinear partonic interactions, strongly suppresses this contribution.[21]
- *SSA in Drell–Yan processes*
Data on unpolarized Drell–Yan processes show a strong $\cos(2\varphi)$ dependence, which might be explained [22] by the so-called Boer–Mulders function,
$$\Delta^N f_{q^\uparrow/p} = -\frac{k_\perp}{M}\, f_1^{\perp q}\,, \tag{14}$$
correlating the transverse spin of quarks inside an unpolarized proton to their transverse momentum $\boldsymbol{k}_\perp$.[10] In such a case, even inside unpolarized protons one can find transversely polarized quarks and the elementary $q\,\bar{q} \to \ell^- \ell^+$ scattering of the Drell–Yan process shows the typical $\cos(2\varphi)$ dependence of double transverse spin asymmetries, Eq. (5). Then, one can learn about $\Delta^N f_{q^\uparrow/p}$.
Drell–Yan processes may exhibit also single transverse spin asymmetries, $d\sigma^\uparrow \neq d\sigma^\downarrow$, which originate from the convolution of trans-

versity and the Boer–Mulders function:

$$d\sigma^{\uparrow} - d\sigma^{\downarrow} \sim \sum_q h_1^q \otimes \Delta^N f_{q^{\uparrow}/p}. \tag{15}$$

Thus, a combined study of the $\cos(2\varphi)$ dependence of unpolarized Drell–Yan processes (which yields information on $\Delta^N f_{q^{\uparrow}/p}$) and of transverse SSA, could allow to extract information on h_1^q.[22,23]

2. Transverse Single Spin Asymmetries (SSA)

Transverse Single Spin Asymmetries are intriguing observables which test the underlying theory at a deep quantum mechanical level and, experimentally, very often result in unexpected and interesting data.[24]

Let us consider, for example, the elastic scattering of two protons in their center of mass frame. If $\boldsymbol{p}$, $\boldsymbol{S}$ and $\boldsymbol{P}_T$ denote respectively the momentum of the incoming proton, its polarization vector and the transverse (with respect to the collision axis) momentum of the scattered proton, the only SSA allowed by parity invariance must be of the form:

$$A_S \equiv \frac{d\sigma^{\boldsymbol{S}} - d\sigma^{-\boldsymbol{S}}}{d\sigma^{\boldsymbol{S}} + d\sigma^{-\boldsymbol{S}}} \sim \boldsymbol{S} \cdot (\boldsymbol{p} \times \boldsymbol{P}_T) \sim P_T \sin(\Phi_S - \Phi)\,, \tag{16}$$

where Φ_S and Φ are the azimuthal angles, around the collision axis, of the spin vector and the scattered proton, respectively. Notice that A_S is zero if $\boldsymbol{S}$ lies in the scattering plane and if $P_T = 0$.

SSA are then possible in principle, with the azimuthal angular dependence of Eq. (16). Do they occur at the partonic, pQCD, level? Let us consider the example of the elastic scattering of two identical quarks, in their center of mass frame, with the scattering taking place in the (xz) plane. All observables can be expressed in terms of 5 independent helicity amplitudes, $M_{\lambda_1',\lambda_2';\lambda_1,\lambda_2}$, which are usually denoted as:

$$\begin{aligned} &M_{++;++} \equiv \Phi_1\,, \qquad M_{--;++} \equiv \Phi_2\,, \qquad M_{+-;+-} \equiv \Phi_3\,, \\ &M_{-+;+-} \equiv \Phi_4\,, \qquad M_{-+;++} \equiv \Phi_5\,. \end{aligned}$$

If only one of the two initial quarks is polarized along the y-direction with spin $\uparrow$ or $\downarrow$ and we do not look at the final quark polarizations, we can have $d\sigma^{\uparrow} \neq d\sigma^{\downarrow}$. In terms of the helicity amplitudes:

$$A_N \equiv \frac{d\sigma^{\uparrow} - d\sigma^{\downarrow}}{d\sigma^{\uparrow} + d\sigma^{\downarrow}} \sim \mathrm{Im}\,\left[\Phi_5\,(\Phi_1 + \Phi_3 + \Phi_2 - \Phi_4)^*\right]\,, \tag{17}$$

which can be non zero only if the single helicity flip amplitude Φ_5 is non zero and if different amplitudes have relative phases. The first condition implies a factor m_q/E_q (quark mass over its energy) and the second requires considering higher order contributions (as lowest order there are no relative phases). Altogether, one has

$$A_N^q \simeq \frac{m_q}{E_q}\,\alpha_s\,, \tag{18}$$

where α_s is the strong coupling constant. This makes single spin asymmetries in the partonic interactions entirely negligible.[25] However, sizeable SSA are observed at the hadronic level in elastic[26] and inclusive processes.[12,27–29]

2.1. *SSA in SIDIS processes, data and extraction of the Sivers function*

Transverse SSA can be understood, even within perturbative QCD or QED dynamics, by taking into account spin effects in the non perturbative components of the hard scattering factorized schemes. One such example has been discussed in Subsection 1.2, where the non perturbative spin–$\boldsymbol{p}_\perp$ dependent contribution is given by the Collins mechanism, Eq. (6), which leads to the SSA given in Eq. (9).

Another spin–intrinsic motion correlation can be introduced in the parton distribution functions, as originally suggested by Sivers:[30]

$$\hat{f}_{a/p^\uparrow}(x,\boldsymbol{k}_\perp) = \hat{f}_{a/p}(x,k_\perp) + \tfrac{1}{2}\,\Delta^N\hat{f}_{a/q^\uparrow}(x,k_\perp)\,\boldsymbol{S}\cdot(\hat{\boldsymbol{p}}\times\hat{\boldsymbol{k}}_\perp)\,, \tag{19}$$

$$\hat{f}_{a/p^\uparrow}(x,\boldsymbol{k}_\perp) - \hat{f}_{a/p^\downarrow}(x,\boldsymbol{k}_\perp) = \Delta^N\hat{f}_{a/q^\uparrow}(x,k_\perp)\,\sin(\Phi_S-\varphi)\,, \tag{20}$$

which shows how the number density of partons a, inside a transversely polarized proton with polarization $\boldsymbol{S}$, may depend on the relative azimuthal angle between $\boldsymbol{S}$ and the transverse momentum $\boldsymbol{k}_\perp$, $(\Phi_S-\varphi)$.

In SIDIS processes, $\gamma^*\,p \to h\,X$ (Fig. 2), the virtual photon interacts with a parton inside the transversely polarized proton, and the scattered parton then fragments into the observed hadron h, which has a transverse momentum $\boldsymbol{P}_T$, with respect to the γ^*-direction. This P_T can originate (at lowest order) from the initial intrinsic motion of the parton or from the fragmentation process. The latter case was considered in Subsection 1.2, and leads to a $\sin(\Phi_h+\Phi_S)$ dependence. The former case instead, explores the $\boldsymbol{k}_\perp$-distribution of partons inside a transversely polarized proton, and is then related to the Sivers mechanism (20). A model for such a mechanism was introduced in Ref. 31.

The analogue of Eq. (9) now reads (see Ref. 32 for further details):

$$(A_N^h)_{\rm Siv} \simeq \frac{\displaystyle\sum_q \int d^2\boldsymbol{k}_\perp \, \Delta^N f_{q/p^\uparrow}(x, k_\perp) \, \sin(\varphi - \Phi_S) \, \frac{d\hat\sigma^{\ell q\to\ell q}}{dQ^2} \, D_{h/q}(z, p_\perp)}{\displaystyle 2\sum_q \int d^2\boldsymbol{k}_\perp \, f_{q/p}(x, k_\perp) \, \frac{d\hat\sigma^{\ell q\to\ell q}}{dQ^2} \, D_{h/q}(z, p_\perp)} \,, \tag{21}$$

where we have neglected some terms of order $(k_\perp^2/Q^2)$. Notice that, following Ref. 32, we have now taken into account all the intrinsic motion effects: in the distribution functions, in the elementary interactions

$$\frac{d\hat\sigma^{\ell q\to\ell q}}{dQ^2} = e_q^2 \, \frac{2\pi\alpha^2}{\hat s^2} \, \frac{\hat s^2 + \hat u^2}{Q^4} \,, \tag{22}$$

and in the fragmentation functions.

This Sivers asymmetry is extracted from the data on A_N^h, Eq. (10), by selecting the $\sin(\Phi_h - \Phi_S)$ averaged asymmetry:

$$A_{UT}^{\sin(\phi_h-\phi_S)} \simeq \tag{23}$$

$$\frac{\displaystyle\sum_q \int d\phi_S \, d\phi_h \, d^2\boldsymbol{k}_\perp \, \Delta^N f_{q/p^\uparrow}(x, k_\perp) \sin(\varphi - \phi_S) \frac{d\hat\sigma}{dQ^2} \, D_q^h(z, p_\perp) \sin(\phi_h - \phi_S)}{\displaystyle\sum_q \int d\phi_S \, d\phi_h \, d^2\boldsymbol{k}_\perp \, f_q(x, \boldsymbol{k}_\perp) \, \frac{d\hat\sigma^{\ell q\to\ell q}}{dQ^2} \, D_{h/q}(z, p_\perp)} \,.$$

Several groups [33] have exploited the above expression and the HERMES [12] and COMPASS [13] data to extract information, or check models, on the Sivers functions for u and d quarks. The fits to the $\pi^\pm$ HERMES data obtained in Ref. 34 are shown in Fig. 3, together with predictions for π^0 production.

The resulting Sivers functions for u and d quarks turn out to be approximately opposite. This explains why the COMPASS data,[13] taken on a deuterium target, show almost negligible values of $A_{UT}^{\sin(\phi_h-\phi_S)}$. In fact, for a deuterium target one has:

$$A_{UT}^{\sin(\phi_h-\phi_S)} \sim \left(\Delta^N f_{u/p^\uparrow} + \Delta^N f_{d/p^\uparrow}\right)\left(4\, D_{h/u} + D_{h/d}\right) \,. \tag{24}$$

2.2. *SSA in $p^\uparrow \; p \to \pi \, X$ processes*

Let us conclude by mentioning SSA in $p^\uparrow p \to \pi\, X$ processes. Some recent papers have discussed the problem, in the context of QCD with a possible factorization scheme [35] and/or with higher-twist partonic correlations.[36]

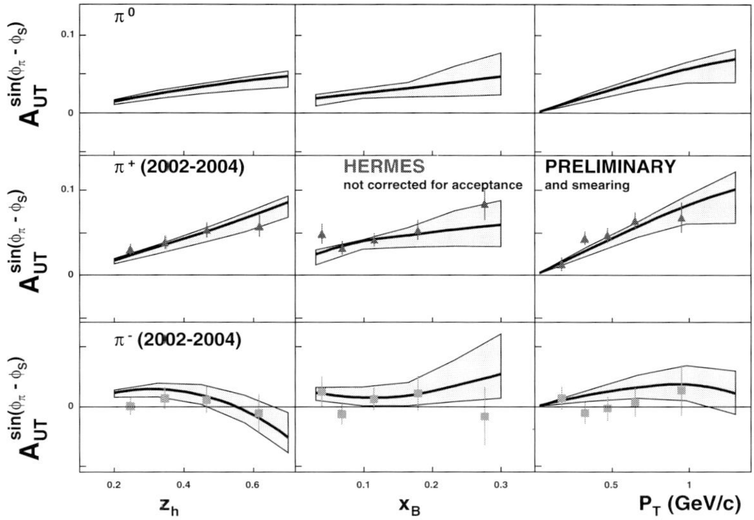

Fig. 3. HERMES data [12] on $A_{UT}^{\sin(\phi_\pi - \phi_S)}$ for scattering off a transversely polarized proton target and pion production; the curves are the results of the fit of Ref. 33. The shaded area spans a region corresponding to one-sigma deviation at 90% CL. The curves for π^0 are predictions based on the extracted Sivers functions.

We only mention here that both Sivers and Collins mechanism, assuming a QCD factorization scheme with parton intrinsic motions, might contribute to a non vanishing A_N; however, it was recently shown that the correct treatment of the elementary dynamics, with non collinear partonic processes and the proper spinor phases taken into account, strongly suppresses the contribution of the Collins mechanism.[21] The Sivers mechanism, instead, is not suppressed [37] and can well explain the observed SSA;[28,29] the Sivers functions active in $\ell\, p$ and $p\, p$ inclusive processes might be the same.

References

1. For a review on the transverse spin structure of the proton, see: V. Barone, A. Drago and P.G. Ratcliffe, *Phys. Rep.* **359** (2002) 1.
2. J. Ralston and D.E. Soper, *Nucl. Phys.* **B152** (1979) 109; J.L. Cortes, B. Pire and J.P. Ralston, *Z. Phys.* **C55**, 409 (1992); R.L. Jaffe and X. Ji, *Nucl. Phys.* **B375** (1992) 527.
3. J.C. Collins, *Nucl. Phys.* **B396** (1993) 161.
4. V. Barone, T. Calarco and A. Drago, *Phys. Rev.* **D56** (1997) 527.
5. O. Martin, A. Schäfer, M. Stratmann and W. Vogelsang, *Phys. Rev.* **D57** (1998) 3084; *Phys. Rev.* **D60** (1999) 117502.

6. PAX Collaboration, e-print archive: hep-ex/0505054.
7. M. Anselmino, V. Barone, A. Drago and N.N. Nikolaev, *Phys. Lett.* **B594** (2004) 97.
8. A.V. Efremov, K. Goeke and P. Schweitzer, *Eur. Phys. J.* **C35** (2004) 207.
9. H. Shimizu, G. Sterman, W. Vogelsang and H. Yokoya, *Phys. Rev.* **D71** (2005) 114007.
10. P.J. Mulders and R.D. Tangerman, *Nucl. Phys.* **B461** (1996) 197; *Erratum-ibid.* **B484** (1997) 538; D. Boer and P.J. Mulders, *Phys. Rev.* **D57** (1998) 5780; D. Boer, P.J. Mulders and F. Pijlman, *Nucl. Phys.* **B667** (2003) 201.
11. A. Bacchetta, U. D'Alesio, M. Diehl and C.A. Miller, *Phys. Rev.* **D70** (2004) 117504.
12. HERMES Collaboration, A. Airapetian *et al.*, *Phys. Rev. Lett.* **94** (2005) 012002; M. Diefenthaler (on behalf of the HERMES collaboration), e-print archive: hep-ex/0507013.
13. COMPASS Collaboration, V.Yu. Alexakhin *et al.*, *Phys. Rev. Lett.* **94** (2005) 202002.
14. W. Vogelsang and F. Yuan, *Phys. Rev.* **D72** (2005) 054028.
15. Belle Collaboration (K. Abe *et al.*), e-print archive: hep-ex/0507063
16. N.N. Nikolaev, contribution to these proceedings.
17. M. Anselmino, M. Boglione, J. Hansson and F. Murgia, *Phys. Rev.* **D54** (1996) 828.
18. R.L. Jaffe, X. Jin and J. Tang, *Phys. Rev. Lett.* **80** (1998) 1166.
19. M. Radici, R. Jakob and A. Bianconi, *Phys. Rev.* **D65** (2002) 074031.
20. P. van der Nat, contribution to these proceedings.
21. M. Anselmino, M. Boglione, U. D'Alesio E. Leader and F. Murgia, *Phys. Rev.* **D71** (2005) 014002.
22. D. Boer, *Phys. Rev.* **D60** (1999) 014012.
23. A. Bianconi and M. Radici, *J. Phys.* **G31** (2005) 645.
24. For a recent review, see A.D. Krisch, e-print archive: hep-ex/0511040
25. G.L. Kane, J. Pumplin and W. Repko, *Phys. Rev. Lett.* **41** (1978) 1689.
26. D.G. Crabb *et al.*, *Phys. Rev. Lett.* **65** (1990) 3241.
27. K. Heller *et al.*, *Phys. Rev. Lett.* **51** (1983) 2025.
28. D.L. Adams *et al.* (E704 Collaboration), *Z. Phys.* **C56** (1992) 181; *Phys. Lett.* **B345** (1995) 569.
29. J. Adams *et al.* (STAR Collaboration), *Phys. Rev. Lett.* **92** (2004) 171801.
30. D. Sivers, *Phys. Rev.* **D41** (1990) 83; **D43** (1991) 261.
31. S.J. Brodsky, D.S. Hwang and I. Schmidt, *Phys. Lett.* **B530** (2002) 99.
32. M. Anselmino, M. Boglione, U. D'Alesio, A. Kotzinian, F. Murgia and A. Prokudin, *Phys. Rev.* **D71** (2005) 074006.
33. M. Anselmino *et al.*, e-print archive: hep-ph/0511017
34. M. Anselmino, M. Boglione, U. D'Alesio, A. Kotzinian, F. Murgia and A. Prokudin, *Phys. Rev.* **D72** (2005) 094007.
35. M. Anselmino, M. Boglione, U. D'Alesio, E. Leader, S. Melis and F. Murgia, e-print archive: hep-ph/0509035; F. Murgia, these proceedings.
36. A. Bacchetta, C.J. Bomhof, P.J. Mulders and F. Pijlman, *Phys. Rev.* **D72** (2005) 034030; A. Bacchetta, e-print archive: hep-ph/0511085.
37. U. D'Alesio and F. Murgia, *Phys. Rev.* **D70** (2004) 074009.

EXPERIMENTAL LECTURES

AZIMUTHAL SINGLE-SPIN ASYMMETRIES FROM POLARIZED AND UNPOLARIZED HYDROGEN TARGETS AT HERMES

G. Schnell
[on behalf of the HERMES Collaboration]

Subatomaire en stralingsfysica,
Universiteit Gent,
9000 Gent, Belgium
E-mail: gunar.schnell@desy.de

Azimuthal single-spin asymmetries for semi-inclusive electro-production of charged pions in deep-inelastic scattering of positrons off transversely polarized and unpolarized protons are presented. In the case of transversely polarized protons, different combinations of the azimuthal angles of the hadron momentum and the proton-spin direction around the virtual-photon direction allow a separation of the so-called Collins and Sivers asymmetries. The asymmetries on longitudinally polarized or unpolarized protons involve subleading-twist distribution or fragmentation functions and thus constitute direct measurements of subleading-twist effects.

1. Introduction

Single-spin asymmetries (SSA) in the azimuthal distribution of lepto-produced hadrons around the virtual photon direction are a valuable tool to explore transverse spin and momentum degrees of freedom in the nucleon. Such SSA's in semi-inclusive deep-inelastic scattering (DIS) have been observed already with unpolarized beam and longitudinally polarized targets.[1] Recently,[2] the HERMES experiment at DESY measured SSA's also on a transversely polarized target. While the interpretation of the data from longitudinally polarized targets is hampered by the fact that the various contributions can not be disentangled without using information from other sources, a transversely polarized target allows a simultaneous measurement of the so-called Collins and Sivers asymmetries. The Collins asymmetry involves the as-yet unmeasured transversity distribution h_1.[3] It describes the imbalance in the number of quarks with their spin (anti)aligned to the spin of a transversely polarized nucleon. It can not be measured in inclusive processes because of its chiral-odd character. However, in semi-inclusive DIS it can appear in conjunction with a chiral-odd fragmentation function. One example is the "Collins function"[4] $H_1^\perp$ – also odd under naive[a] time reversal (T-odd). The combination of h_1 and $H_1^\perp$ results in a sinusoidal distribution of produced hadrons around the virtual photon direction. The Sivers asymmetry, on the other hand, involves a T-odd distribution function. This "Sivers function"[5] $f_{1T}^\perp$ was believed to vanish because of this property but in the recent work initiated by Ref. 6 it was realized that final-state interactions via soft gluons offer a mechanism at leading twist to create the necessary interference of amplitudes for its naive T-odd nature.[7] An interesting property of the Sivers function is its link to orbital angular momentum L_z – the Sivers function vanishes for zero L_z. It thus has the potential to provide one way of measuring L_z. SSA's involving longitudinally polarized targets and unpolarized beam, or unpolarized targets and longitudinally polarized beam are by nature subleading, *i.e.*, at twist-3, in the $1/Q$-expansion of the semi-inclusive DIS cross section. They thus constitute a direct way to measure subleading-twist effects. In particular, the beam SSA involving a longitudinally polarized beam is of interest as it involves the twist-3 distribution e, which can be linked to the pion–nucleon σ-term.

2. Semi-Inclusive DIS on Transversely Polarized Protons

2.1. *Azimuthal Single Spin Asymmetries*

Both the Sivers and the Collins functions belong to the group of unintegrated distribution (DF) and fragmentation functions (FF). They explicitly depend on intrinsic parton transverse momenta and do not survive on integration over the latter. Hence a measurement of either needs to be sensitive to transverse momenta. One possibility is to study the azimuthal distribution of hadrons around the virtual-photon direction. In the case of a transversely polarized target this distribution depends on two azimuthal angles ϕ and ϕ_S as illustrated in Fig. 1.

The luminosity-normalized count-rate asymmetries between opposite target-spin states $(\uparrow, \downarrow)$ can be written as a sum of various sine and co-

[a]Naive time reversal is time reversal without interchange of initial and final states.

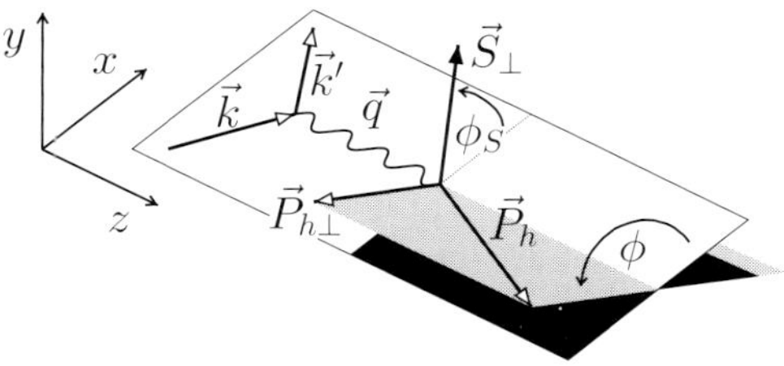

Fig. 1. The definitions of the azimuthal angles.

sine modulations:[8,b]

$$\frac{1}{S_\perp}\frac{N^\uparrow(\phi,\phi_S)-N^\downarrow(\phi,\phi_S)}{N^\uparrow(\phi,\phi_S)+N^\downarrow(\phi,\phi_S)} = \frac{B(\langle y\rangle)}{A(\langle x\rangle,\langle y\rangle)} A_{UT}^{\sin(\phi+\phi_S)}\sin(\phi+\phi_S) + A_{UT}^{\sin(\phi-\phi_S)}\sin(\phi-\phi_S). \quad (1)$$

Here $S_\perp$ is the transverse polarization of the target, the subscript UT denotes unpolarized beam and transverse target polarization, x (Bjorken scaling variable) as well as y (fractional energy of the virtual photon) are the usual DIS Lorentz invariants, and $B(\langle y\rangle) \equiv (1-y)$, $A(\langle x\rangle,\langle y\rangle) \equiv \frac{y^2}{2}+(1-y)(1+R(x,y))/(1+\gamma(x,y)^2)$. The amplitudes of each sine term are proportional in leading order to a convolution integral over transverse momenta of both a DF and a FF:

$$A_{UT}^{\sin(\phi+\phi_S)} \propto \sum_q e_q^2\, \boldsymbol{\mathcal{I}}\left[\frac{\boldsymbol{k_T}\cdot\hat{\boldsymbol{P}}_{\boldsymbol{h}\perp}}{M_h}\, h_1^q(x,p_T^2)\, H_1^{\perp,q}(z,z^2k_T^2)\right], \quad (2)$$

$$A_{UT}^{\sin(\phi-\phi_S)} \propto -\sum_q e_q^2\, \boldsymbol{\mathcal{I}}\left[\frac{\boldsymbol{p_T}\cdot\hat{\boldsymbol{P}}_{\boldsymbol{h}\perp}}{M}\, f_{1T}^{\perp,q}(x,p_T^2)\, D_1^q(z,z^2k_T^2)\right], \quad (3)$$

where $\boldsymbol{p_T}$ ($\boldsymbol{k_T}$) are the intrinsic quark transverse momenta, $\hat{\boldsymbol{P}}_{\boldsymbol{h}\perp}$ is the unit vector in the hadron's transverse momentum direction, and z is the fractional energy of the hadron. These virtual-photon asymmetries were extracted by performing a two-dimensional fit of Eq. (1) in order to minimize uncertainties from systematic correlations.

---

[b]For convenience only the Collins and Sivers sine modulations are taken into account here.

2.2. *Results*

The preliminary results of the measured asymmetries for charged π mesons are plotted in Fig. 2 as functions of x, z, and $P_{h\perp}$. The Collins moments[c] $\langle \sin(\phi + \phi_S) \rangle$ are positive for π^+ and negative for π^-. This is not unexpected as the two valence quark flavors of the proton are predicted to have transversity distributions of opposite sign and they contribute to π^+/π^- production with different strength. However, the large moments for the π^- are somewhat unexpected but could be explained by a large disfavored Collins FF (*e.g.*, the fragmentation of *up* quarks into π^- mesons) which is also opposite in sign compared to the favored one. In fact, recent fits [10] to both the HERMES and the recent COMPASS [11] data give just such a result.

The Sivers moments are positive for π^+ and consistent with zero for π^-. This agrees with, *e.g.*, the prediction in Ref. 12 and is the first direct sign of a T-odd DF in DIS. Following the arguments in Ref. 12, the measured Sivers moments correspond to the orbital angular momentum of the *up*-quarks parallel to the spin of the nucleon, *i.e.*, $L_z^u > 0$. However, one has to keep in mind that at present no formal way leads from the Sivers function to the total quark orbital angular momentum. This is a similar situation to what one has for the anomalous magnetic moment of the proton: although a non-vanishing anomalous magnetic moment requires wave function components with non-vanishing quark orbital angular momentum, no statement can be done about the total orbital angular momentum carried by quarks.[6,13] In fact, the Sivers effect involves the same overlap integrals between the same wave function components that also appear in the expressions for the anomalous magnetic moment.

3. SSA using Longitudinally Polarized Beam or Targets

3.1. *Longitudinally Polarized Target Data Revisited*

Before data on transversely polarized targets became available, SSA's had been measured already in semi-inclusive DIS using unpolarized positrons/electrons and longitudinally polarized targets. The non-vanishing SSA's that had been found for π^+ and π^0 were interpreted quickly as evidence for a non-vanishing Collins function and for transversity. The reason for this is a subtlety in the measurement: In experiments, the target polarization is always defined with respect to the incoming beam direction. How-

[c]Following the *Trento Conventions* [9] azimuthal moments are denoted by either $A_{UT}^{\sin(\phi+\phi_S)}$ or $\langle \sin(\phi \pm \phi_S) \rangle \equiv \frac{1}{2} A_{UT}^{\sin(\phi\pm\phi_S)}$.

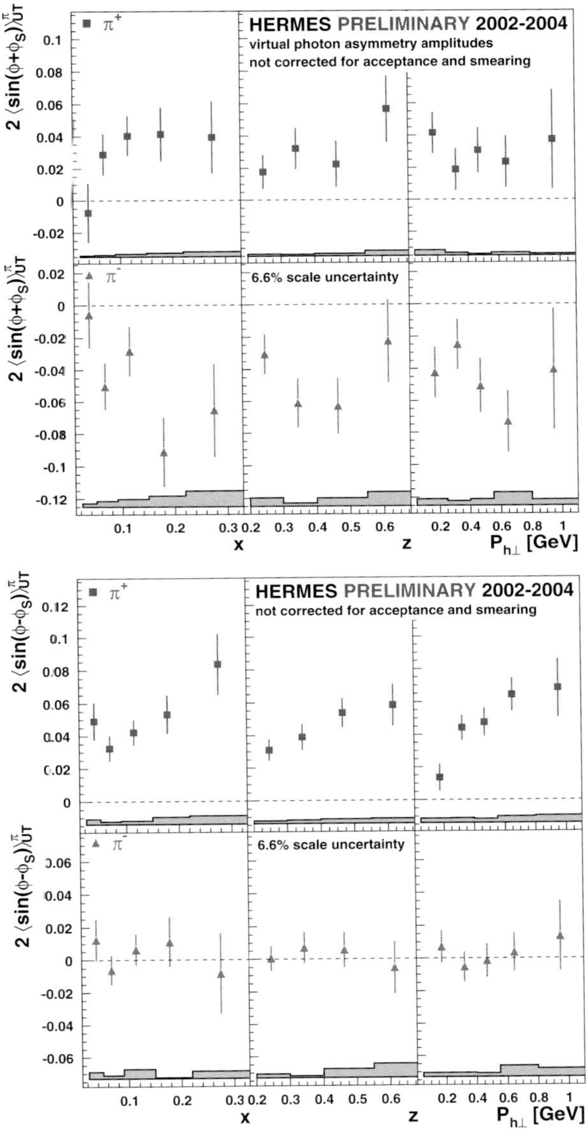

Fig. 2. Collins (top) and Sivers (bottom) azimuthal moments for charged π mesons as labeled in the left panels, as functions of x, z, and $P_{h\perp}$. Error bars are statistical uncertainties only. There is a common 6.6% scale uncertainty due to the uncertainty in the target polarization value. The grey error band is the systematic uncertainty.

ever, in the theoretical treatment of the scattering cross-section, the virtual-photon direction acts as the reference axis. Since the lepton-beam and virtual-photon axes do not coincide there is always a non-vanishing transverse component of the nucleon polarization with respect to the virtual-photon axis (*cf.* Fig. 3). The size of the transverse component depends on the event kinematics, and at HERMES it can be up to 15%.

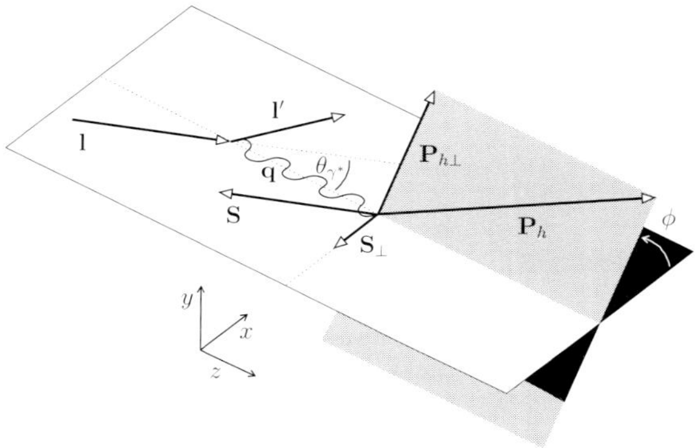

Fig. 3. The azimuthal angles for a target polarized longitudinally with respect to the incoming beam direction. The transverse component $S_\perp$ of the nucleon spin is proportional to $\sin\theta_{\gamma^*}$, where θ_{γ^*} denotes the angle between the beam direction and the virtual-photon axis.

In the case of scattering off a longitudinally polarized target the measured asymmetries thus receive contributions from the longitudinal component of the target polarization as well as from the transverse one. Since the target polarization vector lies in the scattering plane the Sivers and Collins moments show up as a $\sin\phi$ modulation of the measured asymmetry. The longitudinal component of the target polarization can also exhibit a $\sin\phi$ modulation. However, in contrast to the Sivers and Collins moments, which are at leading order in a $1/Q$ expansion of the cross section, the $\sin\phi$ modulation due to the longitudinal component is $1/Q$ suppressed, *i.e.*, subleading twist. As the transverse component of the target polarization with respect to the virtual-photon axis is itself small, it is *a priori* not clear what is the dominant contribution to the measured $\sin\phi$ modulation: the leading-twist Collins and Sivers moments or the subleading-twist contribution stemming from the dominant longitudinal nucleon-polarization component.

The new measurements from a transversely polarized target can now be used to clarify this situation, as the contributions from the Sivers and

Collins moments can be subtracted. The subleading-twist $\sin\phi$ moment, $\langle \sin\phi \rangle^{q}_{UL}$, can be extracted via:[14]

$$\langle \sin\phi \rangle^{q}_{UL} \simeq \langle \sin\phi \rangle^{l}_{UL} + \sin\theta_{\gamma^*} \left(\langle \sin(\phi+\phi_S) \rangle^{l}_{UT} + \langle \sin(\phi-\phi_S) \rangle^{l}_{UT} \right), \quad (4)$$

where the superscripts denote nucleons polarized either with respect to the lepton-beam direction (l) (*i.e.*, the measured Sivers, Collins, and longitudinal $\sin\phi$ moments) or to the virtual-photon direction (q). Hence combining the results from transversely and longitudinally polarized protons one obtains, using Eq. (4), the purely subleading-twist contribution to the cross section.

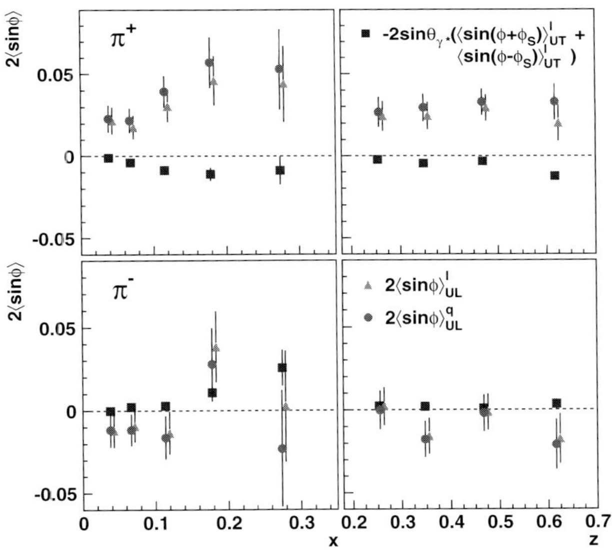

Fig. 4. The various azimuthal moments appearing in the measurement of the $\sin\phi$ modulations of single-spin asymmetries on a longitudinally polarized hydrogen target for charged pions as functions of x and z. The open symbols are the measured lepton-axis moments. The ones from a transversely polarized target are multiplied by $-\sin\theta_{\gamma^*}$ according to their appearance in the longitudinal lepton-axis moments. The closed symbol is the subleading-twist contribution to the measured lepton-axis asymmetries on a longitudinally polarized target. The triangles are slightly shifted horizontally for distinction. An overall systematic error of 0.003 is not included in the figure.

In Fig. 4 the resulting subleading $\langle \sin\phi \rangle^{q}_{UL}$ moments for charged π's are shown as functions of x and z, in addition to the measured $\langle \sin\phi \rangle^{l}_{UL}$ moments, and the contributions to these moments from the Sivers and

Collins asymmetries.[15] The measured $\langle \sin\phi \rangle_{UL}^{\mathrm{l}}$ moments are dominated by the subleading-twist contribution, at least for the π^+. The subleading-twist contribution is significantly positive for π^+, while it is consistent with zero for π^-. Therefore it becomes clear that the measured $\langle \sin\phi \rangle_{UL}^{\mathrm{l}}$ moments cannot be interpreted in terms of the Sivers or Collins effect alone. In fact, the subleading-twist contribution includes various terms, containing either subleading-twist distribution or fragmentation functions, that have to be considered as well.[16]

3.2. *Beam-Spin Asymmetry*

The semi-inclusive cross section also exhibits an azimuthal modulation for longitudinally polarized beam and unpolarized targets. Similar to $\langle \sin\phi \rangle_{UL}^{\mathrm{q}}$, the $\langle \sin\phi \rangle_{LU}$ moment originates from subleading-twist expressions. The interest in $\langle \sin\phi \rangle_{LU}$ is quite high as it involves the subleading-twist DF $e(x)$. The latter is related to the pion–nucleon σ-term, which in turn is related to the strangeness content of the nucleon.[17] As $e(x)$ is a chiral-odd object, it has to couple to another chiral-odd object. In analogy to the situation of the chiral-odd transversity DF, one can couple $e(x)$ to the Collins FF, leading to a $\sin\phi$ modulation of the cross section dependent on the beam helicity. In order to isolate this $\sin\phi$ term, one again can look at SSA's. But this time, in contrast to the target-spin asymmetries presented earlier, it is the helicity of the beam that is flipped. By means of spin rotators in the HERA electron ring in front of and behind the spectrometer, this is easily accomplished at HERMES. As one is not interested in polarized protons as a target, the data from an unpolarized hydrogen target as well as the polarization-averaged data from the longitudinally polarized hydrogen target is used in this analysis.

In contrast to the target-spin asymmetries, the $\langle \sin\phi \rangle_{LU}$ moment was not extracted by a fit to the count-rate asymmetry (which could be formed similar to Eq. (1)) but by the so-called *moment method*:

$$A_{LU}^{\sin\phi} = \frac{1}{P_B} \frac{\sum_{i=1}^{N^+} \sin\phi_i - \sum_{i=1}^{N^-} \sin\phi_i}{\frac{1}{2}(N^+ + N^-)}, \qquad (5)$$

where $\pm$ denotes positive/negative helicity of the beam, $N^{\pm}$ is the number of selected events with a detected hadron for each beam-spin state and $P_B = 0.533 \pm 0.029$ is the mean absolute beam polarization measured by the two HERA polarimeters.

Preliminary results of the $A_{LU}^{\sin\phi}$ moments for charged and neutral π's

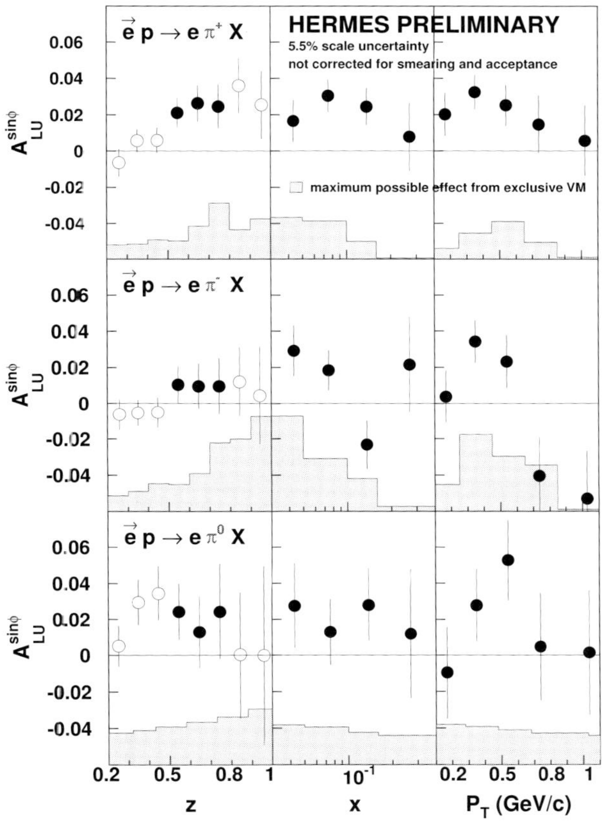

Fig. 5. The beam SSA on a hydrogen target as functions of z, x, and $P_{h\perp}$ for π mesons as indicated in the left panels. The preliminary results for the x and $P_{h\perp}$ dependences are obtained for the range $0.5 < z < 0.8$ (indicated by full circles in the left column). Error bars represent the statistical uncertainty only and the shaded bands represent upper limits for a possible uncertainty from the contribution of exclusive vector meson production (and subsequent decay) to the semi-inclusive pion sample. An additional 5.5% scale uncertainty is due to the beam polarization measurement.

are shown as functions of z, x, and $P_{h\perp}$ in Fig. 5. To reduce the contamination of events from the target fragmentation region as well as contributions from exclusive processes, a range of $0.5 < z < 0.8$ was chosen in the measurements of the x and $P_{h\perp}$ dependences (full circles in Fig. 5). In the above-mentioned z-range the measured moments are positive for π^+ and consistent with zero for the other two π's. This measurement extends a former CLAS measurement of $A_{LU}^{\sin\phi}$ at JLAB [18] to lower x and higher Q^2

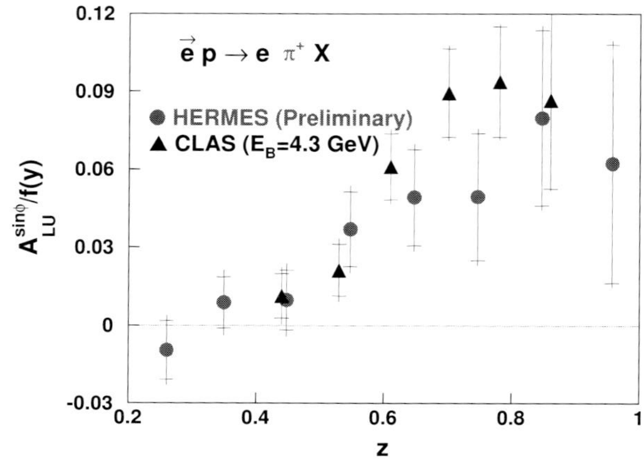

Fig. 6. Comparison of BSA results by HERMES and CLAS, after rescaling the HERMES results by the y-dependent function $f(y) = \frac{2y\sqrt{1-y}}{1-y+y^2/2}$ to take into account the different kinematic regions of the two experiments.

regions. The z dependence from both the HERMES and the CLAS measurements are in good agreement within statistical accuracy as shown in Fig. 6, after accounting for differences in the kinematics (as described in the caption).

A theoretical interpretation of these results in terms of $e(x)$ and the Collins FF is hampered by the possibility of many subleading-twist contributions to $A_{LU}^{\sin\phi}$. Similar to the subleading-twist contribution in the case of a longitudinally polarized target, various subleading-twist DF and FF contribute to the measured beam-spin asymmetry.[16] The disentanglement of these contributions requires further theoretical and experimental studies.

Another uncertainty comes through the contribution to the analyzed π sample from the decay of exclusively produced vector mesons. A sizeable fraction of these vector mesons cannot be reconstructed due to the limited detector acceptance. Therefore the π's into which they decay cannot be distinguished from "ordinarily produced" π's. The contribution from the decay of vector mesons in the semi-inclusive pion sample has been investigated using a PYTHIA Monte Carlo simulation. To estimate the maximal contribution to the asymmetry, positivity limits on the vector meson asymmetries, the asymmetry transfer to the decay pion, and the fraction of π's from vector mesons in the analyzed data sample have been used. The resulting uncertainties are indicated in Fig. 5 as an error band.

4. Outlook

The data presented here are only a first glimpse of the total data set which HERMES will take with a transversely polarized target. The complete data set will not only allow a more precise measurement of the presented Collins and Sivers moments but also the extraction of the involved DF's and FF's. More data will also be collected for the analysis of the beam-spin azimuthal asymmetries. Here, not only proton data can be analyzed but also data from the deuteron (and/or heavier targets).

Acknowledgments

I would like to thank Harut Avakian, Eduard Avetisyan, Alessandro Bacchetta, Matthias Burkardt and Markus Diehl for many useful and stimulating discussions. Special thanks goes also to the organizers of Transversity 2005, in particular, to Vincenzo Barone and Philip G. Ratcliffe, for this interesting workshop and the generous support.

References

1. A. Airapetian *et al.*, *Phys. Rev. Lett.* **84** (2000) 4047.
   A. Airapetian *et al.*, *Phys. Rev.* **D64** (2001) 097101.
   A. Airapetian *et al.*, *Phys. Lett.* **B562** (2003) 182.
2. A. Airapetian *et al.*, *Phys. Rev. Lett.* **94** (2005) 012002.
3. J.P. Ralston and D.E. Soper, *Nucl. Phys.* **B152** (1979) 109.
4. J. Collins, *Nucl. Phys.* **B396** (1993) 161.
5. D.W. Sivers, *Phys. Rev.* **D41** (1990) 83.
6. S.J. Brodsky, D.S. Hwang, and I. Schmidt, *Phys. Lett.* **B530** (2002) 99.
7. J.C. Collins, *Phys. Lett.* **B536** (2002) 43;
   X. Ji and F. Yuan, *Phys. Lett.* **B543** (2002) 66;
   A.V. Belitsky, X. Ji, and F. Yuan, *Nucl. Phys.* **B656** (2003) 165.
8. P.J. Mulders, R.D. Tangerman, *Nucl. Phys.* **B461** (1996) 197.
   D. Boer, P.J. Mulders, *Phys. Rev.* **D57** (1998) 5780.
9. A. Bacchetta, U. D'Alesio, M. Diehl, and C.A. Miller, *Phys. Rev.* **D70** (2004) 117504.
10. W. Vogelsang and F. Yuan, *Phys. Rev.* **D72** (2005) 054028.
11. V. Yu. Alexakhin *et al.*, *Phys. Rev. Lett.* **94** (2005) 202002.
12. M. Burkardt, *Nucl. Phys.* **A735** (2004) 185.
13. M. Burkardt and G. Schnell, arXiv:hep-ph/0510249.
14. M. Diehl and S. Sapeta, *Eur. Phys. J.* **C41** (2005) 515.
15. A. Airapetian *et al.*, *Phys. Lett.* **B622** (2005) 14.
16. A. Bacchetta, P.J. Mulders, and F. Pijlman, *Phys. Lett.* **B595** (2004) 309.
17. A.V. Efremov, K. Goeke, P. Schweitzer, *Phys. Rev.* **D67** (2003) 114014.
18. H. Avakian *et al.*, *Phys. Rev.* **D69** (2004) 112004.

COLLINS AND SIVERS ASYMMETRIES ON THE DEUTERON FROM COMPASS DATA

I. Horn
(on behalf of the COMPASS Collaboration)
Helmholtz-Institut für Strahlen- und Kernphysik
Nussallee 14–16
53115 Bonn, Germany
E-mail: igor.horn@cern.ch

COMPASS is a fixed polarised target experiment presently running at CERN. In 2002, 2003, and 2004 a 160 GeV/c polarised μ^+ beam was utilized coming from SPS and scattered off a ^{6}LiD (deuteron) target. The nucleons in the target can be polarised either longitudinally or transversely with respect to the μ^+ beam. Around 20% of the running time has been dedicated to transverse polarisation measurements on the deuteron target. The final results for the Collins and the Sivers asymmetries extracted from the 2002 data are presented. Error estimates are shown for the data taken in 2003 and 2004. The COMPASS collaboration plans a run with a NH_3 (proton) target in 2006. Estimated projections for the statistical accuracy on the proton are also given.

1. Introduction

The transversity distribution $\Delta_T q(x) = q(x)_\uparrow - q(x)_\downarrow$ supplies together with the unpolarised distribution functions $q(x)$ and the helicity distribution $\Delta q(x)$ a description of the quark structure of the nucleon at the twist-two (leading twist) level. The inclusive cross-section for deep inelastic scattering of longitudinally polarised leptons on polarised nucleons, at the leading twist, can be written as a function of $q(x)$, $\Delta q(x)$ and $\Delta_T q(x)$.[1] $\Delta_T q(x)$ is chiral-odd and can be extracted from the Drell–Yan process in polarised nucleon–nucleon scattering or from semi-inclusive deep inelastic scattering (SIDIS) on a transversely polarised target, where a part of the hadronic system is detected in final state. In SIDIS the convolution of $\Delta_T q(x)$ with the chiral-odd Collins fragmentation function $\Delta D_q^h(z, p_T^h)$ is measured via azimuthal single spin asymmetries (SSA).[2] Another mechanism of generating SSA was proposed by Sivers [3] which takes into account the dependence of the nucleon structure on the intrinsic quark transverse momentum k_T.

This effect is described by Sivers distribution function, $\Delta_T^0 q(x, k_T)$. Collins and Sivers effects are functions of linearly independent kinematic variables and, therefore, decouple from each other in SIDIS on transversely polarised nucleons. In general, there are several methods to study transversity at COMPASS. These are the azimuthal distributions of single hadrons (discussed here), the azimuthal dependence of the plane containing hadron pairs [4] and the measurements of transverse polarisation of Λ hyperons.[5]

2. Collins and Sivers angles

The Collins fragmentation function of a quark of flavour q in a hadron h can be written [6]

$$\Delta D_{\mathbf{q}}^h(z, \mathbf{p_T^h}) = D_q^h(z, p_T^h) + \Delta D_q^h(z, p_T^h) \cdot \sin \Phi_C, \tag{1}$$

where $\mathbf{p_T^h}$ is the hadron transverse momentum with respect to the virtual photon direction and $z = E_h/(E_l - E_{l'})$ is the fraction of available energy carried by the hadron, E_h is the energy of the hadron, E_l is the energy of the incoming lepton and $E_{l'}$ is the energy of the scattered lepton. Φ_C is so-called Collins angle. The latter is usually defined in the system where the z-axis is the virtual photon direction and the x–z plane is the muon scattering plane, see Fig. 1. In this frame $\Phi_C = \phi_h - \phi'_s$, where ϕ_h is the hadron

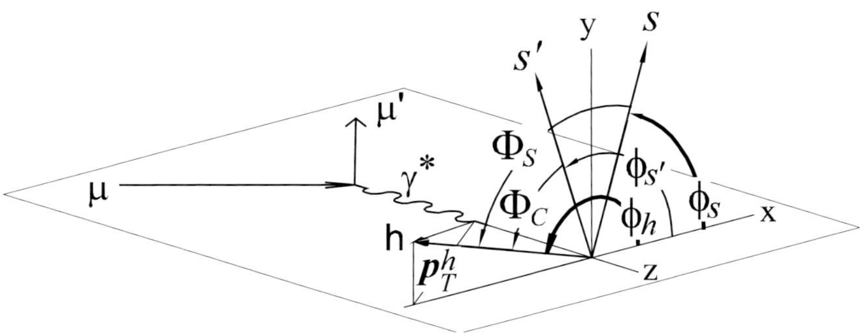

Fig. 1. Definition of Collins and Sivers angles in COMPASS.

azimuthal angle, and ϕ'_s is the azimuthal angle of the transverse spin of the struck quark. QED calculations are used to show that the components of the quark spin in initial and final state are related [7] $\phi'_s = \pi - \phi_s$, therefore $\Phi_C = \phi_h + \phi_s - \pi$. The fragmentation function $\Delta D_q^h(z, p_T^h)$ couples to transverse spin distribution function $\Delta_T q(x)$ and generates a SSA, A_{Coll}, which depends on x, z and p_T^h.

According to Sivers the difference in the probability of finding an unpolarised quark of transverse momentum $\mathbf{k_T}$ and -$\mathbf{k_T}$ inside a polarised nucleon can be expressed in the following way:[8]

$$P_{q/p^\uparrow}(x, \mathbf{k_T}) - P_{q/p^\uparrow}(x, -\mathbf{k_T}) = \sin\Phi_S \cdot \Delta_0^T q(x, k_T^2), \tag{2}$$

where $\Phi_S = \phi_h - \phi_s$ is the azimuthal angle of the quark with respect to the nucleon transverse spin orientation. The SSA A_{Siv} has a non-vanishing contribution at leading twist [9,10] due to coupling of the Sivers function to the unpolarised fragmentation function $D_q^h(z, p_T^h)$.

3. COMPASS experiment and data analysis

COMPASS [11,12] is a fixed target experiment located in CERN. A 160 GeV μ^+ beam, naturally polarised by the π^+ decay, with an intensity of $2{\cdot}10^8$/spill is used, with a 4.8 s spill every 16.2 s. The layout of the COMPASS experiment is shown in Fig. 2. Main components of the COMPASS

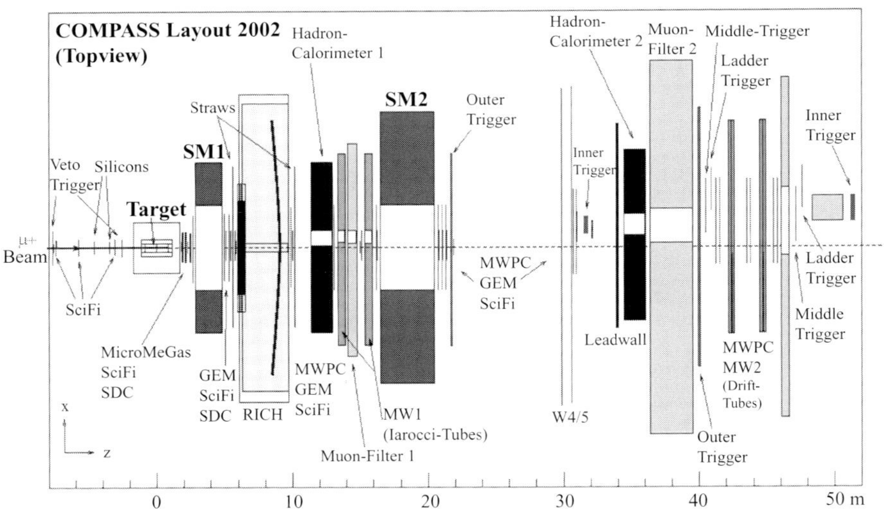

Fig. 2. Schematic view of the COMPASS experiment .

spectrometer are the polarised ^{6}LiD target and two spectrometer magnets: SM1 and SM2. The polarised target consists of two ^{6}LiD cells, each 60 cm long, located along the beam one after the other in two separate RF cavities. Data are taken simultaneously on the two target cells which are oppositely polarised. The polarisation is reversed once a week when operating in the transverse polarisation mode. SM1, the large-angle spectrometer magnet (up to 180 mrad), has an integrated field-strength of 1.0 Tm and enables the measurement of particles of lower momentum. SM2, the small-angle spectrometer magnet (up to 30 mrad), has an integrated field-strength of 4.4 Tm and measures particles with higher momenta. Both spectrometer stages are equipped with detectors for track reconstruction and particle identification.

20% of the total beam-time in 2002, 2003 and 2004 was devoted to the run with the transversely polarised deuteron target. The accumulated sample of 2002 data with transverse polarisation of the target comprises $6 \cdot 10^9$ events. The analysis discussed below refers to this sample. The sample was taken during two separate periods. In each period a polarisation was reversed after 4–5 days by changing the RF frequencies in two target cells.

To select semi-inclusive events, an incoming and scattered muon (primary vertex) plus at least one hadron from this vertex were required. Muon identification was performed with a muon filters, consisting of a large amount of material to pass. To ensure a DIS event, the kinematic cuts $Q^2 > 1\,(\mathrm{GeV}/c)^2$, $W > 5\,\mathrm{GeV}/c^2$ and $0.1 < y < 0.9$ were applied to the data. The upper limit on y was applied to keep radiative corrections small. In addition the transverse momentum cut $p_T^h > 0.1\,\mathrm{GeV}/c$ was applied to unambiguously calculate angles. Asymmetries have been extracted for the hadron with the highest energy (leading, $z > 0.25$) as well as for all detected hadrons with $z > 0.2$.

Collins and Sivers asymmetries were fitted separately. The number of events for each cell is given by

$$N(\Phi_{C/S}) = N_0 \cdot \alpha(\Phi_{C/S})(1 + \epsilon_{C/S} \sin \Phi_{C/S}), \tag{3}$$

where $\epsilon_{C/S}$ is the amplitude of the experimental asymmetry, α is a function which depends on the detector acceptance, and N_0 is product of the muon flux, the number of particles in the target and the spin averaged cross-section. The amplitude can be written as a function of the Collins and Sivers asymmetries:

$$\epsilon_C = A_{\mathrm{Coll}} \cdot P_T \cdot f \cdot D_{NN} \qquad \epsilon_S = A_{\mathrm{Siv}} \cdot P_T \cdot f, \tag{4}$$

where P_T (≈ 0.5) is the target polarisation, $D_{NN} = (1-y)/(1-y+y^2/2)$ is the transverse spin transfer coefficient (for Sivers $D_{NN} = 1$), and f (≈ 0.4)

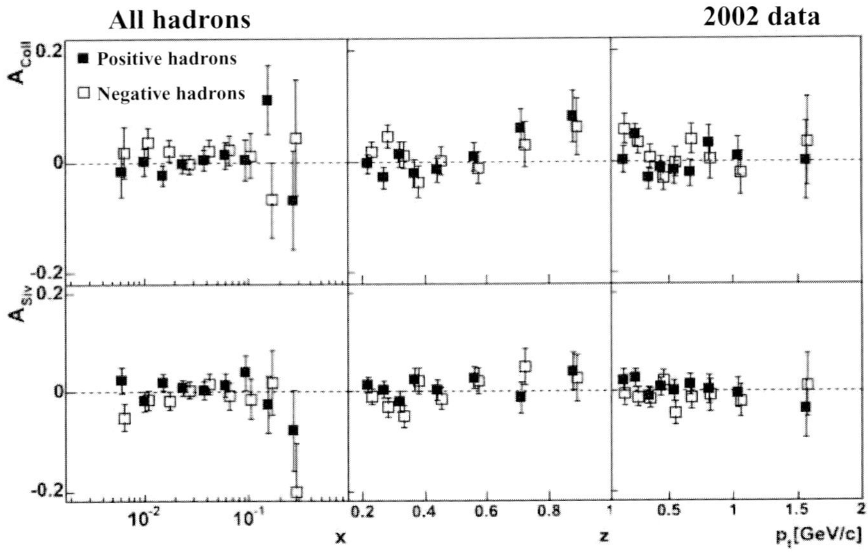

Fig. 3. Collins (upper row) and Sivers (lower row) asymmetry for positive (filled squares) and negative (open squares) all hadrons as a function of x, z and p_T^h (left to right). Only statistical errors are shown.

is the target dilution factor. The asymmetry $\epsilon_{C/S}$ is fitted separately for the two target cells with two opposite spin orientations:

$$\epsilon_{C/S} \sin \Phi_{C/S} = \frac{N_h^{\uparrow}(\Phi_{C/S}) - R \cdot N_h^{\downarrow}(\Phi_{C/S})}{N_h^{\uparrow}(\Phi_{C/S}) - R \cdot N_h^{\downarrow}(\Phi_{C/S})}. \tag{5}$$

$R = N_{h,\text{tot}}^{\uparrow}/N_{h,\text{tot}}^{\downarrow}$ is a normalisation factor and corresponds to the ratio of the total number of events with two target polarisation orientations. The obtained results for the asymmetries are plotted against the kinematic variables x, z and p_T^h (see Fig. 3 for all hadrons and for leading hadrons see Fig. 4).[13] Only statistical errors are shown. Filled squares correspond to the positively charged hadrons and unfilled squares points correspond to the negatively charged hadrons.

Possible sources for systematic errors have been thoroughly investigated and various tests were performed. It was concluded that systematic errors are smaller than the quoted statistical errors.[13]

Extensive Monte Carlo (MC) studies were performed and a good agreement between MC and data has been achieved. MC investigations have

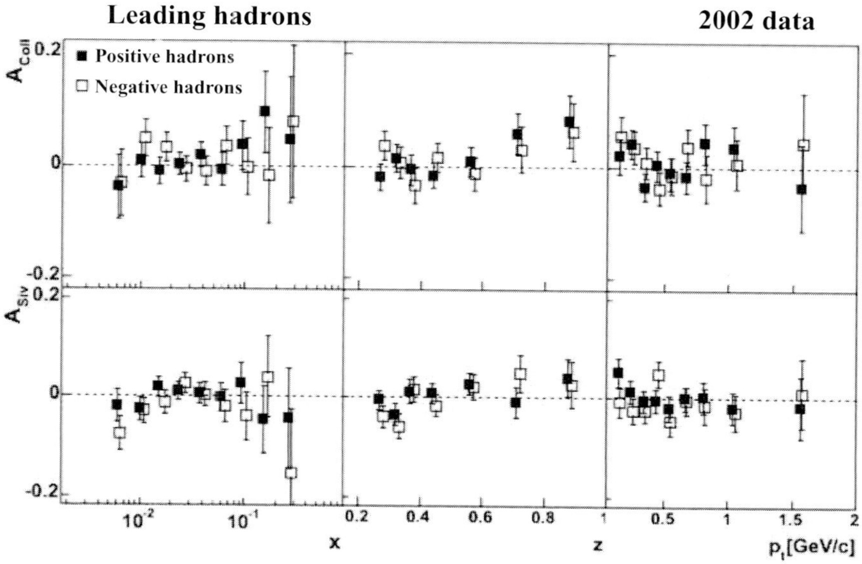

Fig. 4. Collins (upper row) and Sivers (lower row) asymmetry for positive (filled squares) and negative (open squares) leading hadrons as a function of x, z and p_T^h (left to right). Only statistical errors are shown.

shown that around 20% of correctly reconstructed leading hadrons are not pions but mainly kaons and protons.

4. Results and discussions

As we can see in Figs. 3 and 4 both the Collins and Sivers asymmetries are small and compatible with zero. This may hint to a cancellation between proton and neutron or the Collins mechanism is too small. However, if the Collins function $\Delta D_q^h(z, p_T^h)$ is not zero and as large as indicated from preliminary results by the BELLE [14,15] collaboration, then this is an evidence for the cancellation in the isoscalar target.

Recent phenomenological fits were performed on HERMES and COMPASS data [16,17] for Sivers asymmetries and for Collins asymmetries.[18] The compatibility of HERMES results [19] for protons and COMPASS results on the deuteron target was demonstrated.

In 2003 and 2004 COMPASS has collected data on a transversely polarised deuteron target. The estimated statistical errors for the Collins asymmetries from all 2002–2004 data are compared with the statistical errors of the published data in Fig. 5 (top plots). Apart from that, it is possible to separate π, K and p employing the information from RICH (Ring Imaging Čerenkov Detector) available for 2003 and 2004 data. The COMPASS collaboration plans to take data on a transversely polarised proton target (NH_3) in 2006. The statistical accuracy for the measurements of $A_{\rm Coll}$ on both the deuteron (2002–2004 data) and proton targets, assuming 30 days of data taking, is shown in Fig. 5 as a function of x. The current setup was implied.

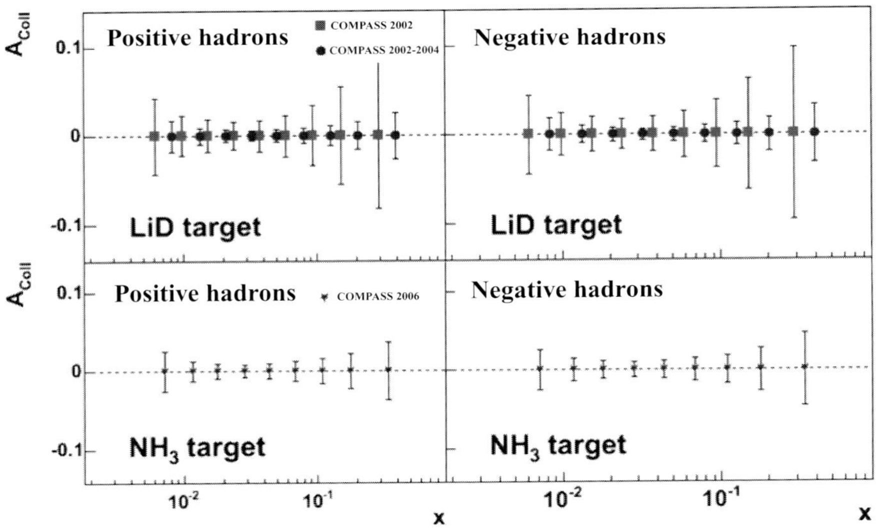

Fig. 5. Estimate of the statistical errors for $A_{\rm Coll}$ as a function of x for deuterium (top) and proton (bottom) targets for positive (left) and negative hadrons (right).

COMPASS 2002 measurements have shown that $A_{\rm Coll}$ and $A_{\rm Siv}$ are small and almost vanishing. The COMPASS data on the proton target (NH_3) will be available after the 2006 run. These data with the new measurements of the Collins function reported by BELLE collaboration [14,15] will allow a flavour separation, giving a precise information about a transverse spin structure of the nucleon.

References

1. V. Barone, A. Drago and P.G. Ratcliffe, *Phys. Rep.* **359** (2002) 1.

2. J.C. Collins, *Nucl. Phys.* **B396** (1993) 161.
3. D.W. Sivers, *Phys. Rev.* **D41** (1990) 83.
4. A. Mielech, *Two-hadron asymmetries*, these proceedings.
5. A. Ferrero, *Lambda asymmetries*, these proceedings.
6. G. Baum *et al.*, CERN-SPSLC-96-14.
7. V. Barone and F.G. Ratcliffe, *Transverse spin physics*, World Scientific, 2003.
8. M. Anselmino, V. Barone, A. Drago and F. Murgia, arXiv:hep-ph/0209073.
9. S.J. Brodsky, D S. Hwang and I. Schmidt, *Phys. Lett.* **B530** (2002) 99.
10. J.C. Collins, *Phys. Lett.* **B536** (2002) 43.
11. G.K. Mallot, *Nucl. Instrum. Meth.* **A518** (2004) 121.
12. F. Bradamante [COMPASS Collaboration], *Prog. Part. Nucl. Phys.* **55** (2005) 270.
13. V.Yu. Alexakhin *et al.* [COMPASS Collaboration], *Phys. Rev. Lett.* **94** (2005) 202002.
14. R. Seidl *et al.*, *Measurements of chiral-odd fragmentation functions at BELLE*, these proceedings.
15. K. Abe *et al.* [Belle Collaboration], arXiv:hep-ex/0507063.
16. M. Anselmino *et al.*, *Phys. Rev.* **D71** (2005) 074006.
17. M. Anselmino *et al.*, arXiv:hep-ph/0507181.
18. W. Vogelsang and F. Yuan, *Phys. Rev.* **D72** (2005) 054028.
19. A. Airapetian *et al.* [HERMES Collaboration], *Phys. Rev. Lett.* **94** (2005) 012002.

FIRST MEASUREMENT OF INTERFERENCE FRAGMENTATION ON A TRANSVERSELY POLARIZED HYDROGEN TARGET

P.B. van der Nat
(on behalf of the HERMES collaboration)

Nationaal Instituut voor Kernfysica en Hoge-Energiefysica (NIKHEF), P.O. Box 41882, 1009 DB Amsterdam, The Netherlands
E-mail: natp@nikhef.nl

The HERMES experiment has measured for the first time single target-spin asymmetries in semi-inclusive two-pion production using a transversely polarized hydrogen target. These asymmetries are related to the product of two unknowns, the transversity distribution function and the interference fragmentation function. In the invariant mass range $0.51\,\text{GeV} < M_{\pi^+\pi^-} < 0.97\,\text{GeV}$ the measured asymmetry deviates significantly from zero, indicating that two-pion semi-inclusive deep-inelastic scattering can be used to probe transversity.

1. Introduction

An important missing piece in our understanding of the spin structure of the nucleon is the transversity distribution $h_1(x)$. It is the only one of the three leading-twist quark distribution functions, $f_1(x)$, $g_1(x)$ and $h_1(x)$, that so-far remains unmeasured. The function $h_1(x)$ describes the distribution of transversely polarized quarks in a transversely polarized nucleon. It is quite difficult to measure $h_1(x)$, since it is a chiral-odd function, which can only be probed in combination with another chiral-odd function. This can be done in semi-inclusive DIS, where the second chiral-odd object is a fragmentation function, describing the fragmentation of the struck quark into one or more final-state hadrons.

HERMES is one of the pioneering experiments on this subject. The structure function $h_1(x)$ is probed by measuring various single-spin asymmetries. First, a longitudinally polarized target [1] was used and more recently a transversely polarized target was used.[2] In these experiments, single spin asymmetries (SSA's) were only measured for *single-hadron* semi-inclusive DIS (SIDIS). However, already in 1993 Collins *et al.*[3] and in 1998

Jaffe *et al.*[4] suggested to study transversity in two-hadron SIDIS. Although this comes at the expense of a larger statistical uncertainty, there is a good reason for looking at SSA's in two-hadron SIDIS: the measured SSA's relate directly to the product of $h_1(x)$ and the fragmentation function, whereas in single-hadron SIDIS this product is convoluted with the transverse momentum of the hadron. Also measuring SSA's in two-hadron SIDIS provides an independent method of measuring $h_1(x)$, since it involves a different fragmentation function as compared to single-hadron SIDIS.

In order to finally extract the structure function $h_1(x)$, one needs to know the value of the involved fragmentation function. Although this function is also still unknown, it can be cleanly measured in e^+e^- experiments, such as Belle and Babar.

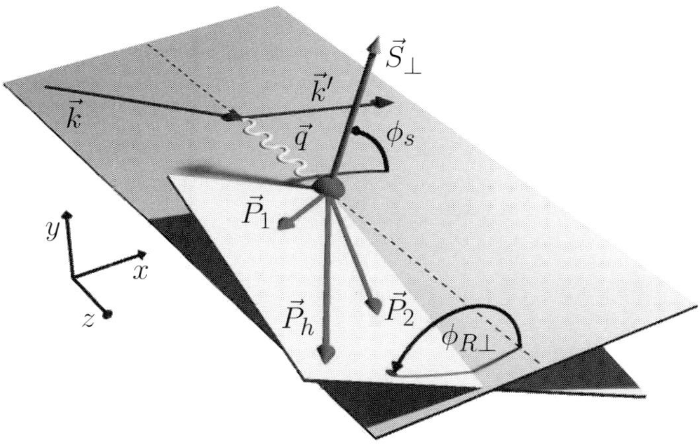

Fig. 1. Kinematic planes, where $\phi_{R\perp}$ is the angle between the plane spanned by the incident ($\vec{k}$) and scattered lepton ($\vec{k}'$) and the plane spanned by the two detected pions $\vec{P}_1$ (π^+) and $\vec{P}_2$ (π^-) with $\vec{P}_h \equiv \vec{P}_1 + \vec{P}_2$.

2. Single-Spin Asymmetry

The transversity distribution can be accessed experimentally by measuring the single target-spin asymmetry, defined as:

$$A_{UT}(\phi_{R\perp}, \phi_S, \theta) = \frac{1}{|S_T|} \frac{N^\uparrow(\phi_{R\perp}, \phi_S, \theta)/N^\uparrow_{\rm DIS} - N^\downarrow(\phi_{R\perp}, \phi_S, \theta)/N^\downarrow_{\rm DIS}}{N^\uparrow(\phi_{R\perp}, \phi_S, \theta)/N^\uparrow_{\rm DIS} + N^\downarrow(\phi_{R\perp}, \phi_S, \theta)/N^\downarrow_{\rm DIS}} = \frac{\sigma_{UT}}{\sigma_{UU}}, \tag{1}$$

where UT refers to Unpolarized beam and Transversely polarized target. The asymmetry is evaluated as a function of the angles $\phi_{R\perp}$, ϕ_S and θ which are defined in Fig. 1[a] and 2. Explicitly:

$$\phi_{R\perp} = \frac{\vec{q} \times \vec{k} \cdot \vec{R_T}}{|\vec{q} \times \vec{k} \cdot \vec{R_T}|} \cos^{-1} \frac{\vec{q} \times \vec{k} \cdot \vec{q} \times \vec{R_T}}{|\vec{q} \times \vec{k}||\vec{q} \times \vec{R_T}|} \tag{2}$$

and

$$\phi_S = \frac{\vec{q} \times \vec{k} \cdot \vec{S}_\perp}{|\vec{q} \times \vec{k} \cdot \vec{S}_\perp|} \cos^{-1} \frac{\vec{q} \times \vec{k} \cdot \vec{q} \times \vec{S}_\perp}{|\vec{q} \times \vec{k}||\vec{q} \times \vec{S}_\perp|}, \tag{3}$$

where R_T is the component of R ($\vec{R} \equiv (\vec{P_1} - \vec{P_2})/2$) perpendicular to P_h ($\vec{P_h} \equiv \vec{P_1} + \vec{P_2}$), i.e. $\vec{R_T} = R - (R \cdot \hat{P_h})\hat{P_h}$.

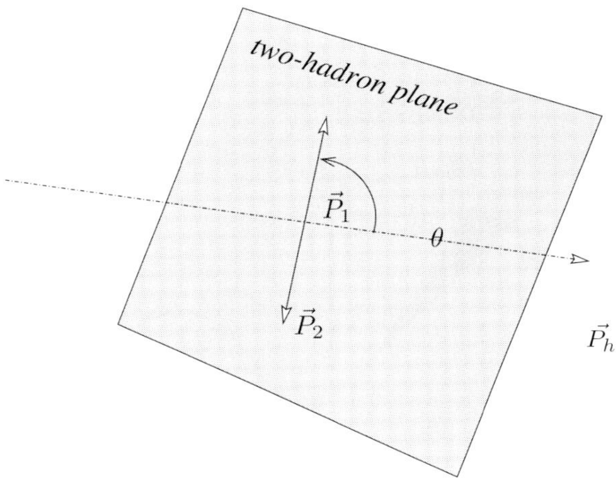

Fig. 2. Description of the polar angle θ, in the center-of-mass frame of the two pions. The vector $\vec{P_h}$ is evaluated in the hadronic center-of-mass system.

The azimuthal angle ϕ_S represents the spin direction of the target "↑" state and $N^{\uparrow(\downarrow)}(\phi_{R\perp}, \phi_S, \theta)$ is the number of semi-inclusive $\pi^+\pi^-$-pairs in the target ↑(↓) spin state. These numbers are normalized to the corresponding number of DIS events, $N^{\uparrow}_{\mathrm{DIS}}$ and $N^{\downarrow}_{\mathrm{DIS}}$, respectively. The quantity $|S_T|$ indicates the average target polarization. The asymmetry is equal to the ratio of σ_{UT} and σ_{UU}, which are the polarized and unpolarized cross sections, respectively. According to Bacchetta *et al.*[6] σ_{UT} can be written at

[a] The angle definitions are consistent with the "Trento Conventions".[5]

leading-twist[b] as:

$$\sigma_{UT} = -\sum_q \frac{\alpha^2 e_q^2}{2\pi Q^2 y}(1-y)|\vec{S}_\perp|\frac{|\vec{R}|}{M_{\pi\pi}} \sin(\phi_{R\perp}+\phi_S)\sin\theta\, h_{1,q}(x)$$
$$\times \left[H_{1,q}^{\sphericalangle,sp}(z,M_{\pi\pi}^2) + \cos\theta\, H_{1,q}^{\sphericalangle,pp}(z,M_{\pi\pi}^2)\right] , \quad (4)$$

where $|\vec{R}| = \frac{1}{2}\sqrt{M_{\pi\pi}^2 - 4M_\pi^2}$, with $M_{\pi\pi}$ the invariant mass of the pion pair, M_π the pion mass and x, y and z the standard scaling variables used in semi-inclusive DIS. The transversity distribution $h_1(x)$ couples to a combination of two-hadron interference fragmentation functions, $H_1^{\sphericalangle,sp}$ and $H_1^{\sphericalangle,pp}$. These functions describe the interference between different production channels of the pion pair; $H_1^{\sphericalangle,sp}$ relates to the interference between s- and p-wave states and $H_1^{\sphericalangle,pp}$ to the interference between two p-wave states.

A two-dimensional fit function of the form

$$f(\phi_{R\perp}+\phi_S,\theta) = p_0 + p_1 \sin(\phi_{R\perp}+\phi_S)\sin\theta \quad (5)$$

was used to extract from the measured asymmetry the part related to the product $h_1 H_1^{\sphericalangle,sp}$, where $p_1 \equiv A_{UT}^{\sin(\phi_{R\perp}+\phi_S)\sin\theta}$.

3. Results

The present results are based on data taken in the period from 2002 until 2004 using a transversely polarized hydrogen target in the HERMES experiment at DESY. The average target polarization, $|S_T|$, was $75.4 \pm 5.0\,\%$.

In Fig. 3 the data for $A_{UT}^{\sin(\phi_{R\perp}+\phi_S)\sin\theta}$ are shown versus the invariant mass of the $\pi^+\pi^-$-pair. The asymmetry is clearly positive over the entire invariant mass range and largest in the region of the ρ^0 mass. The corresponding invariant mass distribution is shown in the left plot of Fig. 4. Whereas the results on SSA's in two-hadron fragmentation using a *longitudinally* polarized deuterium target[8] gave a hint of a sign change of the asymmetry at the ρ^0 mass (0.770 GeV) as predicted in Ref. 4 the new results presented here are clearly inconsistent with such behavior. However, a good description of the data is given by a refined version[9] of a prediction which uses a spectator model for the fragmentation functions.[10]

In Fig. 5 the raw asymmetry is shown in bins of $\phi_{R\perp}+\phi_S$, integrated over the invariant mass range $0.51\,\text{GeV} < M_{\pi\pi} < 0.97\,\text{GeV}$. This plot shows that a clear $\sin(\phi_{R\perp}+\phi_S)$ behavior is present in the data. The plot includes

[b]See Ref. 7 for the sub-leading twist expression.

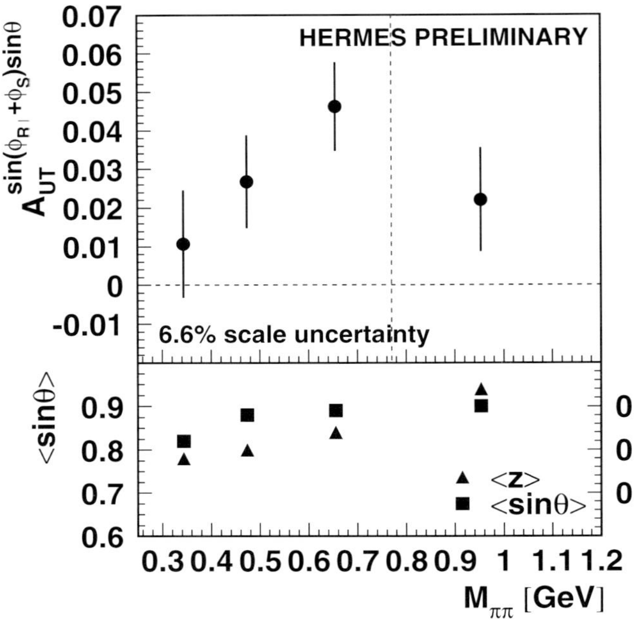

Fig. 3. The asymmetry $A_{UT}^{\sin(\phi_{R\perp}+\phi_S)\sin\theta}$ versus the invariant mass of the $\pi^+\pi^-$-pair (using mass binning, with the bin boundaries at 0.25, 0.40, 0.55, 0.77, 2.0 GeV).

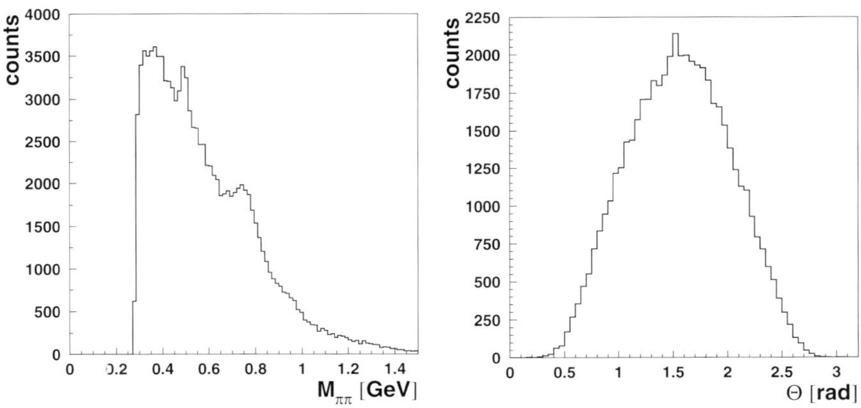

Fig. 4. The left plot shows the distribution of the invariant mass of the $\pi^+\pi^-$-pairs and the right plot shows the distribution of the angle θ (for the invariant mass range $0.51\,\text{GeV} < M_{\pi\pi} < 0.97\,\text{GeV}$.

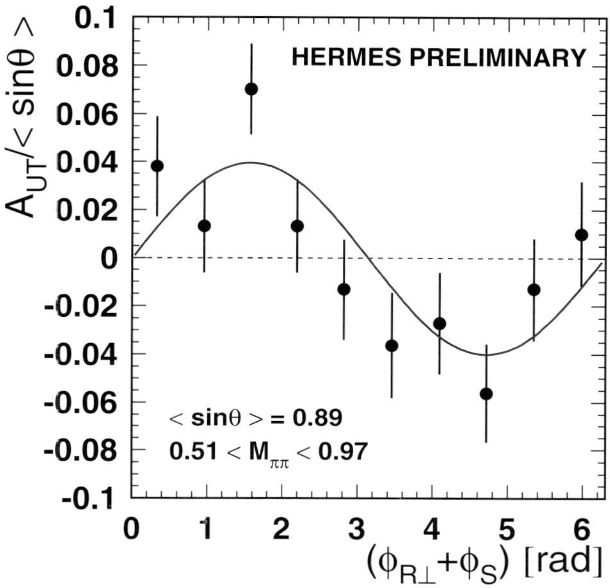

Fig. 5. The asymmetry A_{UT} divided by the average $\langle \sin\theta \rangle$ versus the angle combination $(\phi_{R\perp} + \phi_S)$.

a curve resulting from fitting the data with

$$f(\phi_{R\perp} + \phi_S) = p_0 + p_1 \sin(\phi_{R\perp} + \phi_S), \tag{6}$$

where

$$p_1 \equiv A_{UT}^{\sin(\phi_{R\perp}+\phi_S)\sin\theta} = 0.040 \pm 0.009\,(\mathrm{stat}) \pm 0.003\,(\mathrm{syst}). \tag{7}$$

Due to the peaked shape of the θ-distribution (right plot in Fig. 4) the asymmetry is mostly evaluated around $\theta = \frac{\pi}{2}$. Therefore the value of $A_{UT}^{\sin(\phi_{R\perp}+\phi_S)\sin\theta}$ is insensitive to whether one uses this one-dimensional fit function, integrating over θ, or a two-dimensional one, like Eq. (5).

Data taking with a transversely polarized hydrogen target will continue until November 2005 after which the analysis of the full data sample is expected to lead to a decrease of the uncertainty on the asymmetry with approximately a factor of $\sqrt{2}$. Further steps in the analysis include looking at the part of the asymmetry coupling to $H_1^{\sphericalangle,pp}$ and studying the x and z dependence of the asymmetries.

Acknowledgments

We acknowledge the support of the Dutch Foundation for Fundamenteel Onderzoek der Materie (FOM) and the European Community-Research Infrastructure Activity under the FP6 "Structuring the European Research Area" programme (Hadron Physics, contract number RII3-CT-2004-506078).

References

1. A. Airapetian *et al.* (HERMES), *Phys. Lett.* **B562** (2003) 182.
2. A. Airapetian *et al.* (HERMES), *Phys. Rev. Lett.* **94** (2005) 012002.
3. J.C. Collins, S.F. Heppelmann and G.A. Ladinsky, *Nucl. Phys.* **B420** (1994) 565.
4. R.L. Jaffe, X. Jin, and J. Tang. *Phys. Rev. Lett.* **80** (1998) 1166.
5. A. Bacchetta *et al.*, *Phys. Rev.* **D70** (2004) 117504.
6. A. Bacchetta and M. Radici, *Phys. Rev.* **D67** (2003) 094002.
7. A. Bacchetta and M. Radici, *Phys. Rev.* **D69** (2004) 074026.
8. P.B. van der Nat and K. Griffioen, *in proceedings of SPIN 2004* (2004), hep-ex/0501009.
9. M. Radici, these proceedings, hep-ph/0510165.
10. M. Radici, R. Jakob and A. Bianconi *Phys. Rev.* **D65** (2002) 074031.

TWO-HADRON ASYMMETRIES IN THE COMPASS EXPERIMENT

Adam Mielech
on behalf of the COMPASS Collaboration
INFN, Padriciano 99, 34100 Trieste, Italy
E-mail: mielech@ts.infn.it

The COMPASS experiment covers a broad physics program with muon and hadrons beams. The muon beam can be scattered on a longitudinally or transversely polarised target. In this article the production of hadron pairs in deep inelastic scattering of muons on transversely polarised ^{6}LiD target is described. Data selection cuts and results on the spin asymmetries are presented. The results are based on the data collected by the experiment in the years 2002 and 2003. Plans and prospects for the future are discussed, in particular in connection with hadron identification.

1. Two hadron production and transversity

The transverse spin distribution function $\Delta_T q(x)$, called "transversity", is a chiral-odd object. It is not accessible in inclusive DIS, but it can be measured together with another chiral-odd function. First attempts to address transversity have been made by measuring single spin asymmetries of hadrons produced in DIS processes on transversely polarised nucleons, where the asymmetry is expected to be the product of transversity and a spin-dependent fragmentation function, the "Collins function".[1] Our results obtained for the 2002 data are already published.[2]

Another process is semi-inclusive DIS where at least two hadrons are observed in the final state. Here the cross section at leading twist can be parametrised in terms of the convolution of transversity with an interference fragmentation function $H_1^{\triangleleft}(z, M_{\rm inv}^2)$. If a pair of hadrons is the result of the fragmentation of transversely polarised quark, an asymmetry $A_{\phi_{RS}}$ depending on the angle between the scattering plane and the 2 hadron plane is expected:

$$A_{\phi_{RS}} \propto \frac{\Sigma_i e_i^2 \Delta_T q_i(x) H_1^{\triangleleft}(z, M_{\rm inv}^2)}{\Sigma_i e_i^2 q_i(x) D_i^h(z, M_{\rm inv}^2)}. \qquad (1)$$

Here D_i^h is the spin independent fragmentation function, z is the sum of scaled hadron energies ($z = z_{h1} + z_{h2}$), M_{inv}^2 is the invariant mass of two hadrons and $q_i(x)$ is the unpolarised distribution function. The sum is over the quark flavours i. The expected properties of the interference fragmentation function and suggested experimental access to it are described in several publications.[1,3–8]

The angle ϕ_{RS} is defined as $\phi_{RS} = \phi_R + \phi_S - \pi$, where ϕ_R is the azimuthal angle of the $\mathbf{R_T}$ vector, $\mathbf{R_T}$ is the component of the vector of the difference of the two hadron momenta ($\mathbf{P_{h1}} - \mathbf{P_{h2}}$) perpendicular to their sum ($\mathbf{P_{h1}} + \mathbf{P_{h2}}$), ϕ_S is the azimuthal angle of initial quark spin.

The reference system for the measurement is defined by the scattering plane of the lepton beam ($\mathbf{l}, \mathbf{l}'$) and the virtual photon ($\mathbf{q}$) direction as shown in Fig. 1.

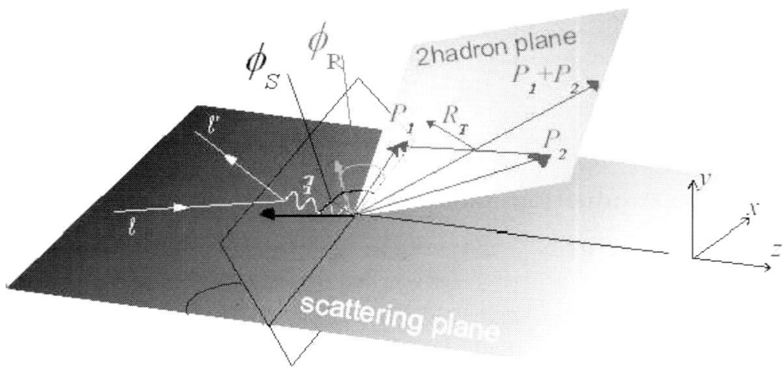

Fig. 1. Reference system and angles definitions.

The asymmetry (1) is related to the experimentally measured counting rate asymmetry, which is defined in the following way:

$$\begin{aligned} A_m(\phi_{RS}) &= \frac{N^\uparrow(\phi_{RS}) - rN^\downarrow(\phi_{RS})}{N^\uparrow(\phi_{RS}) + rN^\downarrow(\phi_{RS})} \\ &= A_{UT}^{\sin\phi_{RS}} \sin\phi_{RS} \\ &= DPfA_{\phi_{RS}} \sin\phi_{RS}, \end{aligned} \tag{2}$$

where $N^{\uparrow(\downarrow)}$ is the number of events for the target with the up (down) polarisation orientation in the laboratory system, $r = \frac{N_{\text{tot}}^\uparrow}{N_{\text{tot}}^\downarrow}$ is the ratio of the total number of events (i.e. integrated over the angle ϕ_{RS}) for the two polarisation orientations, $D = \frac{(1-y)}{(1-y+y^2/2)}$ is the depolarisation factor, P is the target polarisation, f is the dilution factor of the polarised target.

The asymmetries $A_{UT}^{\sin\phi_{RS}}$ are obtained from the fit to the $A_m(\phi_{RS})$ distributions.

2. COMPASS experiment

The COMPASS experiment [9,10] covers a broad physics program with muon and hadrons beams. The experimental setup of COMPASS for the muon program consists of the transversely (or longitudinally) polarised target and of a two-stage magnetic spectrometer with particle identification and hadron calorimeters.

The muon beam has the energy of 160 GeV. The target consists of two 60 cm long cells filled with ^{6}LiD which has a dilution factor of 40%. The target material is polarised to about 50%. The polarisation of the material in the first cell is opposite to the one in the second cell, and periodically (once a week) reversed in both cells. This configuration allows cancellation of the acceptance effects in the measured asymmetry.

The first stage of the COMPASS spectrometer is used for the low momenta and large angle tracks reconstruction. High momentum tracks are reconstructed in the second stage. The tracking system consists of various type and size detectors and include small ($50 \times 70\,\mathrm{mm}^2$) silicon and scintillating fiber planes in the beam region, medium size drift, proportional and novel micromesh chambers and large ($3.2 \times 2.8\,\mathrm{m}^2$) straw tube trackers. A set of hodoscopes allows to trigger on events satisfying a suitable correlation between the scattering angle and the momentum loss of the scattered muon. Large angle events are triggered on the basis of the energy deposited in the hadron calorimeter.

Particle identification (not used for the presented results on asymmetry), is provided by a RICH detector, which allows to identify pion, kaon and proton tracks of momenta above 3, 10 and 17 GeV respectively.

3. Data selection

3.1. *Event selection*

In this article the data collected on the transversely polarised ^{6}LiD target during the years 2002 and 2003 are presented. The following event selections are applied: Properly reconstructed events have an incident and a scattered muon and a primary vertex inside the target cells.

DIS events are defined by selection $Q^2 > 1\,\mathrm{GeV}^2$. Further cuts are performed on the scaled photon energy: $0.1 < y < 0.9$. The final requirement is that the events contain at least one reconstructed hadron pair with oppositely charged hadrons.

3.2. *Muon and hadron selection*

There is a necessity to distinguish muons and hadrons among the tracks outgoing from the primary vertex. The muon identification procedure uses muon filter detectors and the hadron calorimeter signal. Moreover, the condition that tracks should have more than 30 radiation length in the spectrometer is imposed.

Hadrons are all particles not identified as muons. Current fragmentation region is assured by cuts on $z_h > 0.1$ and $x_F > 0.1$ for each hadron. In order to remove exclusive mesons a cut on $z < 0.9$ is applied. An additional selection on the $\mathbf{R_T} > 0.05\,\mathrm{GeV}$ is performed in order to have well defined angles.

4. Results

The resulting asymmetries are presented in Fig. 2 as functions of the three variables: x, z, and M_{inv}. For the invariant mass calculation, all hadrons are assumed to be pions.

The results are compatible with zero. The indicated errors are statistical. The size of the systematics errors were estimated by evaluating "false asymmetries" of the data. They were obtained by scrambling the data with opposite polarisations into fake configurations. The extracted false asymmetries are compatible with zero and their statistical errors are of the same size as the statistical uncertainties of the physics result.

5. Future prospects – hadron identification

In the nearest future we plan to investigate other two-hadron asymmetries, mainly for identified charged hadron pairs: $\pi\pi$, $K\pi$ and KK. So far some investigations were performed, concerning the possibility separating different hadron flavours using the COMPASS RICH,[11] which has been fully operational for the COMPASS transversal data taking starting from the year 2003.

The effect of the RICH identification of hadrons is illustrated on Fig. 3, which refers to the data collected in the year 2003: the left plot shows the

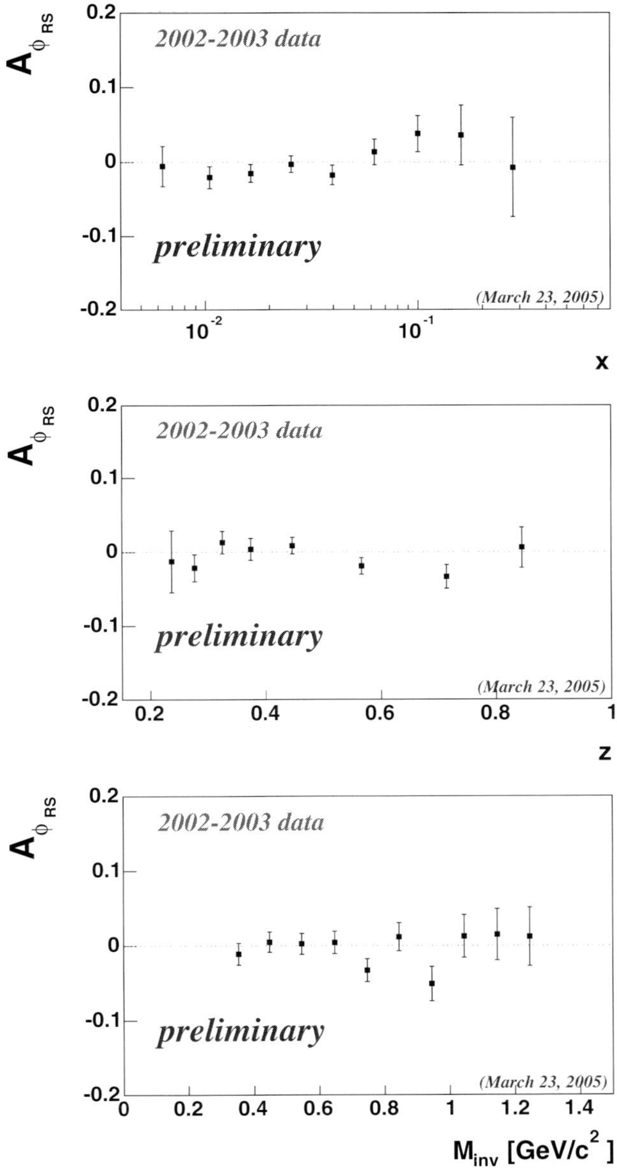

Fig. 2. Asymmetry as a function of x_{Bjorken} (upper plot), $z = z_1 + z_2$ (middle plot), and two-hadron invariant mass m_{inv} (lower plot).

invariant mass distribution comparison for all reconstructed hadron combinations and for the combinations where both particles have an identification answer of RICH. The fraction of the combinations with such answer is 74% of all reconstructed pairs. In the plot to the right one can see the effect of hadron identification: in the three histograms the invariant mass is evaluated following the mass assignments as given by the RICH.

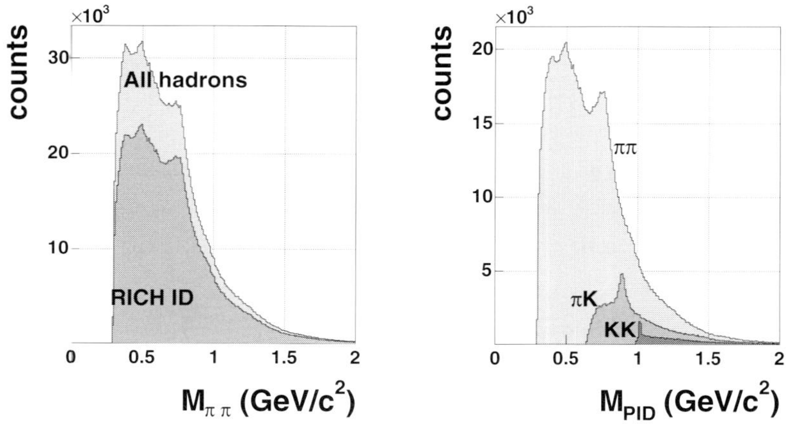

Fig. 3. Invariant mass distribution of two hadrons for the COMPASS 2003 data with transverse target polarisation. The light histogram on the left plot shows the invariant mass of all reconstructed hadron pairs. The dark histogram on the left plot shows only combinations where both hadrons are identified by RICH. In both cases for both hadrons π mass is assumed. On the right plot: histograms of the invariant masses of hadron pairs with mass hypothesis for each hadron as given by the RICH.

References

1. J.R. Collins, S.F. Heppemann and G.A. Ladinsky, *Nucl. Phys.* **B420** (1994) 565.
2. V.Yu. Alexakhin *et al.* (COMPASS Collaboration), *Phys. Rev. Lett.* **94** (2005) 202002 and I. Horn, these proceedings.
3. X. Artru and J.R. Collins, *Z. Phys.* **C69** (1996) 277.
4. R.L. Jaffe, X. Jin and J. Tang, *Phys. Rev. Lett.* **80** (1998) 1166.
5. M. Radici, R. Jakob, A. Bianconi, *Phys. Rev.* **D65** (2002) 074031.
6. A. Bianconi, S. Boffi, R. Jakob and M. Radici, *Phys. Rev.* **D62** (2000) 034008.
7. A. Bacchetta and M. Radici, *Phys. Rev.* **D69** (2004) 074026.
8. A. Bacchetta and M. Radici, Proc. of the *DIS2004*, hep-ph/0407354 (2004).
9. G. Baum *et al.* (COMPASS Collaboration), CERN-SPSLC-96-14.
10. G.K. Mallot, *Nucl. Instrum. Meth.* **A518** (2004) 121.
11. E. Albrecht *et al.*, accepted for publication in *Nucl. Instrum. Meth.* **A553** (2005) 215 and references therein.

MEASUREMENTS OF CHIRAL-ODD FRAGMENTATION FUNCTIONS AT BELLE

R. Seidl, M. Grosse Perdekamp, D. Gabbert

University of Illinois at Urbana-Champaign
1100 W. Green Street,
Urbana, IL 61801, USA
RIKEN BNL Research Center
Upton, NY 11973-5000, USA
E-mail: rseidl@uiuc.edu, mgp@uiuc.edu, dgabbert@uiuc.edu

A. Ogawa

Brookhaven National Laboratory
RIKEN BNL Research Center
Upton, NY 11973-5000, USA
E-mail: akio@bnl.gov

K. Hasuko

RIKEN
Wako, Saitama,351-0198, Japan

Measurements of the so far unknown chiral-odd quark transverse spin distribution in either semi-inclusive Deep Inelastic Scattering(SIDIS) or inclusive measurements in pp collisions at RHIC will require the knowledge of chiral-odd fragmentation functions which serve as analyzer for transverse quark spin. Examples for these chiral-odd fragmentation functions are the so-called Collins fragmentation functions or the two-hadron interference fragmentation functions. The HERMES experiment has provided first evidence that transversity distributions and the related fragmentation functions might be different from zero. However, in order to extract quark transversity distributions from the transverse spin asymmetries observed in HERMES independent measurements of the relevant fragmentation functions will be required. These measurements can be carried our in e^+e^- annihilation into hadrons. We present a first measurement of Collins asymmetries with the Belle experiment using a data sample of $29.0\,\mathrm{fb}^{-1}$.

1. Introduction

At leading twist three quark distribution functions (DF) describe the nucleon structure. While helicity-average and helicity-difference distributions have been studied experimentally the helicity flip transversity distributions remain unknown. The latter cannot be measured in inclusive DIS due to their chiral-odd nature, since all possible interactions are chiral-even for nearly massless quarks. Therefore one needs an additional chiral-odd function in the cross section to access transversity. This can be either achieved by an anti quark transversity DF in double transversely polarized Drell–Yan processes or the combination with a chiral-odd fragmentation function in semi-inclusive processes in either transversely polarized DIS or proton–proton collisions. The most prominent members of chiral-odd fragmentation functions are the so-called Collins [1] fragmentation function and the interference fragmentation function.[2]

2. The Belle experiment

The Belle [3,4] experiment at the asymmetric e^+e^- collider KEK-B [5] in Tsukuba, Japan, has been designed to study CP violation in B meson decays. Its center of mass energy is usually set to the $\Upsilon(4S)$ resonance at $\sqrt{s} = 10.58\,\mathrm{GeV}$. However, part of the data was recorded 60 MeV below the resonance for background studies. We have used this off-resonance data sample to measure spin dependent fragmentation functions. At the present time an integrated luminosity of $47.1\,\mathrm{fb}^{-1}$ has been accumulated in the off-resonance data sample.

The aerogel Čerenkov counter (ACC), time-of-flight (TOF) detector and the central drift chamber (CDC) enable a good particle identification and tracking, which is crucial for these fragmentation function measurements. Using the information from the silicon vertex detector (SVD), tracks directly originating from the interaction vertex are selected. This selection criteria reduced contributions from hadronic decays of heavy mesons.

Radiative events with hard gluons emitted from the primordial quarks are eliminated with a cut on the thrust variable, $T > 0.8$. This enhances the fraction of events with 2-jet topology and the thrust axis is used as approximation of the original quark direction. To ensure that the pions did not originate from the decay of a vector meson and might be mistakenly put in the wrong hemisphere a lower cut on the fractional energy of 0.2 is performed.

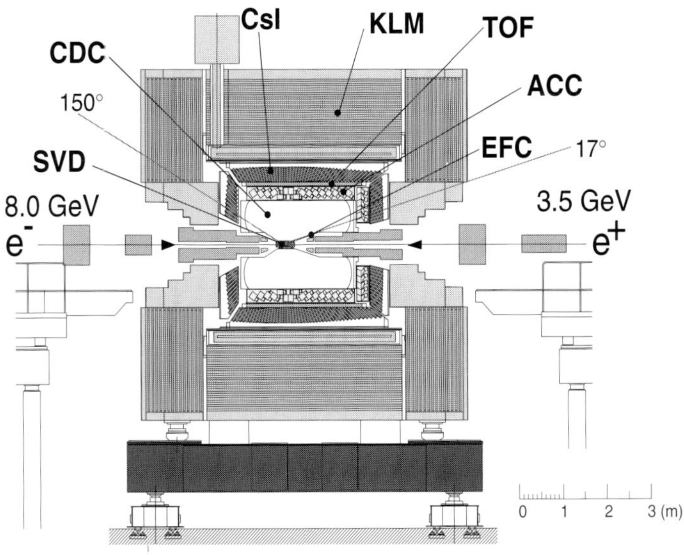

Fig. 1. A schematic side view of the Belle detector.

3. Collins Fragmentation Function

The Collins effect occurs in the fragmentation of a transversely polarized quark with polarization $\mathbf{S_q}$ and 3-momentum $\mathbf{k}$ into an unpolarized hadron of transverse momentum $\mathbf{P}_{h\perp}$ with respect to the original quark direction. According to the Trento convention [6] the number density for finding an unpolarized hadron h produced from a transversely polarized quark q is defined as:

$$D_{hq^\uparrow}(z, P_{h\perp}) = D_1^q(z, P_{h\perp}^2) + H_1^{\perp q}(z, P_{h\perp}^2)\frac{(\hat{\mathbf{k}} \times \mathbf{P}_{h\perp}) \cdot \mathbf{S}_q}{zM_h}, \tag{1}$$

where the first term describes the unpolarized FF $D_1^q(z, P_{h\perp}^2)$, with $z \stackrel{CMS}{=} \frac{2E_h}{Q}$ being the fractional energy the hadron carries relative to half of the CMS energy Q. The second term, containing the Collins function $H_1^{\perp q}(z, P_{h\perp}^2)$, depends on the spin of the quark and thus leads to an asymmetry as it changes sign under flipping the quark spin. The vector product causes a $\sin(\phi)$ modulation in the azimuthal distribution of the hadron yields around the original quark momentum axis. ϕ is the azimuthal angle between the transverse momentum of the hadron and the plane defined by the quark spin and its momentum. In e^+e^- hadron production the

Collins effect can be observed by correlating the fragmentation of quark and anti-quark in opposing hemispheres. Combining two hadrons from different hemispheres in two-jetlike events, with azimuthal angles ϕ_1 and ϕ_2 as defined in Fig. 2, results in a $\cos(\phi_1+\phi_2)$ modulation of the observed di-hadron yield. In the CMS these azimuthal angles are defined between the transverse component of the hadron momenta with regard to the thrust axis $\hat{n}$ and the plane spanned by the lepton momenta and $\hat{n}$. The comparison of the thrust axis calculations using reconstructed and generated tracks in the MC sample shows an average angular separation between the two of 75 mrad with a root mean square of 74 mrad. The discrepancy between true and reconstructed jet axis can lead to biases in one of the two analysis methods used for the Collins measurement. Following Ref. 7 one either computes the azimuthal angles of each pion relative to the thrust axis which results in a $\cos(\phi_1+\phi_2)$ modulation or one calculates the azimuthal angle relative to the axis defined by the 2nd pion which results in a $\cos(2\phi_0)$ modulation (see Fig. 2). While the first method directly accesses moments of the Collins functions the second method also contains a convolution integral of the Collins FF over possible transverse momenta of the hadrons.

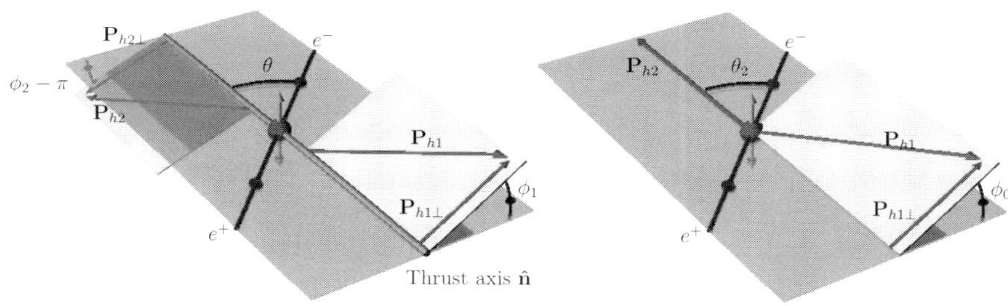

Fig. 2. Description of the azimuthal angles ϕ_0, ϕ_1 and ϕ_2 relative to the scattering plane defined by the lepton axis and either the thrust axis $\hat{n}$ or the momentum of the 2nd hadron P_{h2}.

3.1. *Measured asymmetries*

We measure the normalized yields $N(2\phi)/N_0$, where $N(2\phi)$ denotes the number of hadron pairs in bins of either $2\phi_0$ or $\phi_1+\phi_2$ and N_0 is the average number of hadron pairs in the whole angle interval. The main background, producing similar azimuthal asymmetries as the Collins effect, is the radiation of soft gluons. This gluonic contribution is proportional to the unpolarized FF and is independent of the charge of the hadrons. Consequently taking the ratio of the normalized distributions for unlike-sign over like-sign pairs the gluonic distributions drop out in leading order:

$$\begin{aligned} R &:= \frac{\frac{N(2\phi_0)}{N_0}|_{\text{unlike sign}}}{\frac{N(2\phi_0)}{N_0}|_{\text{like sign}}} \\ &\approx 1 + \frac{\sin^2\theta}{1+\cos^2\theta}\left(F(\frac{H_1^{\perp,\text{fav}}}{D_1^{\text{fav}}}, \frac{H_1^{\perp,\text{disfav}}}{D_1^{\text{disfav}}}) + \mathcal{O}f(Q_T, \alpha_S)^2\right)\cos(2\phi_0)\,, \end{aligned} \tag{2}$$

where θ is the angle between the colliding leptons and the produced hadron. Q_T is the transverse momentum of the virtual photon as seen in the two-hadron center of mass frame. Favored and disfavored FF describe the fragmentation of a light quark into a pion of same (i.e. $u \to \pi^+, d \to \pi^-$) or opposite charge sign (i.e. $u \to \pi^-, d \to \pi^+$). A similar relation also holds for the $\cos(\phi_1 + \phi_2)$ method. Those double ratios are then fit by the sum of a constant term and a $\cos(2\phi_0)$ or $\cos(\phi_1 + \phi_2)$ modulation. The reconstruction of azimuthal asymmetries was tested in MC, where nonzero asymmetries were created by reweighting the distributions in the generated variables. While the $\cos(2\phi_0)$ reproduces those generated asymmetries without bias, the $\cos(\phi_1 + \phi_2)$ method seems to underestimate the asymmetries by $(17.6 \pm 1.1)\%$. This behavior can be understood since this method takes the thrust axis into account which only approximates the quark-antiquark axis. The asymmetry reconstruction was also tested in $\tau^+\tau^- \to \pi\pi\nu\bar{\nu}$ MC where the azimuthal asymmetries due to parity violating weak decays are known. Preliminary results for the cosine fits to the double ratios can be seen in Fig. 3 for charged pion pairs, where the combined z-bins are obtained by adding the symmetric bins of the 4×4 $z \in [0.2, 0.3, 0.5, 0.7, 1.0]$ bins. A clear nonzero asymmetry is visible. Additionally the data shows a rising behavior with rising fractional energy z. The systematic errors are obtained by taking the differences of the double ratio results compared with results obtained by subtracting the unlike from the like sign asymmetries. Also the constant fit to the double ratios obtained in MC (without a Collins contribution) together with its statistical error and a similar fit to double ratios of positively charged over negatively charged pion pair data were assigned as systematic error. Also the differences to the results when fitting the double ratios also with higher order azimuthal modulations were added to the systematic errors. The possible systematic error due to charm quark contribution in the data sample was estimated by the contribution of charm events to the data sample in each of the bins in conjunction with a raw asymmetry due to radiative effects as seen in MC.

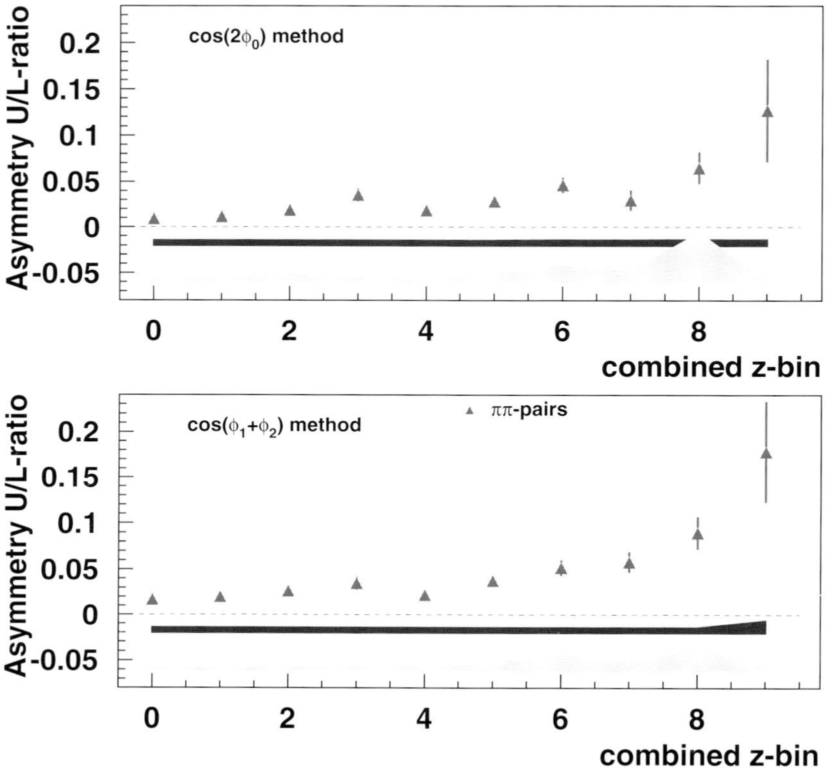

Fig. 3. Double ratio results for the $\cos(2\phi_0)$ and the $\cos(\phi_1 + \phi_2)$ method. The upper error bars correspond to systematic errors, the lower error bars to possible contributions by charm quarks.

References

1. J.C. Collins, *Nucl. Phys.* **B396** (1993) 161.
2. R.L. Jaffe, X. Jin and J. Tang, *Phys. Rev.* **D57** (1998) 5920.
3. A. Abashian *et al.* (Belle), *Nucl. Instrum. Meth.* **A479** (2002) 117.
4. Y. Ushiroda *et al.*, *Nucl. Instrum. Meth.* **A511** (2003) 6.
5. S. Kurokawa, E. Kikutani, *Nucl. Instrum. Meth.* **A499** (2003) 1; and other papers included in this volume.
6. A. Bacchetta, U. D'Alesio, M. Diehl, A. Miller, *Phys. Rev.* **D70** (2004) 117504.
7. D. Boer, R. Jakob, P.J. Mulders, *Phys. Lett.* **B424** (1998) 143.

LAMBDA ASYMMETRIES

A. Ferrero
University of Torino and INFN – Torino

on behalf of the COMPASS Collaboration

The measurement of the transverse spin quark distribution functions $\Delta_T q(x)$ is an important part of the physics program of the COMPASS experiment at CERN. The transversity distributions, being chiral-odd objects, are not accessible in inclusive deep inelastic scattering (DIS). The most promising channels for the measurement of the transversity distributions in semi-inclusive DIS (SIDIS) are the Collins effect, the azimuthal asymmetries in two hadrons production and the spin transfer to the Lambda hyperons. In this paper we focus on the semi-inclusive Lambda production mechanism, showing the connection between the measured polarization and the $\Delta_T q(x)$ functions. We derive an expression for the experimental $\Lambda \to p\pi^-$ angular distribution that is at first order independent of the experimental acceptance, and we present the preliminary results for the Lambda polarization as a function of the x Bjorken variable. The analysis is based on the 2002 and 2003 COMPASS data with transverse spin target configuration.

1. Introduction

The quark structure of the nucleon at twist-two level is fully specified by three distribution functions: the momentum distributions $q(x)$, the helicity distributions $\Delta q(x)$ and the transverse spin distributions $\Delta_T q(x)$. The transverse spin distributions, being chiral-odd objects, are only accessible in semi-inclusive deep inelastic scattering (SIDIS) or in hadron-hadron collisions, and are therefore difficult to be measured experimentally.

The "golden channels" for the measurement of $\Delta_T q(x)$ in SIDIS are the azimuthal asymmetries in pion production [1] and the azimuthal asymmetries in two hadrons production.[2] A possible alternative approach, based on the measurement of the spin transfer to the Lambda hyperons produced in the deep inelastic scattering on transversely polarized targets, has been originally suggested in Refs. 3–5, and more recently by Anselmino *et al.*[6] The measurement is based on the following idea: when a lepton interacts with one of the valence quarks of a transversely polarized nucleon, the scattered quark leaves the nucleon in a polarization state that is completely determined by its transverse spin distribution function inside the nucleon ($\Delta_T q(x)$) and the kinematics of the lepton–photon vertex. In particular, if the struck quark was initially in the same polarization state as the parent nucleon, then after the scattering the quark will be polarized along an axis that is obtained by reflecting the target polarization axis with respect to the normal to the lepton scattering plane. The final quark polarization is reduced by the so-called virtual photon depolarization factor, originating from the lepton–photon QED vertex and given by $D(y) = 2(1-y)/[1+(1-y)^2]$, where is y is the fraction of the incoming lepton energy carried by the exchanged virtual photon.

The struck quark has a certain probability to fragment into a Lambda hyperon. If at least part of its polarization is transferred in the fragmentation process, the angular distribution in the weak $\Lambda \to p\pi^-$ decay can provide information on the initial polarization state of the quark in the nucleon. The Lambda polarization measured experimentally is therefore given by:

$$\begin{aligned} P_T^\Lambda &= \frac{d\sigma^{lp^\uparrow \to \Lambda^\uparrow X} - d\sigma^{lp^\uparrow \to \Lambda^\downarrow X}}{d\sigma^{lp^\uparrow \to \Lambda^\uparrow X} + d\sigma^{lp^\uparrow \to \Lambda^\downarrow X}} = \\ &= f P_S D(y) \cdot \frac{\sum_q e_q^2 \Delta_T q(x) \Delta_T D_{\Lambda/q}(z)}{\sum_q e_q^2 q(x) D_{\Lambda/q}(z)}, \end{aligned} \tag{1}$$

where the T-axis is the polarization vector of the struck quark as described before, P_S and f are the target polarization and dilution factor respectively, and $\Delta_T D_{\Lambda/q}(z)$ is the polarized fragmentation function that describes the spin transfer from the quark to the final state hyperon.

2. Selection of Lambda events

The analysis is based on the data sample with transverse target polarization collected in 2002 and 2003 by the COMPASS [7] experiment at CERN. For a detailed description of the apparatus, please refer to Ref. 8 and references therein.

The event selection is based on the requirement of a scattering with large momentum transfer ($Q^2 > 1\,\mathrm{GeV}^2/c^2$) in the ^{6}LiD target material, together with a two-body charged decay of a neutral particle downstream of the target. The reconstructed position of the primary interaction vertex must be within the geometrical volume occupied by the target material. In order to ensure an equal beam flux in both target cells, the extrapolated beam trajectory must not intersect the cylindrical surfaces of the cell volumes.

The Λ hyperons undergo the decay $\Lambda \rightarrow p\pi^-$ in about 64% of the cases. The decay is detected as a V-shaped vertex in the reconstructed events, with the two oppositely charged decay particles bent in opposite directions by the spectrometer magnets. The typical decay length at the COMPASS energies is about 50 cm.

The main sources of background in the Λ sample come from K^0 decays, photon conversion and fake vertices from accidental track associations. The background is significantly reduced when the longitudinal position of the decay vertex is restricted to a region between the target exit window and the first MicroMega station. The contamination of e^+e^- pairs from photon conversions is significantly reduced by requiring a minimal transverse momentum $p_T > 23\,\mathrm{MeV}/c$ of the decay proton with respect to the decaying hyperon. The background is further reduced when a cut on the collinearity between the Lambda momentum vector and the line connecting the primary and decay vertex is applied. The angle between the two vectors must be lower than 10 mrad.

Only events with a reconstructed Lambda invariant mass between $1.07\,(\mathrm{GeV}/c)^2$ and $1.37\,(\mathrm{GeV}/c)^2$ are kept in the final event sample used for the polarization calculation. The number of Lambda decay events in the sample are estimated by fitting the invariant mass distribution with a Gaussian peak combined with a 3rd degree polynomial for the background parametrization. The mass range includes a significant fraction of the background on both sides of the Lambda peak.

The overall number of detected Lambda decays in the sample used in the analysis, corresponding to the full transversity data collected by the experiment in the years 2002 and 2003, is about 20000.

3. Extraction of the polarization

The angular distribution of the decay proton in the Lambda rest frame, measured in the experiment, is given by

$$\frac{dN}{d\theta_T^*} = N_0 \cdot \left(1 + \alpha P_T^\Lambda \cos(\theta_T^*)\right) \cdot Acc(\theta_T^*), \tag{2}$$

where θ_T^* is the proton emission angle with respect to the T-axis in the Lambda rest frame. The $Acc(\theta_T^*)$ function represents the distortion of the theoretical angular distribution introduced by the experimental apparatus. This distortion is usually corrected by combining real data and Monte-

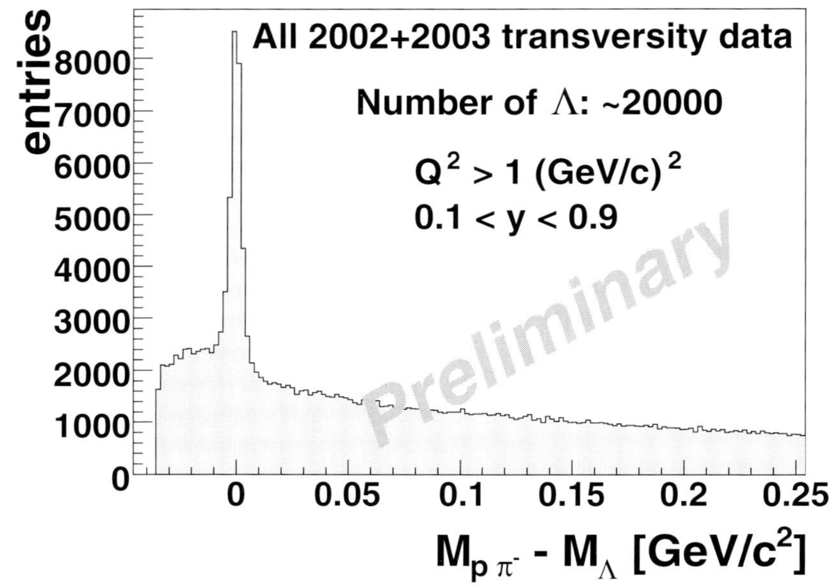

Fig. 1. Invariant mass spectrum of the data sample used in this analysis, after all the event selection cuts. The overall number of detected Λ decays is about 20000.

Carlo (MC) simulations. This approach is however quite sensitive to the accuracy of the MC description of the experiment, and requires huge MC data samples to get a good statistical accuracy.

In this analysis we used a technique based only on real data samples, exploiting some of the symmetries of the experimental apparatus. The technique is based on the combination of two data taking periods, in the same experimental conditions but with opposite target cell polarizations, so that the acceptance functions cancel and only the terms proportional to the true Lambda polarization remain.

We will denote with $Acc^{+(-)}_{1(2)}(\theta^*_T)$ the acceptance for Lambdas coming from the target cell with spin orientation $+(-)$ and data taking period $1(2)$. The number of Lambdas emitting the proton at an angle θ^*_T with respect to the T-axis is therefore given by

$$N^{+(-)}_{1(2)}(\theta^*_T) = \Phi^{+(-)}_{1(2)} \cdot \left(\frac{d\sigma}{d\Omega}\right)^0 \cdot \left(1 + \alpha P^{+(-)}_T \cos(\theta^*_T)\right) \cdot Acc^{+(-)}_{1(2)}(\theta^*_T), \quad (3)$$

where $\Phi^{+(-)}_{1(2)}$ denotes the muon beam flux.

The experimental symmetries and the muon flux normalizations allow to write the following relations:

$$Acc_1^{+(-)}(\theta_T^*) = Acc_2^{-(+)}(\theta_T^*) \tag{4}$$

$$\Phi_1^+ \cdot \Phi_2^+ = \Phi_1^- \cdot \Phi_2^- = \Phi_1 \cdot \Phi_2 \tag{5}$$

Under these assumptions the following counting rate asymmetry:

$$\epsilon_T(\theta_T^*) =$$

$$\frac{\left[\sqrt{\frac{N_1^+(\theta_T^*)}{\Phi_1^+}\frac{N_2^+(\theta_T^*)}{\Phi_2^+}}+\sqrt{\frac{N_1^-(\pi-\theta_T^*)}{\Phi_1^-}\frac{N_2^-(\pi-\theta_T^*)}{\Phi_2^-}}\right]-\left[\sqrt{\frac{N_1^+(\pi-\theta_T^*)}{\Phi_1^+}\frac{N_2^+(\pi-\theta_T^*)}{\Phi_2^+}}+\sqrt{\frac{N_1^-(\theta_T^*)}{\Phi_1^-}\frac{N_2^-(\theta_T^*)}{\Phi_2^-}}\right]}{\left[\sqrt{\frac{N_1^+(\theta_T^*)}{\Phi_1^+}\frac{N_2^+(\theta_T^*)}{\Phi_2^+}}+\sqrt{\frac{N_1^-(\pi-\theta_T^*)}{\Phi_1^-}\frac{N_2^-(\pi-\theta_T^*)}{\Phi_2^-}}\right]+\left[\sqrt{\frac{N_1^+(\pi-\theta_T^*)}{\Phi_1^+}\frac{N_2^+(\pi-\theta_T^*)}{\Phi_2^+}}+\sqrt{\frac{N_1^-(\theta_T^*)}{\Phi_1^-}\frac{N_2^-(\theta_T^*)}{\Phi_2^-}}\right]}. \tag{6}$$

is proportional to the Lambda polarization P_T:

$$\epsilon_T(\theta_T^*) = \alpha P_T^\Lambda \cos\theta_T^*, \tag{7}$$

and the Lambda polarization can be extracted from the slope of the $\epsilon_T(\cos\theta_T^*)$ distribution.

In the present analysis the proton decay angle distributions have been divided into only two bins. In this case the above formula simplifies to $\epsilon_T = \alpha P_T^\Lambda/2$.

4. Results

The multiplicative factors appearing on the right side of Eq. (1) all contribute to reduce the polarization that is measurable experimentally. In the COMPASS case, the average target polarization is about 50% and the dilution factor is ~ 0.45. The $D(y)$ depends on the scattering kinematics; the average value of y in the event sample used in this analysis is $\langle y\rangle \simeq 0.48$, giving a mean depolarization factor of $\langle D(y)\rangle \simeq 0.8$. That means that the experimentally measurable polarization is not significantly reduced by the typical COMPASS trigger acceptance.

In addition, Eq. (1) is only valid in the case that the Lambda originates from the scattered quark (current fragmentation region). In the real case, the reconstructed Lambda sample contains also a fraction of Lambdas originating from the fragmentation of the target remnants (target fragmentation region). Nevertheless, the small acceptance of the COMPASS spectrometer for events at $x_F < 0$ and $z < 0.2$ naturally suppresses the contamination of Lambdas produced in the target fragmentation.

The number of Lambda decay events available in the sample allowed not only to extract the overall transverse polarization, but also to investigate its dependence over the Bjorken x kinematical variable. The measured P_T as a function of x is shown in Fig. 2 for the full data sample and in Fig. 3 for the DIS region ($Q^2 > 1\,\mathrm{GeV}^2/c^2$). The measured values are compatible with zero in all the accessible x range. The data points at $x \sim 0.1$, were the transversity distribution function is expected to be peaked, still needs improvement in statistics, therefore no conclusion can be drawn yet on the spin transfer from the target to the final state Lambdas. The addition of the 2004 data sample is expected to double the available statistics for all the considered x bins.

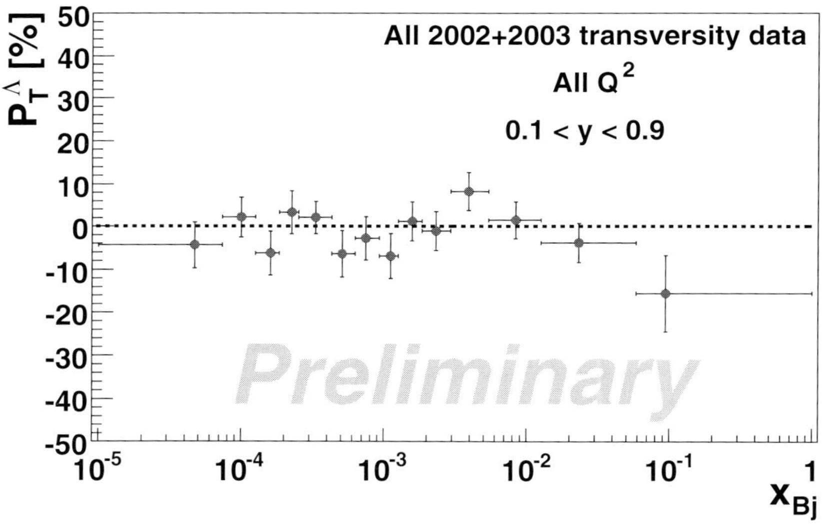

Fig. 2. Measured Λ polarization as a function of x, without the cut $Q^2 > 1\,\mathrm{GeV}^2/c^2$.

Acknowledgements

I would like to thank Prof. Raimondo Bertini and Prof. Aram Kotzinian for many fruitful discussions.
Work supported by INFN and PRIN #200329177_003.

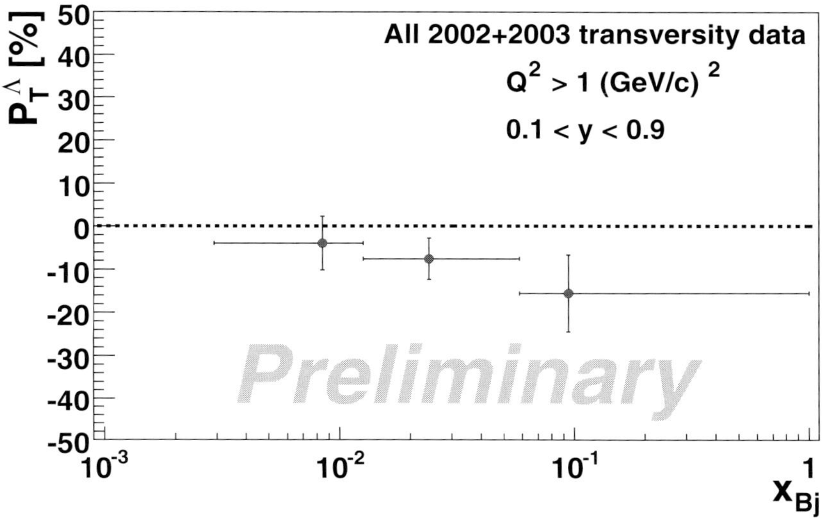

Fig. 3. Measured Λ polarization as a function of x, for the kinematical region of $Q^2 > 1\,\mathrm{GeV}^2/c^2$.

References

1. V.Yu. Alexakhin *et al.*, *Phys. Rev. Lett.* **94** (2005) 202002.
2. A. Mielech, in these proceedings
3. F. Baldracchini *et al.*, *Fortsch. Phys.* **30** (1981) 505.
4. X. Artru and M. Mekhfi, *Nucl. Phys.* **A532** (1991) 351.
5. R.A. Kunne *et al.*, "Electroproduction of polarized Lambdas", Saclay CEN – LNS-Ph-93-01.
6. M. Anselmino, "Transversity and Lambda polarization", proc. of the *Workshop on Future Physics @ COMPASS*, Sept. 26–27, 2002, CERN.
7. COMPASS, A Proposal for a COmmon Muon and Proton Apparatus for Structure and Spectroscopy, (CERN/SPSLC 96-14, SPSLC/P297, 1996).
8. G.K. Mallot, *Nucl. Instrum. Meth.* **A518** (2004) 121; F. Bradamante, hep-ex/0411073, and references therein.

TRANSVERSE SPIN AT PHENIX: RESULTS AND PROSPECTS

C. Aidala,[9] for the PHENIX Collaboration:
A. Adare,[8] S. Afanasiev,[22] N.N. Ajitanand,[48] Y. Akiba,[42,43] H. Al-Bataineh,[37] J. Alexander,[48] K. Aoki,[27,42] L. Aphecetche,[50] R. Armendariz,[37] S.H. Aronson,[3] J. Asai,[43] R. Averbeck,[49] T.C. Awes,[38] B. Azmoun,[3] V. Babintsev,[18] G. Baksay,[14] L. Baksay,[14] A. Baldisseri,[11] K.N. Barish,[4] P.D. Barnes,[30] B. Bassalleck,[36] S. Bathe,[4] S. Batsouli,[38] V. Baublis,[41] A. Bazilevsky,[3] S. Belikov,[3] R. Bennett,[49] Y. Berdnikov,[45] A.A. Bickley,[8] J.G. Boissevain,[30] H. Borel,[11] K. Boyle,[49] M.L. Brooks,[30] H. Buesching,[3] V. Bumazhnov,[18] G. Bunce,[3,43] S. Butsyk,[30,49] S. Campbell,[49] B.S. Chang,[57] J.-L. Charvet,[11] S. Chernichenko,[18] C.Y. Chi,[9] J. Chiba,[23] M. Chiu,[19] I.J. Choi,[57] T. Chujo,[54] P. Chung,[48] A. Churyn,[18] V. Cianciolo,[38] C.R. Cleven,[16] B.A. Cole,[9] M.P. Comets,[39] P. Constantin,[30] M. Csanád,[13] T. Csörgő,[24] D. d'Enterria,[9] T. Dahms,[49] K. Das,[15] G. David,[3] M.B. Deaton,[1] K. Dehmelt,[14] H. Delagrange,[50] A. Denisov,[18] A. Deshpande,[43,49] E.J. Desmond,[3] O. Dietzsch,[46] A. Dion,[49] M. Donadelli,[46] O. Drapier,[28] A. Drees,[49] A.K. Dubey,[56] A. Durum,[18] V. Dzhordzhadze,[4] Y.V. Efremenko,[38] J. Egdemir,[49] F. Ellinghaus,[8] W.S. Emam,[4] A. Enokizono,[29] H. En'yo,[42,43] S. Esumi,[53] K.O. Eyser,[4] D.E. Fields,[36,43] M. Finger,[5,22] F. Fleuret,[28] S.L. Fokin,[26] Z. Fraenkel,[56] A. Franz,[3] J. Franz,[49] A.D. Frawley,[15] K. Fujiwara,[42] Y. Fukao,[27,42] T. Fusayasu,[35] S. Gadrat,[31] I. Garishvili,[51] A. Glenn,[8] H. Gong,[49] M. Gonin,[28] J. Gosset,[11] Y. Goto,[42,43] R. Granier de Cassagnac,[28] N. Grau,[21] S.V. Greene,[54] M. Grosse Perdekamp,[19,43] T. Gunji,[7] H.-Å. Gustafsson,[32] T. Hachiya,[17] A. Hadj Henni,[50] C. Haegemann,[36] J.S. Haggerty,[3] H. Hamagaki,[7] R. Han,[40] H. Harada,[17] E.P. Hartouni,[29] K. Haruna,[17] E. Haslum,[32] R. Hayano,[7] X. He,[16] M. Heffner,[29] T.K. Hemmick,[49] T. Hester,[4] H. Hiejima,[19] J.C. Hill,[21] R. Hobbs,[36] M. Hohlmann,[14] W. Holzmann,[48] K. Homma,[17] B. Hong,[25] T. Horaguchi,[42,52] D. Hornback,[51] T. Ichihara,[42,43] K. Imai,[27,42] M. Inaba,[53] Y. Inoue,[44,42] D. Isenhower,[1] L. Isenhower,[1] M. Ishihara,[42] T. Isobe,[7] M. Issah,[48] A. Isupov,[22] B.V. Jacak,[49] J. Jia,[9] J. Jin,[9] O. Jinnouchi,[43] B.M. Johnson,[3] K.S. Joo,[34] D. Jouan,[39] F. Kajihara,[7] S. Kametani,[7,55] N. Kamihara,[42] J. Kamin,[49] M. Kaneta,[43] J.H. Kang,[57] H. Kano,[42] H. Kanoh,[42,52] D. Kawall,[43] A.V. Kazantsev,[26] A. Khanzadeev,[41] J. Kikuchi,[55] D.H. Kim,[34] D.J. Kim,[57] E. Kim,[47] E. Kinney,[8] A. Kiss,[13] E. Kistenev,[3] A. Kiyomichi,[42] J. Klay,[29] C. Klein-Boesing,[33] L. Kochenda,[41] V. Kochetkov,[18] B. Komkov,[41] M. Konno,[53] D. Kotchetkov,[4] A. Kozlov,[56] A. Král,[10] A. Kravitz,[9] J. Kubart,[5,20] G.J. Kunde,[30] N. Kurihara,[7] K. Kurita,[44,42] M.J. Kweon,[25] Y. Kwon,[25,51] G.S. Kyle,[37] R. Lacey,[48] J.G. Lajoie,[21] A. Lebedev,[21] D.M. Lee,[30] M.K. Lee,[57] T. Lee,[47] M.J. Leitch,[30] M.A.L. Leite,[46] B. Lenzi,[46] X. Li,[6] T. Liška,[10] A. Litvinenko,[22] M.X. Liu,[30] B. Love,[54] D. Lynch,[3] C.F. Maguire,[54] Y.I. Makdisi,[3] A. Malakhov,[22] M.D. Malik,[36] V.I. Manko,[26] Y. Mao,[40,42] L. Masek,[5,20] H. Masui,[53] F. Matathias,[9] M. McCumber,[49] P.L. McGaughey,[30] Y. Miake,[53] P. Mikes,[5,20] K. Miki,[53] T.E. Miller,[54] A. Milov,[49]

S. Mioduszewski,[3] M. Mishra,[2] J.T. Mitchell,[3] M. Mitrovski,[48] A. Morreale,[4] D.P. Morrison,[3] T.V. Moukhanova,[26] D. Mukhopadhyay,[54] J. Murata,[44,42] S. Nagamiya,[23] Y. Nagata,[53] J.L. Nagle,[8] M. Naglis,[56] I. Nakagawa,[42,43] Y. Nakamiya,[17] T. Nakamura,[17] K. Nakano,[42,52] J. Newby,[29] M. Nguyen,[49] B.E. Norman,[30] A.S. Nyanin,[26] E. O'Brien,[3] S. Oda,[7] C.A. Ogilvie,[21] H. Ohnishi,[42] M. Oka,[53] H. Okada,[27,42] K. Okada,[43] O.O. Omiwade,[1] A. Oskarsson,[32] M. Ouchida,[17] K. Ozawa,[7] R. Pak,[3] D. Pal,[54] A.P.T. Palounek,[30] V. Pantuev,[49] V. Papavassiliou,[37] J. Park,[47] W.J. Park,[25] S.F. Pate,[37] H. Pei,[21] J.-C. Peng,[19] H. Pereira,[11] V. Peresedov,[22] D.Yu. Peressounko,[26] C. Pinkenburg,[3] M.L. Purschke,[3] A.K. Purwar,[30] H. Qu,[16] J. Rak,[36] A. Rakotozafindrabe,[28] I. Ravinovich,[56] K.F. Read,[38,51] S. Rembeczki,[14] M. Reuter,[49] K. Reygers,[33] V. Riabov,[41] Y. Riabov,[41] G. Roche,[31] A. Romana,[28] M. Rosati,[21] S.S.E. Rosendahl,[32] P. Rosnet,[31] P. Rukoyatkin,[22] V.L. Rykov,[42] B. Sahlmueller,[33] N. Saito,[27,42,43] T. Sakaguchi,[3] S. Sakai,[53] H. Sakata,[17] V. Samsonov,[41] S. Sato,[23] S. Sawada,[23] J. Seele,[8] R. Seidl,[19] V. Semenov,[18] R. Seto,[4] D. Sharma,[56] I. Shein,[18] A. Shevel,[41,48] T.-A. Shibata,[42,52] K. Shigaki,[17] M. Shimomura,[53] K. Shoji,[27,42] A. Sickles,[49] C.L. Silva,[46] D. Silvermyr,[38] C. Silvestre,[11] K.S. Sim,[25] C.P. Singh,[2] V. Singh,[2] S. Skutnik,[21] M. Slunecka,[5,22] A. Soldatov,[18] R.A. Soltz,[29] W.E. Sondheim,[30] S.P. Sorensen,[51] I.V. Sourikova,[3] F. Staley,[11] P.W. Stankus,[38] E. Stenlund,[32] M. Stepanov,[37] A. Ster,[24] S.P. Stoll,[3] T. Sugitate,[17] C. Suire,[39] J. Sziklai,[24] T. Tabaru,[43] S. Takagi,[53] E.M. Takagui,[46] A. Taketani,[42,43] Y. Tanaka,[35] K. Tanida,[42,43] M.J. Tannenbaum,[3] A. Taranenko,[48] P. Tarján,[12] T.L. Thomas,[36] M. Togawa,[27,42] A. Toia,[49] J. Tojo,[42] L. Tomasek,[20] H. Torii,[42] R.S. Towell,[1] V-N. Tram,[28] I. Tserruya,[56] Y. Tsuchimoto,[17] E. Tujuba,[28] C. Vale,[21] H. Valle,[54] H.W. van Hecke,[30] J. Velkovska,[54] R. Vertesi,[12] A.A. Vinogradov,[26] M. Virius,[10] V. Vrba,[20] E. Vznuzdaev,[41] M. Wagner,[27,42] D. Walker,[49] X.R. Wang,[37] Y. Watanabe,[42,43] J. Wessels,[33] S.N. White,[3] D. Winter,[9] C.L. Woody,[3] M. Wysocki,[8] W. Xie,[43] Y. Yamaguchi,[55] A. Yanovich,[18] Z. Yasin,[4] J. Ying,[16] S. Yokkaichi,[42,43] G.R. Young,[38] I. Younus,[36] I.E. Yushmanov,[26] W.A. Zajc,[9] O. Zaudtke,[33] C. Zhang,[38] S. Zhou,[6] J. Zimányi,[24] and L. Zolin[22]

[1]*Abilene Christian University, Abilene, TX 79699, USA*
[2]*Department of Physics, Banaras Hindu University, Varanasi 221005, India*
[3]*Brookhaven National Laboratory, Upton, NY 11973-5000, USA*
[4]*University of California - Riverside, Riverside, CA 92521, USA*
[5]*Charles University, Ovocny trh 5, Praha 1, 116 36, Prague, Czech Republic*
[6]*China Institute of Atomic Energy (CIAE), Beijing, People's Republic of China*
[7]*Center for Nuclear Study, Graduate School of Science, University of Tokyo, 7-3-1 Hongo, Bunkyo, Tokyo 113-0033, Japan*
[8]*University of Colorado, Boulder, CO 80309, USA*
[9]*Columbia University, New York, NY 10027 and Nevis Laboratories, Irvington, NY 10533, USA*
[10]*Czech Technical University, Zikova 4, 166 36 Prague 6, Czech Republic*
[11]*Dapnia, CEA Saclay, F-91191, Gif-sur-Yvette, France*
[12]*Debrecen University, H-4010 Debrecen, Egyetem tér 1, Hungary*
[13]*ELTE, Eötvös Loránd University, H - 1117 Budapest, Pázmány P. s. 1/A, Hungary*
[14]*Florida Institute of Technology, Melbourne, FL 32901, USA*
[15]*Florida State University, Tallahassee, FL 32306, USA*
[16]*Georgia State University, Atlanta, GA 30303, USA*
[17]*Hiroshima University, Kagamiyama, Higashi-Hiroshima 739-8526, Japan*
[18]*IHEP Protvino, State Research Center of Russian Federation, Institute for High*

Energy Physics, Protvino, 142281, Russia
[19]*University of Illinois at Urbana-Champaign, Urbana, IL 61801, USA*
[20]*Institute of Physics, Academy of Sciences of the Czech Republic, Na Slovance 2, 182 21 Prague 8, Czech Republic*
[21]*Iowa State University, Ames, IA 50011, USA*
[22]*Joint Institute for Nuclear Research, 141980 Dubna, Moscow Region, Russia*
[23]*KEK, High Energy Accelerator Research Organization, Tsukuba, Ibaraki 305-0801, Japan*
[24]*KFKI Research Institute for Particle and Nuclear Physics of the Hungarian Academy of Sciences (MTA KFKI RMKI), H-1525 Budapest 114, PO Box 49, Budapest, Hungary*
[25]*Korea University, Seoul, 136-701, Korea*
[26]*Russian Research Center "Kurchatov Institute", Moscow, Russia*
[27]*Kyoto University, Kyoto 606-8502, Japan*
[28]*Laboratoire Leprince-Ringuet, Ecole Polytechnique, CNRS-IN2P3, Route de Saclay, F-91128, Palaiseau, France*
[29]*Lawrence Livermore National Laboratory, Livermore, CA 94550, USA*
[30]*Los Alamos National Laboratory, Los Alamos, NM 87545, USA*
[31]*LPC, Université Blaise Pascal, CNRS-IN2P3, Clermont-Fd, 63177 Aubiere Cedex, France*
[32]*Department of Physics, Lund University, Box 118, SE-221 00 Lund, Sweden*
[33]*Institut für Kernphysik, University of Muenster, D-48149 Muenster, Germany*
[34]*Myongji University, Yongin, Kyonggido 449-728, Korea*
[35]*Nagasaki Institute of Applied Science, Nagasaki-shi, Nagasaki 851-0193, Japan*
[36]*University of New Mexico, Albuquerque, NM 87131, USA*
[37]*New Mexico State University, Las Cruces, NM 88003, USA*
[38]*Oak Ridge National Laboratory, Oak Ridge, TN 37831, USA*
[39]*IPN-Orsay, Universite Paris Sud, CNRS-IN2P3, BP1, F-91406, Orsay, France*
[40]*Peking University, Beijing, People's Republic of China*
[41]*PNPI, Petersburg Nuclear Physics Institute, Gatchina, Leningrad region, 188300, Russia*
[42]*RIKEN, The Institute of Physical and Chemical Research, Wako, Saitama 351-0198, Japan*
[43]*RIKEN BNL Research Center, Brookhaven National Laboratory, Upton, NY 11973-5000, USA*
[44]*Physics Department, Rikkyo University, 3-34-1 Nishi-Ikebukuro, Toshima, Tokyo 171-8501, Japan*
[45]*Saint Petersburg State Polytechnic University, St. Petersburg, Russia*
[46]*Universidade de São Paulo, Instituto de Física, Caixa Postal 66318, São Paulo CEP05315-970, Brazil*
[47]*System Electronics Laboratory, Seoul National University, Seoul, South Korea*
[48]*Chemistry Department, Stony Brook University, Stony Brook, SUNY, NY 11794-3400, USA*
[49]*Department of Physics and Astronomy, Stony Brook University, SUNY, Stony Brook, NY 11794, USA*
[50]*SUBATECH (Ecole des Mines de Nantes, CNRS-IN2P3, Université de Nantes) BP 20722 - 44307, Nantes, France*
[51]*University of Tennessee, Knoxville, TN 37996, USA*
[52]*Department of Physics, Tokyo Institute of Technology, Oh-okayama, Meguro, Tokyo 152-8551, Japan*

[53] *Institute of Physics, University of Tsukuba, Tsukuba, Ibaraki 305, Japan*
[54] *Vanderbilt University, Nashville, TN 37235, USA*
[55] *Waseda University, Advanced Research Institute for Science and Engineering, 17 Kikui-cho, Shinjuku-ku, Tokyo 162-0044, Japan*
[56] *Weizmann Institute, Rehovot 76100, Israel*
[57] *Yonsei University, IPAP, Seoul 120-749, Korea*

The Relativistic Heavy Ion Collider (RHIC), as the world's first and only polarized proton collider, offers a unique environment in which to study the spin structure of the proton. In order to study the proton's transverse spin structure, the PHENIX experiment at RHIC took data with transversely polarized beams in 2001–02 and 2005, and it has plans for further running with transverse polarization in 2006 and beyond. Results from early running as well as prospective measurements for the future will be discussed.

1. Introduction

The Relativistic Heavy Ion Collider (RHIC) has opened up a new energy regime in which to study the spin structure of the proton. Polarization of more than 50% has so far been achieved for 100-GeV proton beams, with expectations that this value will rise to 70% in 2006 or 2007.

The PHENIX experiment, one of two large experiments at RHIC, specializes in the measurement of photons, electrons, and muons as well as high-transverse-momentum (p_T) probes in general over a limited acceptance, with good particle identification capabilities. It has a high rate capability and sophisticated trigger systems, allowing measurement of rare processes. The PHENIX detector [1] consists of two mid-rapidity ($|\eta| < 0.35$) spectrometers, primarily for identifying and tracking charged particles as well as measuring electromagnetic probes, forward spectrometers for identifying and tracking muons ($1.2 < |\eta| < 2.4$), and interaction detectors.

Several polarization-averaged cross sections have been measured for 200-GeV collisions at RHIC and found to be in good agreement with next-to-leading-order (NLO) pQCD calculations.[2–5] The ability of NLO pQCD to describe RHIC cross section data well and with little scale dependence provides a solid foundation for using it to interpret polarized data in a similar kinematic regime.

2. Current Results

Large transverse single-spin asymmetries (SSAs) have been observed in spin-dependent proton–proton scattering experiments spanning a wide range of energies, as well as in semi-inclusive deep-inelastic scattering. The origin of these asymmetries remains unclear, but several different mechanisms have been proposed, as described for example in Refs. 6–9.

From data collected in 2001–02 ($0.15\,\text{pb}^{-1}$, $\langle P_{\text{beam}} \rangle \sim 15\%$), PHENIX measured the left–right transverse single-spin asymmetry (A_N) for neutral pion and charged hadron production at $x_F \sim 0.0$ up to a transverse momentum of $5\,\text{GeV}/c$ from polarized proton–proton interactions at $\sqrt{s} = 200\,\text{GeV}$.[4] As can be seen in Fig. 1, the asymmetries observed for production of both neutral pions and inclusive charged hadrons are consistent with zero within a few percent over the measured p_T range.

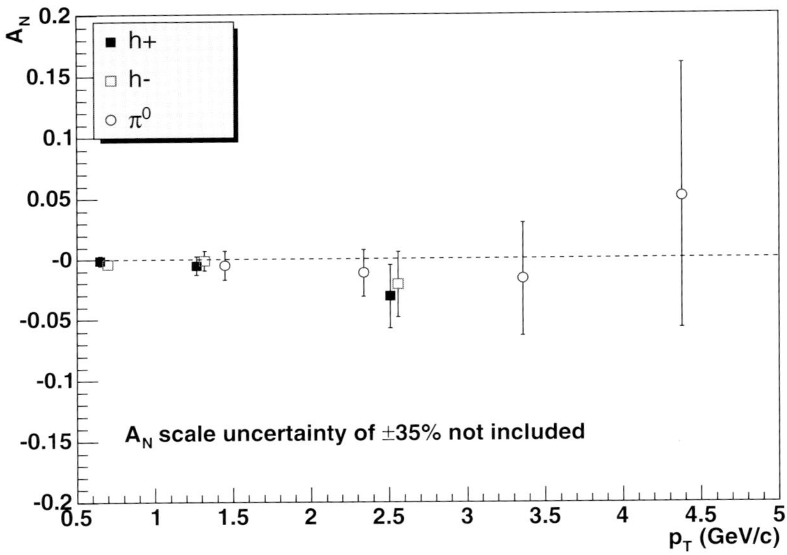

Fig. 1. Mid-rapidity neutral pion and charged hadron transverse single-spin asymmetries. Points for positive hadrons have been shifted down by $50\,\text{MeV}/c$ to improve readability. The error bars represent statistical uncertainties.

The result is consistent with mid-rapidity results for neutral pions at $\sqrt{s} = 19.4\,\text{GeV}$.[10] The present measurement is complementary to that of Ref. 3, which observed large asymmetries for forward neutral pions at $\sqrt{s} = 200\,\text{GeV}$. Neutral pion production at forward rapidity is expected to originate from processes involving valence quarks, whereas particle production at mid-rapidity is dominated by gluon–gluon and quark–gluon processes. As evident from Fig. 2, π^0 production in the p_T range covered by the recent PHENIX measurement is nearly half from gluon–gluon scattering and half from gluon–quark scattering. As such, the asymmetry is not very sensitive to mechanisms involving quarks.

In the forward direction at PHENIX, a large negative transverse SSA of approximately -11% in the production of neutrons from 200-GeV $p+p$ collisions has been observed.[11] This measurement was made using the RHIC zero-degree calorimeters (ZDCs), hadronic calorimeters covering 2π in azimuth and $4.7 < |\eta| < 5.6$. In 2005 there was a brief period of accelerator commissioning with polarized proton collisions at 410 GeV, and the large negative asymmetry in forward neutron production was found to persist.

The azimuthal asymmetry of forward charged particles was also measured at PHENIX [12] using beam–beam counters (BBCs), which are quartz Čerenkov counters that cover 2π in azimuth and $3.0 < |\eta| < 3.9$. The asymmetry for inclusive forward charged particles was consistent with zero. However, non-zero asymmetries were found in charged particle production from events in which a forward neutron was also detected in the ZDC. A significant negative asymmetry was observed for forward charged particles in neutron-tagged events, with a preliminary value of $(-4.50 \pm 0.50 \pm 0.22) \times 10^{-2}$. A smaller positive asymmetry was found for backward charged particles produced in neutron-tagged events, with a preliminary value of $(2.28 \pm 0.55 \pm 0.10) \times 10^{-2}$. The observed asymmetries for forward and backward charged particles in events with a forward neutron may suggest a diffractive process.

3. Prospective Future Measurements

Despite great theoretical progress in recent years, no single, clear formalism has emerged in which to interpret the currently available data. Further theoretical work and a variety of additional experimental measurements are necessary to understand current results and elucidate the transverse spin structure of the proton.

From a modest transverse-spin data sample taken in 2005 ($0.16\,\mathrm{pb}^{-1}$, $\langle P_{\mathrm{beam}} \rangle \sim 48\%$), PHENIX has begun analysis to obtain improved mid-rapidity A_N results for neutral pions and charged hadrons, expected to provide tighter constraints on the gluon Sivers function. Future higher-statistics samples for these particles at mid-rapidity will reach higher p_T and provide greater sensitivity to transversity and the Collins effect.

There is also analysis underway to obtain first results for A_N of single muons, largely from open charm decay but with significant contributions from light-hadron decays. The current x_F reach for this measurement is

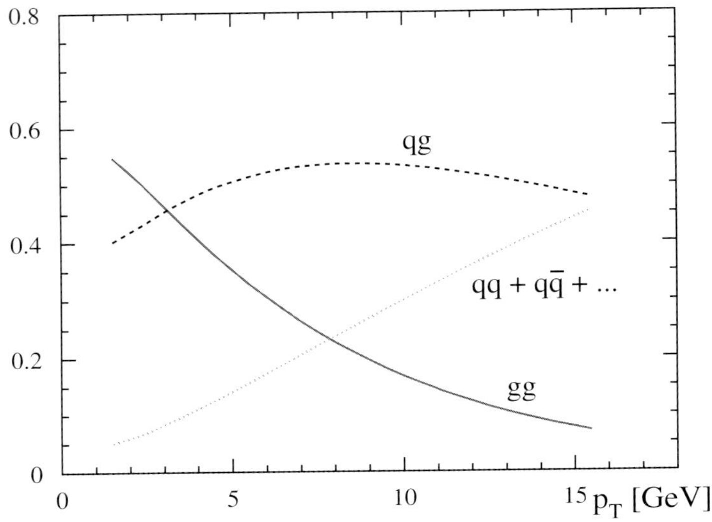

Fig. 2. Relative fractional contributions of partonic processes to mid-rapidity π^0 production at $\sqrt{s} = 200\,\text{GeV}$, calculated by W. Vogelsang.

up to ~ 0.15; higher x_F values would become accessible with lower-energy running. A forward hadron A_N measurement using the PHENIX muon spectrometers may be possible using decay muons and the charged hadrons that punch through the absorber in front of the muon tracker. Careful studies will be needed to understand the particle ratios in this sample.

In 2003 Boer and Vogelsang proposed a single transverse-spin di-jet measurement that could probe the gluon Sivers function.[13] A non-zero Sivers function implies a spin-dependence in the k_T distributions of the partons within the proton, which would lead to an observable spin-dependent asymmetry in $\Delta\varphi$ of back-to-back jets. In 2006, PHENIX intends to perform a measurement similar to the one proposed, using di-hadrons instead of di-jets because of the limited detector acceptance. This analysis will study the spin-dependence of the azimuthal angle between nearly back-to-back π^0-hadron pairs, triggering on a decay photon from the π^0 in order to obtain a higher-statistics sample. Although dilution of the effect is anticipated for hadron rather than jet pairs, studies have shown that it should still be measurable. Fragmentation to the final-state hadrons must also be considered, and some contribution from the Collins mechanism may be present; however, as shown above in Fig. 2, for $p_T \lesssim 5\,\text{GeV}/c$ there is a large contribution to mid-rapidity π^0 production from gluon fragmentation, to which the Collins mechanism does not apply.

Measurement of A_N for direct photons has also been proposed to probe the gluon Sivers function.[14] Direct photon production is dominated by quark–gluon Compton scattering ($q + g \rightarrow \gamma + X$) over a wide range in photon p_T at RHIC. Transverse SSAs of photons and jets in events with correlated photon–jet pairs would access the gluon and quark Sivers functions, respectively, with some ability to identify the x values at which these functions were probed. PHENIX can currently measure A_N of mid-rapidity direct photons. Future upgrades extending the azimuthal coverage for tracking to 2π in the inner region and adding forward electromagnetic calorimetry ($0.9 < |\eta| < 3.0$) are expected to expand the coverage for this measurement as well as make γ–jet and jet–jet measurements feasible.

Yet another proposal has been made to access the gluon Sivers function via mid- to moderate rapidity ($-0.2 < x_F < 0.6$) D meson production at RHIC.[15] PHENIX is currently capable of measuring open charm decays statistically via inclusive single electrons and muons. In the future, a silicon vertex detector upgrade will make it possible to identify D mesons event by event. Note that A_N measurements for charmonium production, also sensitive to the gluon, are already possible at PHENIX. However, the charmonium production mechanism is not as well understood.

The flavor separation of the Sivers function for u, d, $\bar{u}$, and $\bar{d}$ quarks via A_N of forward or backward W boson production, possible once RHIC achieves 500-GeV collisions, has been suggested by Schmidt.[16] The processes of interest at PHENIX are $u + \bar{d} \rightarrow W^+ \rightarrow \mu^+ + \nu_\mu$ and $d + \bar{u} \rightarrow W^- \rightarrow \mu^- + \bar{\nu}_\mu$. An upgrade to trigger on the high-p_T muons from W decays is expected in 2009. The trigger upgrade will also make open charm, charmonium, and Drell–Yan measurements cleaner.

The double transverse-spin asymmetry, A_{TT}, is another observable sensitive to transverse spin quantities. A_{TT} for the Drell–Yan process would provide direct access to transversity. Although this asymmetry is expected to be at the sub-percent level for $\sqrt{s} = 200\,\text{GeV}$, it could reach several percent for $\sqrt{s} < 100\,\text{GeV}$. PHENIX already has an effective di-muon trigger for measuring Drell–Yan pairs; however, the trigger upgrade will improve backgrounds. To measure A_{TT} it would be necessary to optimize the beam energy to balance luminosity against the size of the predicted asymmetry. A first direct measurement of transversity would be an exciting milestone.

4. Summary

The first transverse-spin results from PHENIX are now available, and further results from the brief transverse-spin run in 2005 are forthcoming. A longer period of running with transversely polarized beams is anticipated for 2006. Looking farther ahead, forward detector upgrades will improve access to the kinematic region where large asymmetries have been observed, and mid-rapidity upgrades will improve jet measurements.

Acknowledgments

PHENIX acknowledges support from the Department of Energy and NSF (U.S.A.), MEXT and JSPS (Japan), CNPq and FAPESP (Brazil), NSFC (China), MSMT (Czech Republic), IN2P3/CNRS, and CEA (France), BMBF, DAAD, and AvH (Germany), OTKA (Hungary), DAE (India), ISF (Israel), KRF, CHEP, and KOSEF (Korea), MES, RAS, and FAAE (Russia), VR and KAW (Sweden), U.S. CRDF for the FSU, US–Hungarian NSF-OTKA-MTA, and US–Israel BSF.

References

1. K. Adcox *et al.*, *Nucl. Instrum. Meth.* **A499** (2003) 469.
2. S.S. Adler *et al.*, *Phys. Rev. Lett.* **91** (2003) 241803.
3. J. Adams *et al.*, *Phys. Rev. Lett.* **92** (2004) 171801.
4. S.S. Adler *et al.*, hep-ex/0507073 (2005).
5. S.S. Adler *et al.*, *Phys. Rev.* **D71** (2005) 071102.
6. D.W. Sivers, *Phys. Rev.* **D41** (1990) 83.
7. J.C. Collins, *Nucl. Phys.* **B396** (1993) 161.
8. J.-W. Qiu and G. Sterman, *Phys. Rev.* **D59** (1999) 014004.
9. Y. Kanazawa and Y. Koike, *Phys. Lett.* **B478** (2000) 121.
10. D.L. Adams *et al.*, *Phys. Rev.* **D53** (1996) 4747.
11. A. Bazilevsky *et al.*, 15th. Int. Spin Physics Symposium (SPIN 2002), *AIP Conf. Proc.* **675** (2003) 584.
12. A. Taketani, proc. of the X Advanced Research Workshop on High Energy Spin Physics (SPIN-03), Dubna, Russia (2004) 421.
13. D. Boer and W. Vogelsang, *Phys. Rev.* **D69** (2004) 094025.
14. I. Schmidt, J. Soffer, and J.-J. Yang, *Phys. Lett.* **B612** (2005) 258.
15. M. Anselmino *et al.*, *Phys. Rev.* **D70** (2004) 074025.
16. I. Schmidt, talk presented at this workshop.

TRANSVERSE SPIN AND RHIC

L.C. Bland

Brookhaven National Laboratory,
Upton, NY 11786, USA
E-mail: bland@bnl.gov

The Relativistic Heavy Ion Collider (RHIC) at Brookhaven National Laboratory is the first accelerator facility that can accelerate, store and collide spin polarized proton beams. This development enables a physics program aimed at understanding how the spin of the proton results from its quark and gluon substructures. Spin states that are either parallel (longitudinal) or perpendicular (transverse) to the proton momentum reveal important insight into the structure of the proton. This talk outlines future plans for further studies of transverse spin physics at RHIC.

1. Introduction

There has been renewed experimental and theoretical interest in transverse spin physics. Large transverse single-spin asymmetries (SSA) observed in elastic proton scattering and particle production experiments (hyperon production and pion production) were often viewed as a challenge to QCD, since the chiral properties of the theory should make transverse single spin asymmetries small for inclusive particle production. Many people believed that transverse SSA would disappear when studying polarized $p + p$ collisions at higher collision energies ($\sqrt{s}$) now possible at the Relativistic Heavy Ion Collider (RHIC) at Brookhaven National Laboratory.

The modern perspective views transverse SSA as a challenge to our understanding of hadrons on long distance scales, possibly providing sensitivity to the transversity structure function or to spin- and transverse-momentum dependent distribution functions that are related to parton orbital motion.

In this contribution, I briefly review transverse SSA and related measurements completed at RHIC to date and describe what new measurements are expected in the near-term future.

2. Recent developments in transverse spin physics

This workshop surveyed ongoing experimental and theoretical work. A deeper understanding of spin- and transverse-momentum dependent distribution functions (embodied in the Sivers effect [1]) and fragmentation functions (one of the keys to the Collins effect [2]) has recently emerged. The former were known to violate "naive" time reversal symmetry. Results from a specific model demonstrate their possible existence.[3] This important theoretical development was essentially concurrent with experimental results from semi-inclusive deep inelastic scattering from a transversely polarized proton target that unambiguously observed a non-zero Sivers effect.[4] Also concurrent was the observation of large spin effects in dihadron correlations from e^+e^- collisions that indicates that spin-dependent fragmentation effects are large. This is one of the keys to the Collins effect.[5] These experimental developments, coupled with the observation of non-zero single spin asymmetries in inclusive pion production from the first collisions at RHIC, have led to a reinvigoration of transverse spin physics.

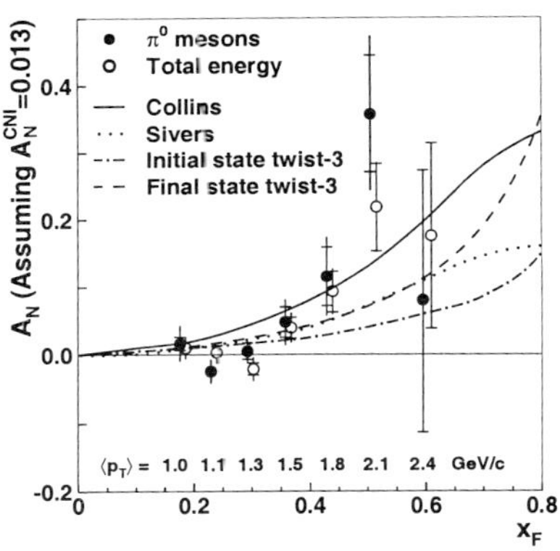

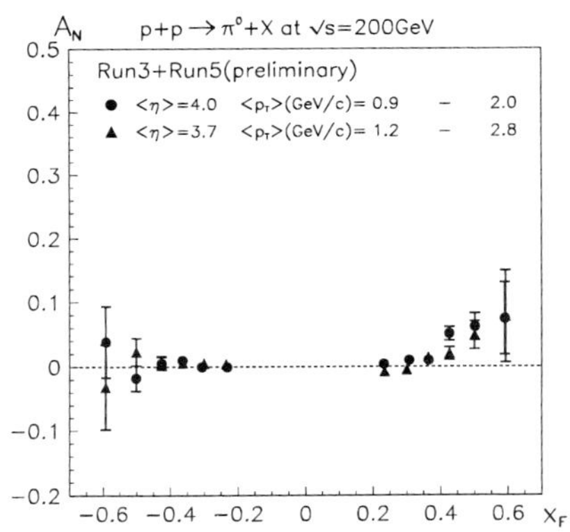

Fig. 1. (left) First results for $p_\uparrow + p \to \pi^0 + X$ analyzing powers (A_N) at $\sqrt{s} = 200\,\text{GeV}$ compared to theoretical predictions available prior to the measurements. (right) More precise preliminary results, using on-line measurements of the beam polarization, for π^0 production A_N obtained in subsequent runs.

Large SSA were observed for $p_\uparrow + p \to \pi^0 + X$ at $\sqrt{s} = 200\,\text{GeV}$ by the STAR collaboration [6] in the first polarized proton collisions at RHIC. They confirmed the expectation,[7–10] not shared by all, that the sizeable SSA observed for pion production at $\sqrt{s} = 20\,\text{GeV}$ [11] would persist at an order of magnitude higher collision energy. These expectations are shown by the theoretical predictions in Fig. 1, available prior to the measurements. Subsequent development of full integration over intrinsic transverse momentum has modified the relative contributions to transverse SSA from different sources.[12] More recent data [13] have improved the statistical precision of the effect and given the first hint of its separated x_F and p_T dependence. Preliminary results from the BRAHMS collaboration indicate that mirror asymmetries $(A_N(\pi^-) \approx -A_N(\pi^+))$ are observed for large rapidity $\pi^{\pm}$ production,[14] similar to the lower-energy results.[11]

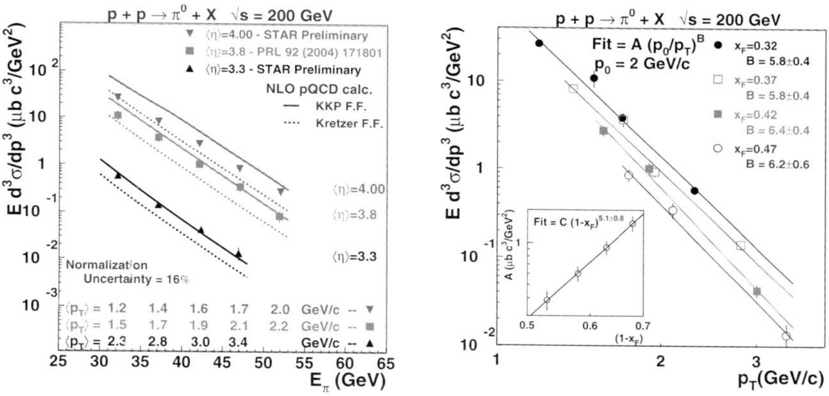

Fig. 2. Results for $p + p \to \pi^0 + X$ cross sections at $\sqrt{s} = 200\,\text{GeV}$ compared to NLO pQCD calculations using conventional parton distribution and fragmentation functions. (right) Parametrized x_F and p_T dependence.

Perhaps most significantly, it has been established that π production cross sections at RHIC collision energies, in the kinematics where single spin effects are observed, are consistent with next-to-leading order perturbative QCD (NLO pQCD) calculations at $\sqrt{s} = 200\,\text{GeV}$ (Fig. 2). This is in marked contrast to the situation at lower $\sqrt{s}$ where measured cross sections far exceed NLO pQCD predictions,[15] apparently consistent with the belief that the transverse SSA in hadroproduction were due to beam fragmentation. The NLO pQCD description at $\sqrt{s} = 200\,\text{GeV}$ describes particle production being due to partons from both beams undergoing a hard scattering prior to fragmenting to the observed hadrons. That description is further supported by experimental data that shows a significant back-to-back peak for hadrons detected at midrapidity for events where a large rapidity π^0 is observed.[16]

Measurements of the cross section for inclusive π^0,[17] charged hadron [18] and jet production[19] at midrapidity have been completed and compared with NLO pQCD calculations.[20] Quantitative agreement with calculations has been found. This agreement is an important basis for the interpretation of spin observables (A_N and the helicity asymmetry, A_{LL}, that is sensitive to gluon polarization). Transverse SSA for midrapidity π^0 and charged hadron production[18] have been measured and are consistent with zero with a precision comparable to the non-zero A_N found at large rapidity at the same p_T of $\sim 2\,\mathrm{GeV}/c$. The midrapidity results may lead to important constraints on the magnitude of the gluon Sivers function.

3. The future

I'll restrict attention to the near-term future since there will be significant data sets obtained with transverse polarization during the upcoming RHIC run following the resolution of budgetary problems. A main objective for midrapidity studies of $p_\uparrow + p$ collisions at $\sqrt{s} = 200\,\mathrm{GeV}$ is to establish if there are spin effects correlated with k_T, a transverse momentum imbalance that is observable if more than one particle, or more than one jet, is observed. Such effects could be a signal of a non-zero Sivers function for

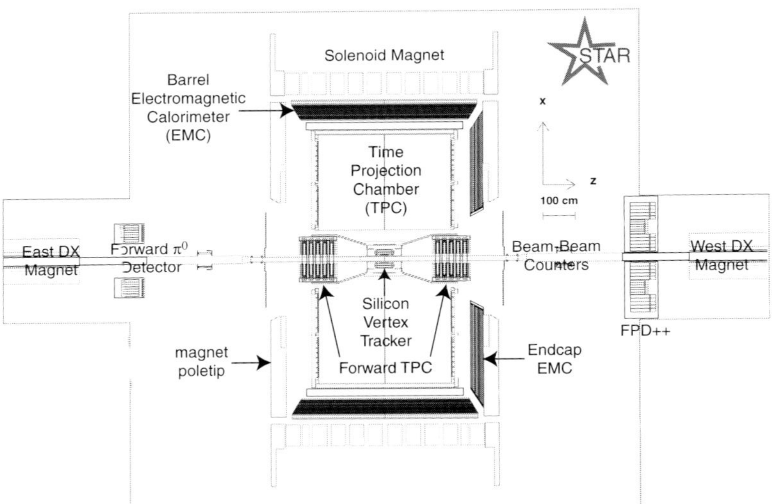

Fig. 3. Layout of the STAR experiment for RHIC run 6. The Forward π^0 Detector (FPD) arrays shown east of the STAR magnet were also present in earlier runs west of the STAR magnet. For run 6, the west FPD has been upgraded to become the FPD++. The Forward Time Projection Chamber provides information about charged hadrons in the angular range spanned by the forward calorimetry.

gluons.[21] STAR (Fig. 3) plans to measure k_T for di-jet events (Fig. 4) and will use vertical polarization. The projected sensitivity is based on existing unpolarized data for the azimuthal angle difference between pairs of midrapidity jets.[22] PHENIX plans to measure k_T by detecting pairs of hadrons in their central arms whose symmetry requires radial polarization to observe a spin effect. Non-zero transverse SSA for midrapidity dihadron production may have contributions from both the Sivers effect and the Collins effect.[23]

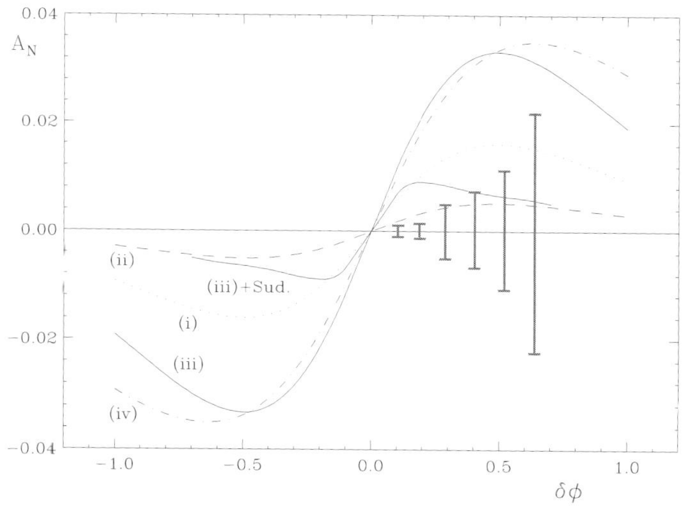

Fig. 4. Projected sensitivity to the analyzing power versus the azimuthal angle difference between pairs of jets detected at midrapidity for a data sample corresponding to $5\,\mathrm{pb}^{-1}$ with beam polarization of 50% compared to theoretical expectations from different models of the gluon Sivers function as discussed in the text.[24]

A portion of the upcoming RHIC run will be devoted to collisions of transversely polarized protons at $\sqrt{s} = 62\,\mathrm{GeV}$. BRAHMS aims to measure transverse SSA for inclusive production of identified charged hadrons at large rapidity ($\eta \approx 3.3$ and 3.9) from these collisions. Their particle identification apparatus will permit measurements up to $x_F \approx 0.6$ at the lower $\sqrt{s}$. The unpolarized cross section systematics discussed earlier [15] would greatly benefit from new forward angle results at $\sqrt{s} = 62\,\mathrm{GeV}$. In the remainder of this section, I'll discuss plans in the upcoming RHIC run for measurements with increased acceptance forward calorimetry in STAR.

An important goal is to address the relative contributions from the

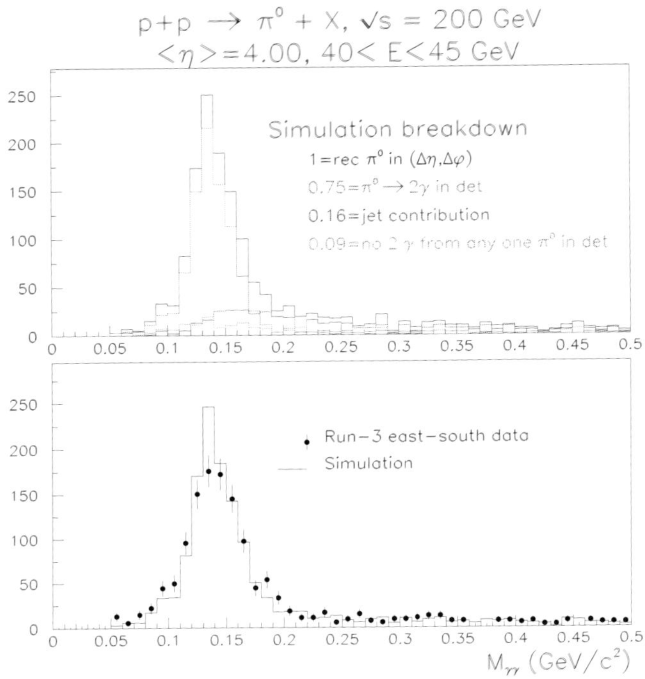

Fig. 5. (bottom) Di-photon invariant mass distribution for events with at least two reconstructed photons. Data compares well to PYTHIA/GEANT simulations. (top) Contributions to simulated di-photon invariant mass distribution for $N_\gamma > 1$ events. Jet-like events are evident.

Collins and Sivers effects to the transverse SSA observed for inclusive forward pion production. One way to disentangle the contributions is to address the question "*is there a significant transverse single-spin asymmetry for jet-like events in $p+p$ collisions?*"

Jet-like events are defined as having three or more photons which are mostly a π^0 and accompanying particles or single photon daughters from two or more π^0. In either case, multiple fragments of the parton scattered through small angles are observed, making the events manifestly jet like. If the detector acceptance for the observed particles is azimuthally symmetric around the thrust axis of the forward scattered parton, then a transverse SSA for jet-like events must be due to the Sivers effect.[1] Integration over all particles detected in an acceptance that is azimuthally symmetric around the thrust axis ensures cancellation of possible contributions from spin- and k_T-dependent fragmentation functions that serve to analyze quark polarizations transferred to the final state (Collins effect [2]). Particularly for events at large x_F, the forward π^0 carries a large fraction of the energy of the forward scattered parton. A precise definition of jet-like behavior is required.

We know that jet-like events are present at large η from results with the STAR Forward Pion Detector (FPD). Fig. 5 shows the reconstructed invariant mass distribution, where $M_{\gamma\gamma} = E_{\mathrm{trig}}\sqrt{1 - z_{\gamma\gamma}^2}\sin(\phi_{\gamma\gamma}/2)$. The total energy ($E_{\mathrm{trig}}$) corresponds to the sum of energy from all towers of one of the FPD modules and is taken as the π^0 energy in the analysis. It is used in conjunction with the opening angle ($\phi_{\gamma\gamma}$) and energy sharing, $z_{\gamma\gamma} = |(E_{\gamma 1} - E_{\gamma 2}|/(E_{\gamma 1} + E_{\gamma 2})$, from the two highest energy photons reconstructed in the event. Jet-like events occur when more than two photons are found, resulting in $E_{\mathrm{trig}} > E_{\pi}$ and therefore $M_{\gamma\gamma} > M_{\pi}$. This is observed in Fig. 5 and is accounted for by simulation, which is decomposed into its various contributions. But, events in the FPD at a given x_F occur primarily in portions of the calorimeter closest to the beam because this minimizes p_T (see right panel of Fig. 2). For such events the FPD does not have azimuthally symmetric acceptance around the thrust axis for additional particles distributed around the reconstructed π^0.

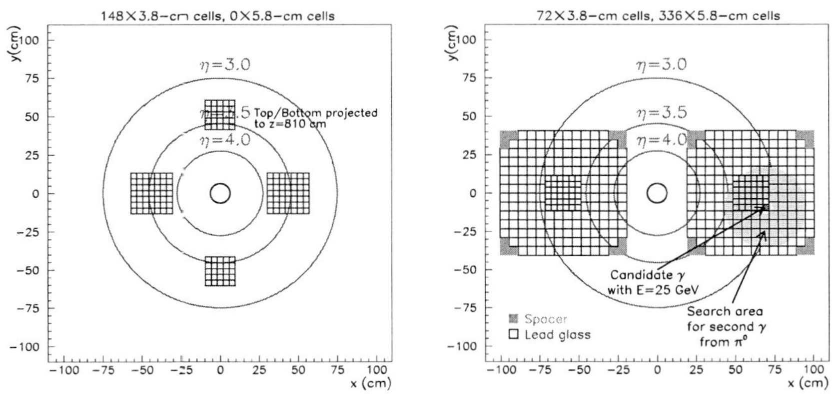

Fig. 6. (left) Layout of STAR forward pion detector used in run 5. (right) Layout of STAR FPD++ that is planned for use in run 6.

The issue of azimuthally symmetric coverage for jet-like events is resolved by an upgrade known as the FPD++ (Fig. 6) that has been built for the upcoming RHIC run as an engineering test of the STAR Forward Meson

Spectrometer (FMS).[25] It consists of two left/right symmetric calorimeters that replace the FPD modules west of the STAR interaction point. The original FPD modules remain on the east side of STAR and are planned to improve the precision of transverse SSA measurements at large x_F. As shown in Fig. 6, the inner portion of each calorimeter module is essentially identical to the FPD. The outer portion of the calorimeter consists of larger cells [26] that are placed with azimuthal symmetry about the inner portion. Events can be selected with the FPD++ in an identical manner as used for the FPD that result in sizeable transverse SSA for π^0 production. The additional detector coverage can be queried for evidence of additional photons that accompany a trigger π^0 thereby signaling jet-like events. Based on the di-photon invariant mass distribution and the photon multiplicity distribution, at least 16% of the π^0 events observed in the FPD with $E_\pi \geq 20\,\text{GeV}$ are accompanied by additional photons in jet-like events.

Fig. 7 shows PYTHIA 6.222 [27] simulations that provide an operational definition of what we mean by jet-like events. PYTHIA is expected to have predictive power in this kinematics because it has been previously shown to agree with measured forward pion cross sections.[28] To explore jet-like events, minimum-bias PYTHIA events are selected when the summed photon energy in the inner portion of a FPD++ module, defined as E_{trig}, exceeds 18 GeV. To facilitate possible reconstruction of forward $\pi^0 + \pi^0$ pairs, events are further required to have more than 3 photons within the full acceptance of an FPD++ module. These requirements mean that selected events with energy from incident photons summed over the entire FPD++ module (E_{sum}) may exhibit jet-like behavior. The upper left panel of Fig. 7 shows the pseudorapidity distribution of the most forward angle hard-scattered parton when $E_{\text{sum}} > 40\,\text{GeV}$. It is peaked at $\eta \approx 3.3$, corresponding to the location of the triggering portion of the FPD++ module. The small background near midrapidity has contributions from large Bjorken x quarks that emit initial-state radiation that subsequently scatters from soft gluons from the other proton. The distribution of the photon energy relative to the thrust axis of the forward scattered parton is shown in the upper right panel of Fig. 7. Jet-like behavior is evident, although evidence for contributions from the underlying event is also present. The summed photon energy within the FPD++ acceptance gives a good representation of the forward scattered parton, albeit shifted in its energy scale. Furthermore, the vector sum of the detected photon momenta faithfully reconstructs (η_{recon}) the direction of the scattered parton. From the middle right panel of Fig. 7, the symmetry of the $\delta\eta = \eta_{\text{recon}} - \eta_{\text{parton}}^{\text{max}}$ distribution

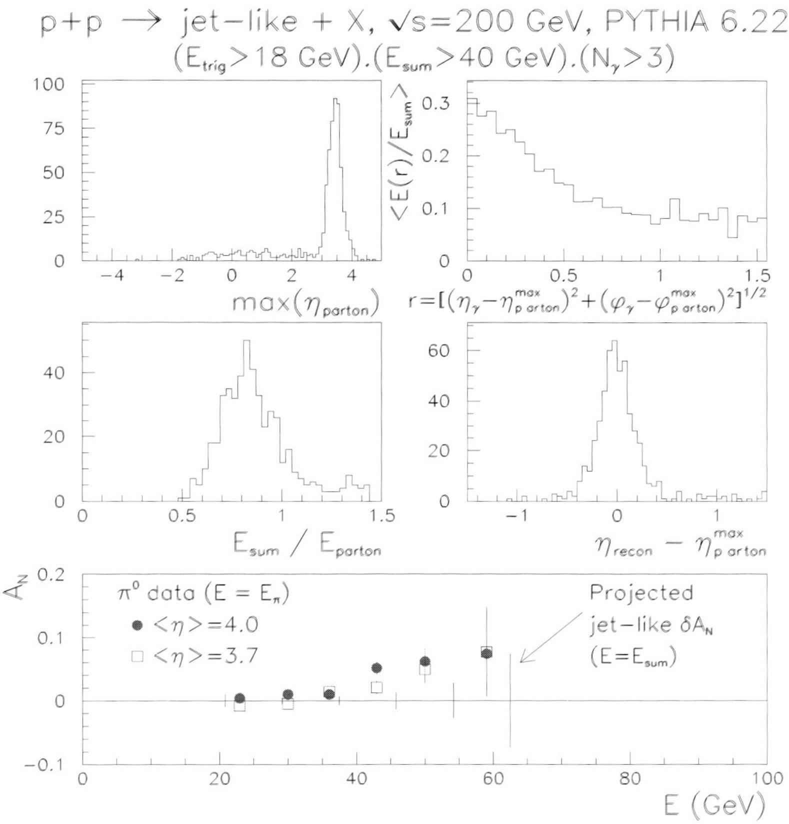

Fig. 7. PYTHIA simulations of FPD++ response for $p+p$ collisions at $\sqrt{s} = 200\,\text{GeV}$. E_{trig} (E_{sum}) is the energy sum in the central section (entire) of the calorimeter (Fig. 6. (upper left) most forward hard-scattered parton pseudorapidity distribution; (upper right) distribution of photon energy relative to the thrust axis showing a standard jet shape atop an underlying event contribution; (middle left) photon energy sum scaled by the most forward hard-scattered parton energy; (middle right) difference distribution of the η reconstructed from the vector sum of the detected photon momenta and the parton η; (bottom) projected statistical precision for jet-like events for $5\,\text{pb}^{-1}$ of polarized proton integrated luminosity.

indicates that the underlying event is not skewed from fragments of the beam jets. Projections for the uncertainties that could be measured on the A_N for these jet-like events with $5\,\text{pb}^{-1}$ of integrated luminosity recorded in a data sample with beam polarization of 50% are shown in the bottom panel of Fig. 7.

In addition, the FPD++ is expected to allow for robust detection of large x_F direct photon production. Rather than searching for coincident photons, the outer portion of the calorimeters can serve to remove neutral meson contributions, to extract events with only a single energetic photon observed in the calorimeter. The resulting yield would be predominantly prompt photon events, including direct photons and fragmentation photons. The calorimetric coverage will allow suppression of most of the latter events. We expect $\approx 70{,}000$ direct photon events with $E \geq 25\,\text{GeV}$ in a data sample from the FPD++ that records $5\,\text{pb}^{-1}$ of $p+p$ collisions at $\sqrt{s} = 200\,\text{GeV}$. These photon energies are larger than the simulated cutoff in the distribution of energy deposition by charged hadrons incident on the calorimeter.

The left/right symmetry of the FPD++ is important for the cancellation of systematic errors. With that symmetry, so-called cross ratio methods can be used for extracting single-spin asymmetries for π^0 production, jet-like events and prompt photon events. Another benefit of the symmetry is for coincident events. The opening angle between the two calorimeters is well matched to that required for large-x_F production of objects with invariant mass of order $3\,\text{GeV}/c^2$ that decay to either photons, neutral mesons or electron–positron pairs.

The end result is that the upcoming RHIC run should provide exciting results for transverse spin physics. Perhaps of greatest interest is the prospect for isolation of the Sivers function both at midrapidity and for forward particle production. Its isolation may establish the dynamical origin of transverse SSA for large-x_F π production and may conclusively demonstrate orbital motion of the constituents of the proton.

4. Acknowledgements

The forward calorimetry work is done with the STAR collaboration. I would like to acknowledge all members of the RHIC spin collaboration (RSC) for making the RHIC-spin program a reality. The RSC is a group including members of the RHIC experiments, members of the BNL Collider-Accelerator Department, experts in the RHIC polarimeters required to determine the beam polarization, and theorists. I would also like to thank Akio Ogawa, Hank Crawford, Jack Engelage, Carl Gagliardi, Steve Heppelmann, Larisa Nogach, Greg Rakness, Gerry Bunce and Werner Vogelsang for their comments and help. This work was supported by DOE Medium Energy Physics.

References

1. D. Sivers, *Phys. Rev.* **D41** (1990) 83; **43** (1991) 261.
2. J. Collins, *Nucl. Phys.* **B396** (1993) 161; J. Collins, S.F. Heppelmann, G.A. Ladinsky, *Nucl. Phys.* **B420** (1994) 565.
3. S.J. Brodsky, D.S. Hwang and I. Schmidt, *Phys. Lett.* **B539** (2002) 99.
4. A. Airapetian *et al.* (HERMES collaboration), *Phys. Rev. Lett.* **94** (2005) 012002.
5. K. Abe *et al.* (Belle collaboration), to be published in the proceedings of the Lepton Photon 2005 Conference [hep-ex/0507063].
6. J. Adams *et al.* (STAR collaboration), *Phys. Rev. Lett.* **92** (2004) 171801.
7. M. Anselmino, M. Boglione, and F. Murgia, *Phys. Rev.* **D60** (1999) 054027; M. Boglione and E. Leader, *Phys. Rev.* **D61** (2000) 114001.
8. M. Anselmino, M. Boglione, and F. Murgia, *Phys. Lett.* **B362** (1995) 164; M. Anselmino and F. Murgia, *ibid.* **B442** (1998) 470; U. D'Alesio and F. Murgia, *AIP Conf. Proc.* **675** (2003) 469.
9. J. Qiu and G. Sterman, *Phys. Rev.* **D59** (1998) 014004.
10. Y. Koike, *AIP Conf. Proc.* **675** (2003) 449.
11. D. L. Adams *et al.*, *Phys. Lett.* **B261** (1991) 201; **B264** (1991) 462.
12. M. Anselmino, M. Boglione, U. D'Alesio, E. Leader and F. Murgia, *Phys. Rev.* **D71** (2005) 014002.
13. D. Morozov (for the STAR collaboration), to be published in the proceedings of XI Workshop on High Energy Spin Physics [hep-ex/0512013].
14. F. Vidabaek (for the BRAHMS collaboration), to be published in the proceedings of the PANIC 2005 Conference [nucl-ex/0601008].
15. C. Bourrely and J. Soffer, *Eur. Phys. J.* **C36** (2004) 371.
16. A. Ogawa (for the STAR collaboration), published in the proceedings of DIS2004 [hep-ex/0408004].
17. S.S. Adler *et al.* (PHENIX collaboration), *Phys. Rev. Lett.* **91** (2003) 241803.
18. S.S. Adler *et al.* (PHENIX collaboration), *Phys. Rev. Lett.* **95** (2005) 202001.
19. M. Miller (for the STAR collaboration), to be published in the proceeding of the PANIC 2005 conference.
20. F. Aversa *et al.*, *Nucl. Phys.* **B327** (1989) 105; B. Jager *et al.*, *Phys. Rev.* **D67** (2003) 054005; D. de Florian, *Phys. Rev.* **D67** (2003) 054004.
21. D. Boer and W. Vogelsang, *Phys. Rev.* **D69** (2004) 094025.
22. T. Henry (for the STAR collaboration), J. Phys. **G30** (2004) S1287.
23. A. Bacchetta, C.J. Bomhof, P.J. Mulders and F. Pijlman, *Phys. Rev.* **D72** (2005) 034030.
24. C. Aidala *et al.*, *Research Plan for Spin Physics at RHIC*, February, 2005 (`http://spin.riken.bnl.gov/rsc/report/masterspin.pdf`).
25. L.C. Bland *et al.*, *Eur. Phys. J.* **C43** (2005) 427.
26. J. Cumalat, *Int. J. Mod. Phys.* **A20** (2005) 3692.
27. T. Sjöstrand, P. Eden, C. Friberg, L. Lönnblad, G. Miu, S. Mrenna, E. Norrbin, *Comput. Phys. Commun.* **135** (2001) 238.
28. L.C. Bland (for the STAR collaboration), contribution to the 10th Workshop on High Energy Spin Physics [hep-ex/0403012].

STUDIES OF TRANSVERSE SPIN EFFECTS AT JLAB

H. Avakian, P. Bosted, V. Burkert and L. Elouadrhiri
(for the CLAS Collaboration)
Jefferson Lab., Newport News, VA 23606, USA

We present ongoing and future studies of single-spin asymmetries in semi-inclusive electroproduction of pions using the CEBAF polarized electron beam. Kinematic dependences of single-spin asymmetries have been measured in a wide kinematic range at CLAS with a polarized NH_3 target. Significant target-spin $\sin 2\phi$ and $\sin\phi$ asymmetries have been observed, indicating a non-zero Collins fragmentation function and supporting future SIDIS measurements with upgraded JLab.

1. Factorization in Semi-Inclusive DIS

In recent years it has become clear that appropriate exclusive and semi-inclusive scattering processes may provide access to parton distributions, generalized to account for not only longitudinal but also transverse degrees of freedom of partons.[1–16] SIDIS cross section at leading twist has eight contributions related to different combinations of polarization states of the incoming lepton and the target nucleon.[17] Corresponding structure functions factorize into transverse momentum dependent (TMD) parton distribution and fragmentation functions (FF), and soft and hard parts.[17] It was shown, that the hard factors in the SIDIS cross section for different contributions are similar at one-loop order [17] and may cancel to large extent in asymmetry observables.

Single-spin asymmetries (SSA) in the distribution of lepto-produced hadrons in the azimuthal angle around the virtual photon direction are a valuable tool for the exploration of transverse spin and momentum degrees of freedom in nucleon structure. A key goal of Semi-Inclusive DIS (SIDIS) studies at moderate energies accessible at JLab is to carefully study the transition between the nonperturbative and perturbative regimes of QCD using simultaneous measurements of the Q^2 and x dependences of cross sections and beam/target spin asymmetries for different final state hadrons and extraction of the corresponding structure functions and separation of the contributions of different distribution and fragmentation functions.

A major issue in studies of semi-inclusive scattering at moderate beam energies is the separation of contributions from current fragmentation (active parton) and target fragmentation (spectators). Systematic studies of factorization of x, z and $P_{h\perp}$ dependences for spin-dependent observables and for different pion flavors will be required to do a flavor decomposition of various PDFs from measured spin and azimuthal asymmetries in SIDIS. It was shown that the CLAS data [18,19] already at existing energies are consistent with factorization and suggest that the high-energy description of the SIDIS process can be extended to the moderate energies of the CLAS measurements.

2. Single-Spin Asymmetries

Large SSAs, observed in hadronic reactions for decades [20,21] have been among the most difficult phenomena to understand from first principles in QCD. Recently, SSA measurements were reported in semi-inclusive DIS (SIDIS) by the HERMES collaboration at HERA [22–24] for longitudinally and transversely polarized targets, by the COMPASS collaboration with a transversely polarized target [25] and by the CLAS collaboration at JLab with a polarized beam and longitudinally polarized target.[18,19] In general, such single-spin asymmetries require a correlation of a particle spin direction and the orientation of the production or scattering plane. In hadronic processes, such correlations can provide a window into the physics of initial and final state interactions at the parton level.[8]

2.1. *Leading order SSA*

For polarized targets, several azimuthal asymmetries already arise at leading order. The following contributions were investigated in Refs. 3, 5, 8, 11, 26 and 27:

$$\sigma_{UL}^{\sin 2\phi} \propto S_L 2(1-y)\sin 2\phi \sum_{q,\bar{q}} e_q^2 x h_{1L}^{\perp q}(x) H_1^{\perp q}(z), \tag{1}$$

$$\begin{aligned}\sigma_{UT}^{\sin\phi} &\propto S_T(1-y)\sin(\phi-\phi_S)\sum_{q,\bar{q}} e_q^2 x h_1(x) H_1^{\perp q}(z),\\ &\quad + S_T(1-y+y^2/2)\sin(\phi+\phi_S)\sum_{q,\bar{q}} e_q^2 x f_{1T}^{\perp q}(x) D_1^q(z),\end{aligned} \tag{2}$$

The ϕ is the azimuthal angle between the scattering plane formed by the initial and final momenta of the electron and the production plane formed by the transverse momentum of the observed hadron and the virtual photon, ϕ_S is the azimuthal angle of the transverse spin in the scattering plane, S_L and S_T are longitudinal and transverse polarizations with respect to virtual photon direction.

The spin-dependent moments $(\sin\phi, \sin 2\phi)$ of the semi-inclusive cross section can be extracted in a fit to the normalized-yield asymmetry

$$A_{UL}(\phi) = \frac{1}{P_T}\frac{N^+ - N^-}{N^+ + N^-}. \tag{3}$$

Here $N^{\pm}$ is the number of events for target polarizations antiparallel/parallel to the incoming beam direction, P_T is the target polarization,

The $\sin\phi$ moment of the SIDIS cross section with a transversely polarized target (σ_{UT})[28] contains contributions both from the Sivers effect (T-odd distribution)[4] and the Collins effect (T-odd fragmentation).[3] Contributions to transverse SSAs from T-odd distributions of initial quarks ($f_{1T}^{\perp q}(x)$ term) and T-odd fragmentation of final quarks ($H_1^{\perp q}(z)$ term) could be separated by their different azimuthal dependence, see Eq. (2).

The HERMES Collaboration has recently measured a transverse spin asymmetry in SIDIS providing the cleanest evidence to date for the existence of a non-zero Collins function,[24] which describes the fragmentation of a transversely polarized quark into pions. This finding is supported by the preliminary data from BELLE[29] indicating a non-zero Collins effect. The large target SSA in semi-inclusive pion production measured at CLAS and analyzed in terms of the Collins fragmentation,[30] also indicate a significant Collins function. To reveal the source of SSA and accomplish the separation of Collins and Sivers contributions, measurements with different target polarizations and with detection of different final state particles may be required.

One important, unique feature of Collins mechanism is the presence of a leading twist $\sin 2\phi$ dependent SSA, so-called "Kotzinian–Mulders asymmetry",[26] the only azimuthal asymmetry arising for the longitudinally polarized target in leading order[5,26,27], see Eq. (1). This contribution involves the TMD Collins fragmentation function $H_1^{\perp}$[3] and the Mulders distribution function $h_{1L}^{\perp}$,[1,5] describing the transverse polarization of quarks in a longitudinally polarized proton.[5,26,27] The distribution function $h_{1L}^{\perp}$ is also accessible in double polarized Drell–Yan, where it gives rise to a $\cos 2\phi$ azimuthal moment in the cross section.[31]

Measurements of the $\sin 2\phi$ SSA allow the study of the Collins effect with no contamination from other mechanisms. A recent measurement of $\sin 2\phi$ moment of σ_{UL} by HERMES[22] is consistent with zero. A measurably

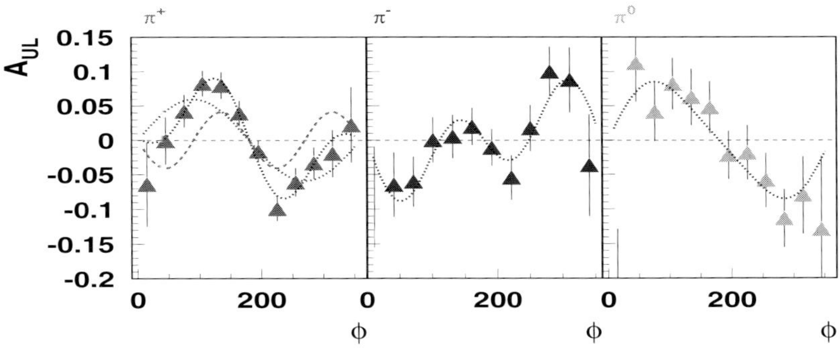

Fig. 1. Preliminary target SSA as a function of azimuthal angle ϕ from data at 5.7 GeV. Dashed and dotted lines correspond to $\sin 2\phi$ and $\sin\phi$ moments respectively

large asymmetry has been predicted only at large x ($x > 0.2$), a region well-covered by JLab.[32] The data for π^+ (Fig. 1) show a clear $\sin\phi$ and $\sin 2\phi$ modulations from which a $\sin\phi$ moment of $0.058 \pm 0.011(stat)$ and $\sin 2\phi$ moment of $-0.041 \pm 0.011(stat)$ have been determined.

Future measurements approved for longitudinal target with a 6 GeV beam will significantly improve the quality of this measurement (see Fig. 2). Upgrade of JLab, enabling measurements at 11 GeV and at much higher luminosity, will help to pin down the corresponding transverse momentum distribution functions.

Projections for transverse target single-spin asymmetry measurements with CLAS at 11 GeV are plotted in Figs. 3 and 4. The curves are calculated by Efremov *et al.* for $z > 0.45$ and the missing mass of the $e'\pi$ system ($M_X(\pi^+) > 1.1$ GeV). The asymmetry is integrated over all hadron transverse momenta. The extraction of the transversity from $A_{UT}^{\sin\phi}$ could be performed using parametrizations for the unpolarized distribution functions $u(x)$ and $\bar{d}(x)$ and certain approximations for the polarized Collins fragmentation function $H_1^{\perp}$.

The same measurement of the transverse asymmetry, provides also access to the Sivers effect. Measurements with proton and neutron targets will enable direct extraction of information on flavor partners of the Sivers function, as the extraction of the Sivers distribution function does not require additional information on the polarized quark (Collins) fragmentation functions. Furthermore, using transverse target asymmetries for π^0 would allow extraction of Sivers function even without any knowledge of the unpolarized fragmentation functions providing a model independent way for Sivers

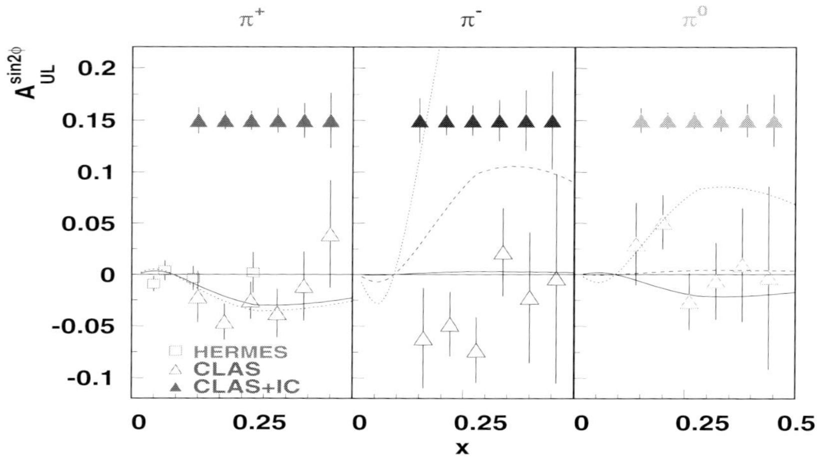

Fig. 2. Dependence on x of longitudinally polarized target SSA, $A_{UL}^{\sin 2\phi}$. Squares are HERMES data for $A_{UL}^{\sin 2\phi}$, open triangles are CLAS measurements with NH_3 target and triangles represent expected statistical errors from JLAB E05-113 proposal at 6 GeV with 1000 hours of data taking. The curves are the theory prediction based on the Collins mechanism[32] for different ratios of $H_{\mathrm{unfav}}/H_{\mathrm{fav}}$. The solid, dashed and dotted lines are respectively for 0, -1.1, and -5.

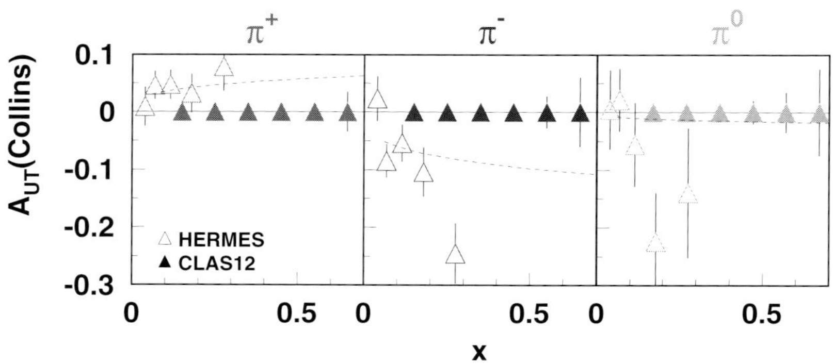

Fig. 3. Projected transverse spin asymmetry from the Collins effect ($A_{UT}^{\sin(\phi+\phi_S)}$) in single π production with CLAS at 11 GeV. The projected error bars have been calculated assuming a luminosity of $10^{35}\,\mathrm{cm}^{-2}\,\mathrm{s}^{-1}$, with a NH_3 target polarization of 85% and a dilution factor 0.176, and 2000 hours of data taking.

function studies. The π^0 studies will also have an important advantage of being less affected by exclusive vector meson background. Projections for transverse asymmetry measurements with upgraded CLAS and transverse target and also for extraction of the Sivers distribution function are shown in Fig. 4.

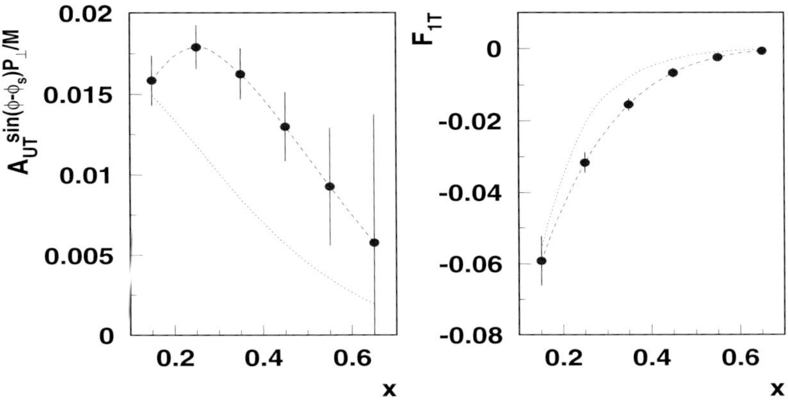

Fig. 4. Projected transverse spin asymmetry from the Sivers effect ($A_{UT}^{\sin(\phi-\phi_S)}$) in single π production with CLAS at 11 GeV. F_{1T} is the sum of Sivers distributions of u and d quarks weighted by their charges squared. The curves are theory predictions for two models of the Sivers function.[32]

2.2. *Sub-leading order SSA*

With better experimental accuracy it may be possible to isolate the higher-twist effects in hard processes, which arise from the quantum mechanical interference of partons in the interacting hadrons. The higher-twist terms are important for understanding long-range quark–gluon dynamics and may be accessible through measurements of certain asymmetries,[2,5,27,33] where they appear as leading contributions.

The complete analysis up to sub-leading-twist and leading order in α_s of longitudinal single-spin asymmetries in semi-inclusive DIS was presented in Ref. 34, completing previous work of Refs. 5 and 28. All contributions involve either combinations of sub-leading-twist distribution functions (e, h_L, $g^\perp$, f_L) with leading-twist fragmentation functions or of leading-twist distribution functions in conjunction with sub-leading-twist fragmentation functions ($G^\perp$, E).[14,34,35]

Distribution functions $g^\perp$ and f_L [14,34] are odd under time reversal (T-odd) and do not have a simple probabilistic interpretation. Their contributions are similar to the leading order Sivers effect, so the same extraction procedure can be used as in case of Sivers functions from measurements of SSAs for π^0s (see Fig. 4). In general both SSAs contain additional contributions depending on the fragmentation function $G^{\perp}$, which is at present unknown. This makes beam and target single-spin asymmetry measurements complimentary and global analysis is required to separate different contributions.

So far factorization has been proven only for leading-twist observables in semi-inclusive deep-inelastic scattering with hadrons in the current fragmentation region. A factorization proof for sub-leading-twist observables is still open. At the moment there are only few model predictions [14,32] and no firm experimental information about any of the sub-leading-twist terms exist.

Measurements of A_{UL} and A_{LU} reported by HERMES [22,23,36] and CLAS [18,19] are first indications about the size of such sub-leading-twist effects in azimuthal target-spin and beam-spin asymmetries (see Fig. 5).

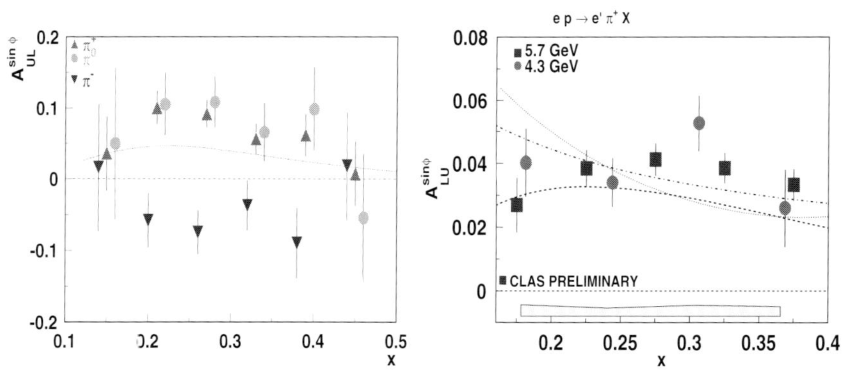

Fig. 5. Sub-leading SSA with polarized target (left) or beam (right). The $\sin\phi$ moment of target SSA for 5.7 GeV beam extracted for π^+, π^-, and π^0. The beam-spin azimuthal asymmetry ($\sin\phi$ moment of the cross section) extracted from hydrogen data at 5.7 GeV (squares) and 4.3 GeV (circles) as a function of x_B in a range $0.5 < z < 0.8$. Curves represent calculations performed assuming only the Sivers effect (dash-dotted lines,[35] dotted [37]) or only the Collins effect (dashed line [32]) contributes to A_{LU}.

3. Conclusions

An experimental investigation of properties of 3D PDFs, requiring precision measurements at upgraded JLab, would serve as an important check of our present day understanding of nuclear structure in terms of quark and gluon properties. Studies at 12 GeV would provide important input for global analysis of observed spin and azimuthal asymmetries, leading to the separation of different contributions and extraction of underlying, essentially unexplored distribution and fragmentation functions.

References

1. J.P. Ralston and D.E. Soper, *Nucl. Phys.* **B152** (1979) 109.
2. R.L. Jaffe and X. Ji, *Nucl. Phys.* **B375** (1992) 527.
3. J.C. Collins, *Nucl. Phys.* **B396** (1993) 161.
4. D.W. Sivers, *Phys. Rev.* **D41** (1990) 83.
5. P.J. Mulders and R.D. Tangerman, *Nucl. Phys.* **B461** (1996) 197; *Erratum-ibid.* **B484** (1996) 538.
6. D. Boer and P.J. Mulders, *Phys. Rev.* **D57** (1998) 5780.
7. M. Anselmino and F. Murgia, *Phys. Lett.* **B442** (1998) 470.
8. S.J. Brodsky, D.S. Hwang and I. Schmidt, *Phys. Lett.* **B530** (2002) 99.
9. V. Barone, A. Drago and P.G. Ratcliffe, *Phys. Rep.* **359** (2002) 1.
10. J.C. Collins, *Phys. Lett.* **B536** (2002) 43.
11. X. Ji and F. Yuan, *Phys. Lett.* **B543** (2002) 66.
12. A.V. Belitsky, X. Ji and F. Yuan, *Nucl. Phys.* **B656** (2003) 165.
13. J. Collins and A. Metz, *Phys. Rev. Lett.* **93** (2004) 252001.
14. A. Metz and M. Schlegel, *Eur. Phys. J.* **A22** (2004) 489.
15. K. Goeke, A. Metz and M. Schlegel, *Phys. Lett.* **B612** (2005) 233.
16. X. Ji, J.-P. Ma and F. Yuan, *Phys. Rev.* **D71** (2005) 034005.
17. X. Ji, J.-P. Ma and F. Yuan, *Phys. Lett.* **B597** (2004) 299.
18. H. Avakian *et al.*, *Phys. Rev.* **D69** (2004) 112004.
19. H. Avakian, P. Bosted, V. Burkert and L. Elouadrhiri, nucl-ex/0509032.
20. K. Heller *et al.*, Proc. of Spin 96, Amsterdam, Sept. 1996, p. 23
21. E704 collaboration (A. Bravar *et al.*), *Phys. Rev. Lett.* **77** (1996) 2626.
22. A. Airapetyan *et al.*, *Phys. Rev. Lett.* **84** (2000) 4047.
23. A. Airapetyan *et al.*, *Phys. Rev.* **D55** (2001) 097101.
24. A. Airapetyan *et al.*, *Phys. Rev.* **D64** (2004) 097101.
25. V.Yu. Alexakhin *et al.*, *Phys. Rev. Lett.* **94** (2005) 202002.
26. A.M. Kotzinian and P.J. Mulders, *Phys. Rev.* **D54** (1996) 1229.
27. A. Kotzinian, *Nucl. Phys.* **B441** (1995) 234.
28. D. Boer and P. Mulders, *Phys. Rev.* **D57** (1998) 5780.
29. Belle Collaboration (K. Abe *et al.*), hep-ex/0507063.
30. A. Efremov, *Annalen Phys.* **13** (2004) 651
31. R.D. Tangerman and P.J. Mulders, *Phys. Rev.* **D51** (1995) 3357.
32. A.V. Efremov *et al.*, *Phys. Rev.* **D67** (2003) 114014; hep-ph/0412420.
33. J. Levelt and P.J. Mulders, *Phys. Lett.* **B338** (1994) 357 [hep-ph/9408257].
34. A. Bacchetta, P.J. Mulders and F. Pijlman, *Phys. Lett.* **B595** (2004) 309.
35. F. Yuan, *Phys. Lett.* **B589** (2004) 28.
36. E. Avetisyan, A. Rostomyan, A. Ivanilov, hep-ex/0408002.
37. A. Afanasev and C. Carlson, hep-ph/0308163.

NEUTRON TRANSVERSITY AT JEFFERSON LAB

J.P. Chen

Jefferson Lab, Newport News, VA 23606, USA
E-mail: jpchen@jlab.org

X. Jiang

Rutgers University, Piscataway, NJ 08855, USA

J.-C. Peng, L. Zhu

University of Illinois, Urbana, IL 61801, USA

for the Jefferson Lab Hall A Collaboration

Nucleon transversity and single transverse spin asymmetries have been the recent focus of large efforts by both theorists and experimentalists. On-going and planned experiments from HERMES, COMPASS and RHIC are mostly on the proton or the deuteron. Presented here is a planned measurement of the neutron transversity and single target spin asymmetries at Jefferson Lab in Hall A using a transversely polarized ^{3}He target. Also presented are the results and plans of other neutron transverse spin experiments at Jefferson Lab. Finally, the factorization for semi-inclusive DIS studies at Jefferson Lab is discussed.

1. Introduction

After forty years of extensive experimental and theoretical efforts, the unpolarized Parton Distribution Functions (PDFs) have been extracted from DIS, Drell–Yan and other processes with excellent precision over a large range of x. The comparison of the structure functions in a large range of Q^2 with QCD evolution equations have provided one of the best tests of QCD. Since the "proton spin crisis" in the 1980s, very active spin-physics program have been carried out at CERN, SLAC and HERA, and, recently, at JLab and RHIC. Longitudinally polarized parton distribution functions have been extracted by a number of groups in recent years, although the precision is not as good as that of the unpolarized PDFs. The other equally important parton distribution functions, the transversity distributions, have only been explored recently.

1.1. *Transversity*

The transversity distributions, $\delta q(x, Q^2)$, are fundamental leading-twist (twist-2) quark distributions, same as the unpolarized and polarized parton distributions, $q(x, Q^2)$ and $\Delta q(x, Q^2)$. In quark–parton models, they describe the net transverse polarization of quarks in a transversely polarized nucleon.[1] There are several special features for the transversity distributions, making them uniquely interesting:

- The difference between the transversity and the longitudinal distributions is purely due to relativistic effects. In the absence of relativistic effects (as in the non-relativistic quark model, where boosts and rotations commute), the transversity distributions are identical to the longitudinally polarized distributions.
- The quark transversity distributions do not mix with gluonic effects[2] and therefore follow a much simpler evolution and have a valence-like behavior.
- The positivity of helicity amplitudes leads to the Soffer's inequality for the transversity:[3] $|h_1^q| \leq \frac{1}{2}(f_1^q + g_1^q)$.
- The lowest moment of h_1^q measures a simple local operator analogous to the axial charge, known as the "tensor charge", which can be calculated from lattice QCD.

Due to the chiral-odd nature of the transversity distribution, it can not be measured in inclusive DIS experiments. In order to measure $\delta q(x, Q^2)$, an additional chiral-odd object is required, such as double-spin asymmetries in Drell–Yan processes, single target spin azimuthal asymmetries in Semi-Inclusive DIS reactions, double-spin asymmetries in Λ production from e-p and p-p reactions and single-spin asymmetries in double pion production from e-p scattering. The first results, from measurements performed by the HERMES[4] and COMPASS[5] collaborations with SIDIS offered the first glimpse of possible effects caused by the transversity distributions.

1.2. *Semi-Inclusive Deep-Inelastic-Scattering*

Semi-inclusive DIS is a powerful tool for probing various parton distributions and fragmentation functions. For producing a spin-zero meson, the SIDIS differential cross section at leading order contains 8 structure functions.[6,7] The single transverse target spin asymmetry term contains the Collins term,[8] which contains the product of the transversity and the Collins fragmentation function and is proportional to $\sin(\phi_h^\ell + \phi_S^\ell)$, the

Sivers term,[9] which contains the product of the Sivers structure function with the regular fragmentation function and is proportional to $\sin(\phi_h^\ell - \phi_S^\ell)$, and a third term proportional to $\sin(3\phi_h^\ell - \phi_S^\ell)$. Azimuthal angles are defined relative to the lepton plane, e.g. $\phi_h^\ell = \phi_h - \phi^\ell$ is the angle between the hadron plane and the lepton plane and $\phi_S^\ell = \phi_S - \phi^\ell$ is the angle between the spin plane and the lepton plane. SIDIS can be used to extract all eight of the twist-2 structure functions by choosing different beam helicity and target polarization configurations, and detecting the azimuthal angular dependences (ϕ_h^l and ϕ_s^l).

1.3. *Experimental Status*

The HERMES collaboration reported two years ago the observation of single-spin azimuthal asymmetries for charged and neutral hadron electroproduction.[10] Using an unpolarized positron beam on longitudinally polarized hydrogen and deuterium targets, the cross section was found to have a $\sin\phi_h^l$ dependence. Although a longitudinally polarized target was used in the HERMES experiment, there is a small (≈ 0.15) nonzero value of polarization transverse to the virtual photon direction. The observed azimuthal asymmetries arise from Collins, Sivers and higher-twist contributions from the longitudinal target polarization. For a longitudinally polarized target ($\phi_s^l = 0$) the Collins and the Sivers mechanisms can not be distinguished.

If the azimuthal asymmetry observed by HERMES is indeed partially caused by the nucleon transversity, a much larger asymmetry is expected for a transversely polarized target. The HERMES[4] and COMPASS[5] collaborations have collected polarized SIDIS data using transversely polarized hydrogen and LiD targets, respectively. A simultaneous fit to $\sin(\phi_h^l + \phi_s^l)$ and $\sin(\phi_h^l - \phi_s^l)$ dependences was applied to the HERMES data to extract the "Collins moments" and the "Sivers moments", respectively. Nonzero Collins moments were clearly observed in the HERMES data with positive azimuthal asymmetry for π^+ and negative azimuthal asymmetry for π^-. The unexpectedly large magnitude for the azimuthal asymmetry of π^- relative to π^+ seems to suggest that the disfavored Collins function is of comparable magnitude, but opposite sign, to the favored Collins function. Furthermore, the observation of the $\sin(\phi_h^l - \phi_s^l)$ moments shows that the Sivers function is nonzero and indeed contributes to the azimuthal asymmetry in SIDIS. Preliminary results[5] from the COMPASS 2002 data show that the $\sin(\phi_h^l + \phi_s^l)$ azimuthal asymmetry is consistent with zero for a transversely polarized LiD target. A factor of 4 increase in statistics is expected for the COMPASS 2002–2004 data.

To be able to extract the transversity from the Collins moments, independent measurements of the Collins fragmentation functions are needed. First results of an extraction of the Collins fragmentation functions from $e^+ + e^-$ measurements by the Belle collaboration at KEK is available now.[11]

2. A planned measurement of neutron transversity at JLab

The Thomas Jefferson National Accelerator Facility (Jefferson Lab, or JLab) produces a continuous-wave electron beam of energy up to 6 GeV. An energy upgrade to 12 GeV is planned in the next few years. The electron beam with a current up to 180 μA is polarized up to 85% by illuminating a strained GaAs cathode with polarized laser light. The electron beam goes into three experimental halls (Halls A, B and C) where the electron scattering off various nuclear targets takes place. The experiments reported here studied semi-inclusive (or inclusive) electron scattering where the scattered electron and one scattered hadron (or only the scattered electron) are detected. The neutron experiments presented here are from Hall A [12] where a polarized ^{3}He target, with in-beam polarization of about 40%, provides an effective polarized neutron target. The polarized luminosity reached is $10^{36}\,\mathrm{s}^{-1}\,\mathrm{cm}^{-2}$. There are two High Resolution Spectrometers (HRS) with momentum up to 4 GeV/c. The HRS detector package consisted of vertical drift chambers (for momentum analysis and vertex reconstruction), scintillation counters (data acquisition trigger) and particle identification detectors: gas Čerenkov counters and lead-glass calorimeters for electron detection or Aerogel and heavy gas Čerenkov detectors and RICH detectors for hadron particle identification. In addition to the HRS's, the BigBite spectrometer with a large acceptance is used for detecting electrons for the semi-inclusive experiments where large acceptance is needed. The BigBite detector package consists of drift chambers, scintillators and a lead-glass calorimeter.

A recently approved JLab proposal [13] plans to measure the single-spin asymmetry of the $\vec{n}(e, e'\pi^-)X$ reaction on a transversely polarized ^{3}He target. The goal of this experiment is to provide the first measurement of the neutron transversity, complementary to the ongoing HERMES and COMPASS measurements on the proton and deuteron. This experiment focuses on the valence quark region, $x = 0.19-0.34$, at $Q^2 = 1.77-2.73\,\mathrm{GeV}^2$. Data from this experiment, when combined with data from HERMES and COMPASS, will provide powerful constraints on the transversity distributions of both u-quarks and d-quarks in the valence region.

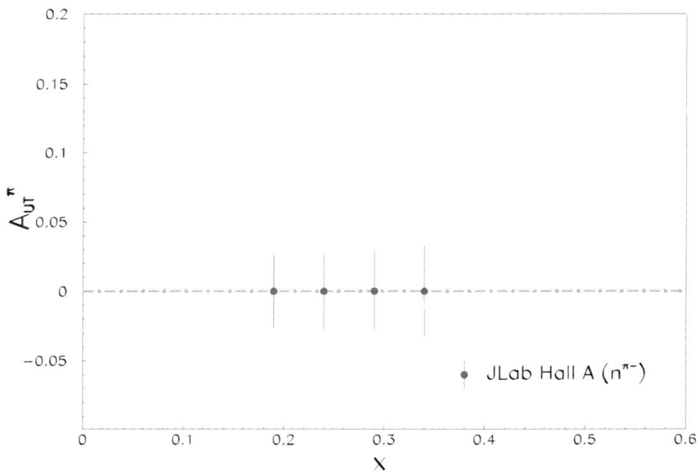

Fig. 1. Expected statistical precision of this experiment.

The experiment will use a 6 GeV electron beam with the Hall A left-side high resolution spectrometer (HRS_L) situated at 16° as the hadron arm, and the BigBite spectrometer located at 30° beam-right as the electron arm. A set of vertical coils will be added to the polarized ^{3}He target to provide tunable polarization directions in all three dimensions. By rotating the target polarization direction in the transverse plane, the coverage in ϕ_s^l is increased, hence facilitating the separation of the Collins and the Sivers effects. Figure 2 shows the expected statistical precision of this experiment with 24 days of beamtime for $\vec{n}(e, e'\pi^-)X$ single-spin asymmetry. Due to the good particle identification in the HRS, K^- data will be collected at the same time, providing a set of precision data to study the transverse spin asymmetries for semi-inclusive K^- production.

The same measurement can be performed with the HRS spectrometer in positive polarity to detect π^+ and K^+. A new proposal is being developed for this measurement.

3. Inclusive neutron transverse spin experiments at JLab

In addition to the transverse spin SIDIS experiments to study transversity, inclusive transverse spin experiments can be used to study higher-twist effects. There are several completed inclusive (double) transverse spin experiments [14–16] which precisely measured the second spin structure function g_2 and its moment d_2. g_2, unlike g_1 and F_1, can not be interpreted in a simple quark–parton model. To understand g_2 properly, it is best to start with the operator product expansion (OPE) method. In the OPE, neglecting quark masses, g_2 can be cleanly separated into a twist-2 and a higher-twist term:

$$g_2(x, Q^2) = g_2^{\mathrm{WW}}(x, Q^2) + g_2^{\mathrm{H.T.}}(x, Q^2) \ . \tag{1}$$

The leading-twist term, g_2^{WW}, can be determined from g_1 [17] and the higher-twist term arises from the quark–gluon correlations. Therefore g_2 provides a clean way to study higher-twist effects. In addition, at high Q^2, the x^2-weighted moment, d_2, is a twist-3 matrix element and is related to the color polarizabilities.[18] Predictions for d_2 exist from various models and lattice QCD.

A precision measurement of g_2^n from JLab E97-103 [14] covered the Q^2 range from 0.58 to 1.36 GeV^2 at $x \approx 0.2$. Results for g_2^n are given on the left panel of Fig. 2. The light-shaded area in the plot gives the leading-twist contribution, obtained by fitting world data [19] and evolving to the Q^2 values of this experiment. The systematic errors are shown as the dark-shaded area near the horizontal axis.

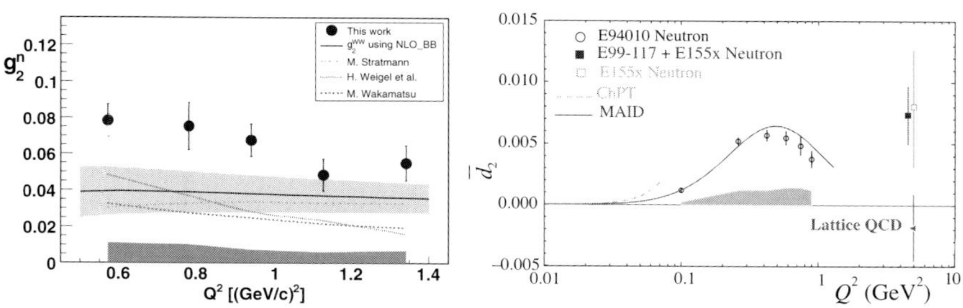

Fig. 2. Results for g_2^n (left) and d_2^n (right) from JLab Hall A.

The precision reached is more than an order of magnitude improvement over that of the best world data.[20] The difference of g_2 from the leading twist part (g_2^{WW}) is due to higher twist effects. The measured g_2^n values are consistently higher than g_2^{WW}. For the first time, there is a clear indication that higher-twist effects become significantly positive at Q^2 below 1 GeV^2, while the bag model [21] and Chiral Soliton model [22,23] predictions of higher-twist effects are negative or close to zero.

The second moment of the spin structure function, d_2, can be extracted from g_1 and g_2 measurements. Due to x^2 weighting, the contributions are dominated by the high-x region and the problem of low-x extrapolation is avoided. The Hall A experiment E99-117 [15] provided data on A_2^n at high-x. Combining these results with the world data,[20] the second moment d_2^n was extracted at an average Q^2 of 5 GeV^2. This result is compared to the previously published result [20] and a calculation by Lattice QCD.[24] While a negative or near-zero value was predicted by Lattice QCD and most models, the new result for d_2^n is positive. Also shown in Fig. 2 are the low Q^2 (0.1–1 GeV^2) results of the inelastic part of d_2^n from another Hall A experiment E94-010,[16,25] which were compared with a Chiral Perturbation Theory calculation [26] and a model prediction.[27]

4. Experimental Tests of Factorization for SIDIS at JLab

Due to the limitation of maximum beam energy (6 GeV now and 12 GeV after the planned energy upgrade), how well factorization works for SIDIS at JLab is an important issue. Due to the high luminosity, it is possible to select kinematical settings keeping Q^2 reasonably large by going to large scattering angles. With an optimal choice of kinematics, the typical SIDIS measurements at JLab will be at Q^2 around 2 GeV^2 with W^2 of 4–10 GeV^2 and W'^2 of around 4 GeV^2 for an x range of 0.1–0.4 and a z range of 0.4–0.6. At what precision level factorization will work can only be answered by experimental tests. First test results are becoming available from $p(e, e'\pi^{+,-})$ experiments at JLab Hall C [28] and Hall B.[29] These results show that at the 10% level, the data are consistent with the factorization assumption.

A JLab Hall A proposal [30] has been conditionally approved to measure unpolarized $(e, e'\pi^{\pm})$ and $(e, e'K^{\pm})$ reactions. The new data will provide further precision tests of factorization. In addition, the pion SIDIS measurement aims to determine $\bar{d} - \bar{u}$ with much better statistical accuracy than the existing HERMES data,[31] which will provide a complementary measurement of the sea asymmetry to the Drell–Yan process.[32]

Acknowledgments

This work is supported by the U.S. Department of Energy (DOE). The Southeastern University Research Association operates the Thomas Jefferson National Accelerator Facility for DOE under contract DE-AC05-84ER40150, Modification No. 175.

References

1. V. Barone, A. Drago and P.G. Ratcliffe, *Phys. Rep.* **359** (2002) 1.
2. C. Bourrely, J. Soffer and O.V. Teryaev, *Phys. Lett.* **B420** (1998) 375.
3. J. Soffer, *Phys. Rev. Lett.* **74** (1995) 1292.
4. M. Diefenthaler, proc. of DIS2005, *AIP* **792** (2005) 933.
5. P. Pagano, proc. of DIS2005, *AIP* **792** (2005) 937.
6. P. Mulders and R.D. Tangerman, *Nucl. Phys.* **B461** (1996) 197.
7. D. Boer and P. Mulders, *Phys. Rev.* **D57** (1998) 5780.
8. J. Collins, *Nucl. Phys.* **B396** (1993) 161.
9. D.W. Sivers, *Phys. Rev.* **D41** (1990) 83.
10. A. Airapetian *et al.*, *Phys. Lett.* **B562** (2003) 182.
11. A. Ogawa *et al.*, proc. of DIS2005, *AIP* **792** (2005) 949.
12. Hall A collaboration: J. Alcorn *et al.*, Nucl. Inst. Meth. **A522** (2004) 294.
13. JLab E03-004, J.P. Chen, X. Jiang and J.-C. Peng, spokespersons; J.-C. Peng *et al.*, proc. of HiX2004, *AIP* **747** (2004) 141.
14. K. Kramer *et al.*, *Phys. Rev. Lett.* **95** (2005) 142002.
15. X. Zheng *et al.*, *Phys. Rev. Lett.* **92** (2004) 012004; X. Zheng *et al.*, *Phys. Rev.* **C70** (2004) 065207.
16. M. Amarian *et al.*, *Phys. Rev. Lett.* **89** (2002) 242301; *ibid.* **92** (2004) 022301; *ibid.* **93** (2004) 152301; Z.E. Meziani *et al.*, *Phys. Lett.* **B613** (2005) 148.
17. S. Wandzura and F. Wilczek, *Phys. Lett.* **B72** (1977) 195.
18. X. Ji and W. Melnitchouk, *Phys. Rev.* **D56** (1997) 1.
19. J. Blümlein and H. Bottcher, *Nucl. Phys.* **B636** (2002) 225.
20. K. Abe *et al.*, E155 collaboration, *Phys. Lett.* **B493** (2000) 19.
21. M. Stratmann, *Z. Phys.* **C60** (1993) 763.
22. H. Weigel, *Pramana* **61** (2003) 921.
23. M. Wakamatsu, *Phys. Lett.* **B487** (2000) 118.
24. M. Gockeler *et al.*, *Phys. Rev.* **D63** (2001) 074506.
25. J.P. Chen, A. Deur and Z.E. Meziani, to appear in *Mod. Phys. Lett.* **A** (2005); nucl-ex/0509007.
26. X. Ji, C. Kao, and J. Osborne, *Phys. Lett.* **B472** (2000) 1; C.W. Kao, T. Spitzenberg and M. Vanderhaeghen, *Phys. Rev.* **D67** (2003) 016001; V. Bernard, T. Hemmert and Ulf-G. Meissner, *Phys. Rev.* **D67** (2003) 076008
27. D. Drechsel, S. Kamalov and L. Tiator, *Phys. Rev.* **D63** (2001) 114010.
28. JLab E00-108, R. Ent, H. Mkrtchyan and G. Niculescu, spokespersons; W. Melnitchouk, R. Ent and C.E. Keppel, *Phys. Rep.* **406** (2005) 127.
29. H. Avagyan, a contribution to these proceedings.
30. JLab E04-114, J.P. Chen, X. Jiang, J.-C. Peng and L. Zhu, spokerspersons (2004).
31. K. Ackerstaff *et al.*, *Phys. Rev. Lett.* **81** (1998) 5519.
32. G.T. Garvey and J.C. Peng, *Prog. Part. Nucl. Phys.* **47** (2001) 203.

$\mathcal{PAX}$: POLARIZED ANTIPROTON EXPERIMENTS

M. Contalbrigo

INFN – Sezione di Ferrara,
Via Saragat 1, 44100 Ferrara, ITALIA
E-mail: contalbrigo@fe.infn.it

Polarized antiprotons, by spin filtering with an internal polarized gas target, provide access to a wealth of single- and double-spin observables. This includes a first direct measurement of the transversity distribution of the valence quarks in the proton, a test of the predicted opposite sign of the Sivers-function, related to the quark distribution inside a transversely polarized nucleon, in Drell–Yan (DY) as compared to semi-inclusive DIS, and a first measurement of the moduli and the relative phase of the time-like electric and magnetic form factors $G_{E,M}$ of the proton. Still open questions in polarized and unpolarized $p\bar{p}$ elastic scattering can be addressed as well. A viable experimental set-up can be realized within the FAIR project for a large European hadron facility, where a low-energy antiproton polarizer ring is used to yield an antiproton beam with sizable polarization. After acceleration, the polarized antiproton beam can be used to collide on a polarized internal hydrogen target (fixed target mode) or with a beam of polarized protons (collider mode). The detector concept for a large-angle apparatus optimized for the detection of lepton pairs of large invariant mass is anticipated.

The polarized antiproton–proton interactions in the High Energy Storage Ring (HESR) at the future Facility for Antiproton and Ion Research (FAIR) at GSI (Darmstadt) will provide unique access to a number of new fundamental physics observables, which can be studied neither at other facilities nor at HESR without transverse polarization of protons and antiprotons. The possibility to achieve polarized proton–antiproton reactions at the HESR ring at GSI has been suggested by the PAX Collaboration last year.[1] Since then, there has been much progress, both in understanding the physics potential of such an experiment [2,3] and in studying the feasibility of building an efficient polarized antiproton facility.[4] They will be briefly reviewed in this report.

1. Physics case

Parton distributions. The spin tomography of the proton would be ever incomplete without the determination of transversity – the quark transverse polarization inside a transversely polarized proton. Unlike the unpolarized quark distribution $q(x, Q^2)$ and the helicity distribution $\Delta q(x, Q^2)$, the transversity $h_1^q(x, Q^2)$ can neither be accessed in inclusive DIS nor can it be reconstructed from the knowledge of $q(x, Q^2)$ and $\Delta q(x, Q^2)$. The transversity may contribute to some single-spin observables in semi-inclusive DIS reactions, but, because of its chiral properties, always in combination with other unknown functions, i.e. the Collins fragmentation function.[5] The transversity distribution is directly accessible uniquely via the double transverse spin asymmetry A_{TT} in the Drell–Yan production of lepton pairs. The theoretical expectations for A_{TT} in the Drell–Yan process with transversely polarized antiprotons interacting with transversely polarized protons at HESR indicate large values in the 0.3–0.4 range.[2] With a significant beam polarization achieved using a dedicated low-energy antiproton polarizer ring (APR) and the luminosity of HESR, the PAX experiment is uniquely suited for the definitive observation of $h_1^q(x, Q^2)$ of the proton for the valence quarks. In addition, other novel parton distributions which are sensible to the transverse spin effects can be studied, i.e. the Sivers functions which correlate the transverse momentum of the quarks to the spin of the parent nucleon.[6] In conjunction with the data on single-spin asymmetries (SSA) in semi-inclusive DIS by the HERMES collaboration,[7] the PAX measurements of the SSA in Drell–Yan reactions on polarized protons can for the first time provide a test of the theoretical prediction of opposite sign of the Sivers functions in the two reactions.[8]

Magnetic and electric form factors. The origin of the unexpected Q^2-dependence of the ratio of the magnetic G_M and electric G_E form factors of the proton as observed at the Jefferson laboratory[9] can be clarified by a measurement of their relative phase in the time-like region, which discriminates strongly between the models for the form factor. This phase can be measured via SSA in the annihilation $\bar{p}p^{\uparrow} \to e^+e^-$ on a transversely polarized target.[10] The double-spin asymmetry will allow independently the $G_E - G_M$ separation and serve as a check of the Rosenbluth separation in the time-like region which has not been carried out so far.

Hard elastic scattering. Arguably, in $p\bar{p}$ elastic scattering the hard scattering mechanism can be checked beyond $|t| = \frac{1}{2}(s - 4m_p^2)$ accessible in the t-u-symmetric pp scattering, because in the $p\bar{p}$ case the u-channel exchange contribution can only originate from the strongly suppressed exotic dibaryon exchange. Consequently, in the $p\bar{p}$ case the polarized hard mechanisms[11] can be tested at t almost twice as large as in pp scattering, at the same center of mass energy s. The double transverse asymmetry in $p\bar{p}$ scattering can be investigated and related with the large effects observed in the pp case.[12]

2. Experimental setup

The PAX collaboration has elaborated a viable scheme which would allow to reach a polarization of the stored antiprotons at HESR–FAIR of $\simeq 0.2-$ -0.3.[4] It is schematically depicted in Fig. 1.

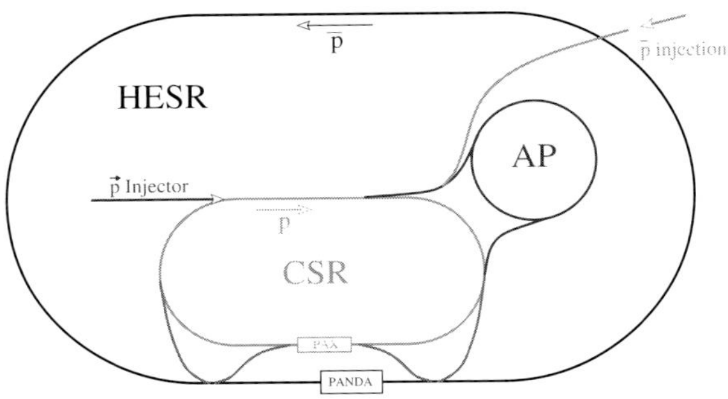

Fig. 1. The proposed accelerator set-up at the HESR, with the equipment used by the PAX collaboration and described in the text: CSR and APR rings, beam transfer lines and polarized proton injector.

(i) The Antiproton Polarizer Ring (APR) has the crucial goal of polarizing antiproton to be accelerated and injected in the other rings. The polarization method is based on spin-filtering by a polarized gas target internal to the beam line. In 1992 an experiment at the Test Storage Ring (TSR) at MPI Heidelberg showed that an initially unpolarized stored 23 MeV proton beam can be polarized by spin-dependent interaction with a polarized hydrogen gas target.[13] The final polarization achieved, of only few %, was limited by the not optimized beam acceptance and momentum. Among the three different mechanisms were initially identified to add up to the measured result,[14] two are now under theoretical debate.[15] The PAX collaboration has planned a set of preparatory test-experiments to investigate these mechanisms and optimize the polarization build up in APR. Preliminary estimates indicate an achievable polarization in the range $\simeq 0.2$–0.3. **(ii)** In the Cooler Synchrotron Ring (CSR, COSY-like) protons or antiprotons can be stored with a momentum up to 3.5 GeV/c for fixed-target experiments or accelerated to be transferred to the HESR ring. **(iii)** The HESR ring, already approved within the FAIR project to serve the PANDA experiment with antiproton up to 15 GeV/c momentum, is upgraded to become a synchrotron. By deflection of the HESR beam into the straight section of the CSR, an asymmetric antiproton–proton collider become feasible at the PAX interaction point. It is worthwhile to stress that, through the employment of the CSR, effectively a second interaction point is formed without additional costs for the laboratory. The proposed solution opens the possibility to run two different experiments at the same time with no interference.

The PAX collaboration proposes an approach that is composed of two phases. During these the major milestones of the project can be tested and optimized before the final goal of a polarized proton–antiproton asymmetric collider, is approached.

(I) A beam of unpolarized or polarized antiprotons with momentum up to 3.5 GeV/c in the CSR ring collides on a polarized hydrogen target in the PAX detector. This first phase, independent of the HESR performance, will allow for the first time the measurement of the time-like proton form factors in single and double polarized $\bar{p}p$ interactions in a wide kinematic range, from close to the threshold up to $Q^2 = 8.5\,\mathrm{GeV}^2$. It would be possible to determine several double spin asymmetries in elastic $\bar{p}^\uparrow p^\uparrow$ scattering. There are no competing facilities at which these topical issues can be addressed.

(II) This phase will allow the first ever direct measurement of the quark transversity distribution h_1, by measuring the double transverse spin asymmetry A_{TT} in Drell–Yan processes $p^\uparrow\bar{p}^\uparrow \to e^+e^-X$ as a function of Bjorken x and Q^2 $(= M^2)$

$$A_{TT} \equiv \frac{d\sigma^{\uparrow\uparrow} - d\sigma^{\uparrow\downarrow}}{d\sigma^{\uparrow\uparrow} + d\sigma^{\uparrow\downarrow}} = \hat{a}_{TT}\frac{\sum_q e_q^2 h_1^q(x_1, M^2) h_1^{\overline{q}}(x_2, M^2)}{\sum_q e_q^2 q(x_1, M^2)\overline{q}(x_2, M^2)} \tag{1}$$

where $q = u, \overline{u}, d, \overline{d} \ldots$, M is the invariant mass of the lepton pair and $\hat{a}_{TT}$ is the calculable double-spin asymmetry of the QED elementary process $q\overline{q} \to e^+e^-$. The most promising scenario foresees a beam of polarized antiprotons from 1.5 GeV/c up to 15 GeV/c circulating in the HESR, colliding on a beam of polarized protons up to 3.5 GeV/c circulating in the CSR. By proper variation of the energy of the two colliding beams, this setup would allow a measurement of the transversity distribution h_1 in the valence region of $x > 0.05$, with corresponding $Q^2 = 4 \ldots 100\,\mathrm{GeV}^2$ (see Fig. 2). The asymmetry A_{TT} is predicted to be larger than 20 % over the full kinematic range, till the

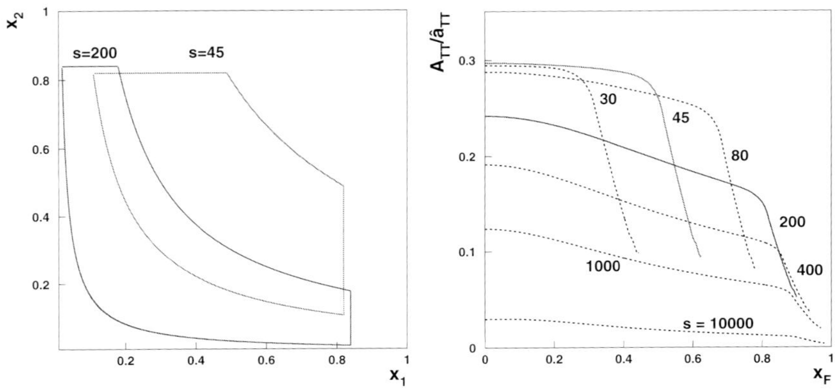

Fig. 2. Left: The kinematic region covered by the h_1 measurement at PAX in phase II. In the asymmetric collider scenario antiprotons of 15 GeV/c impinge on protons of 3.5 GeV/c at c.m. energies of $\sqrt{s} \sim \sqrt{200}$ GeV and $Q^2 > 4\,\text{GeV}^2$. With reduced beam momenta, different c.m. energies (i.e. $\sqrt{s} \sim \sqrt{45}$ GeV) and kinematic domains are explored. Right: the expected asymmetry as a function of Feynman x_F for different values of s and $Q^2 = 16\,\text{GeV}^2$.

highest reachable center-of-mass energy of $\sqrt{s} \sim \sqrt{200}$. The cross section is large as well, with a conservative luminosity value of $2 \cdot 10^{30}\,\text{cm}^{-2}\,s^{-1}$ about $\sim$ 1000 events per day can be expected. The theoretical work on the K-factors for the transversity determination is in progress.[16] For the transversity distribution h_1^q, such an experiment can be considered as the analogue of polarized DIS for the determination of the helicity distribution Δq; the kinematic coverage (x, Q^2) will be similar to that of the HERMES experiment.

3. The PAX detector concept

The detector design proposed in the PAX Technical Proposal (Fig. 3) is optimized for electron pairs detection. The overwhelming hadronic background requires excellent lepton identification. Several detection tools allow one to efficiently identify electrons without an adverse effect on the momentum resolution. High momentum resolution is needed to be sensitive to h_1 dependence on Bjorken x; in addition it opens the interesting possibility to distinguish resonances from continuum and extend the measured range down to 2 GeV dilepton mass, thereby enlarging the Bjorken x coverage of the h_1 measurement and exploiting the larger Drell–Yan cross-section. Alternative detector scenarios, e.g. with $\mu^+\mu^-$ Drell–Yan pair detection capability are also under study.

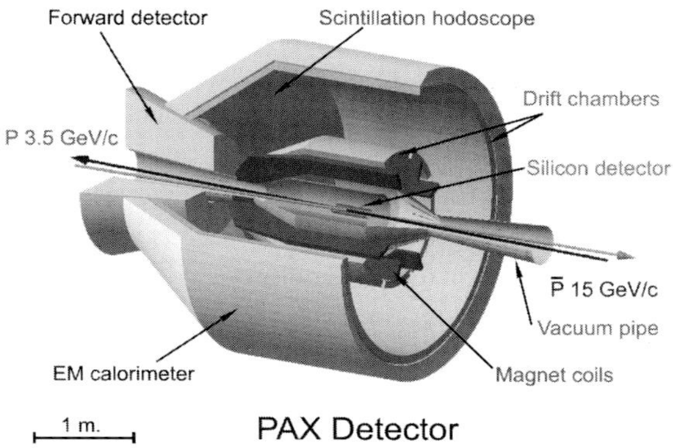

Fig. 3. This conceptual design of the PAX detector is employed to estimate the performance of the detector and to show the feasibility of the transversity measurement in the asymmetric antiproton–proton collider mode at PAX. The artists view is produced by GEANT.

The spectrometer concept comprises a compact silicon vertex detector close to the interaction point, plus a conventional set of drift-chambers external to the toroid magnet, see Fig. 3. The lepton identification is accomplished by a Čerenkov detector, inserted into the free-space of the tracking arm of the drift chambers, plus the electromagnetic calorimeter. The toroid has almost negligible fringe-fields outside its active volume, both internally along the polarized beam line and externally inside the tracking and Čerenkov detector volume. It is compatible with the internal polarized target foreseen for fixed-target mode of Phase-I.

Phase-I: In fixed-target mode the typical luminosity is of the order of 10^{31} cm^{-2}s^{-1}. From the $p\overline{p} \to e^+e^-$ cross-section measured by PS170,[17] it is estimated that the proton time-like form factors can be measured in a relatively short time, from less than 1 day up to few weeks. Only the most challenging measurement of double polarized asymmetry at the highest energy requires few months of data-taking. From the cross-section measured by E838,[18] it is estimated that the most challenging double-polarized measurement in $\overline{p}p$ elastic scattering requires only few hours of data-taking to reach a precision of 0.05 at the maximum transverse momentum achievable in CSR ($t_{\text{PAX}} = 4\,\text{GeV}^2/c^2$).

Phase-II: The major sources of background to the Drell–Yan process are the combinatorial background from meson Dalitz-decays and gamma conversions. Gamma conversions are vetoed requiring a charged hit in the first silicon layer. The residual background can be studied and finally subtracted by investigating wrong-charge candidates (control sample). Charm background can be studied and eventually reduced by reconstructing the secondary vertex of the decay with the silicon detector.[19] The background by particle misidentification can be assumed to be negligible since the PAX detector is designed to provide redundant high-level information about the particle type. Preliminary results based on GEANT simulation show that the signal over background ratio is of the order of one *before* combinatorial background subtraction, and support the view that the background for the e^+e^- Drell–Yan measurement is well under control. Conventional tracking detectors can provide a resolution better than 2 % in dilepton invariant mass: this is sufficient to efficiently distinguish resonance from continuum contributions and to investigate the h_1^q dependence on x. During one year of data-taking, the most interesting valence region can be explored and the h_1^q transverse distribution can be measured with a precision better than 10 %, see Fig. 4.

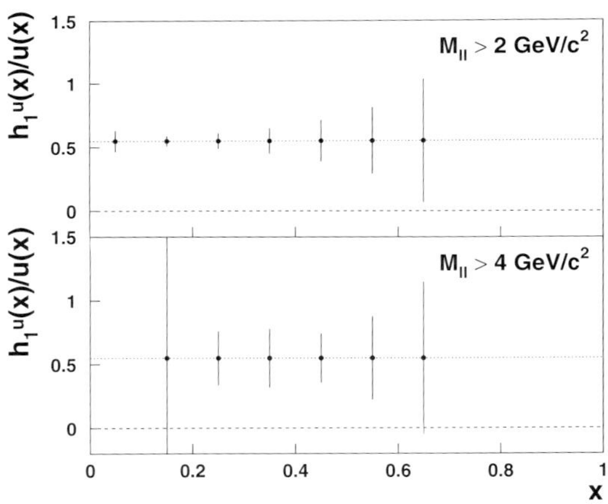

Fig. 4. Expected precision of the $h_1^u(x)$ measurement for one year of data taking in the collider mode at PAX with a conservative luminosity value of $2 \cdot 10^{30}$ cm^{-2}s^{-1}. The data points are plotted along the value of the $h_1^u(x)/u(x)$ ratio corresponding to the $A_{TT}/\hat{a}_{TT} = 0.3$ expectation. The precision achievable within the full $Q^2 > 4\,\text{GeV}^2$ kinematic range is of the order of 10 % (top panel). The bottom panel shows the precision achievable in the restricted $Q^2 > 16\,\text{GeV}^2$ range.

4. Summary

The PAX collaboration has presented a rich and innovative physics program to be realized in the upcoming FAIR hadron facility. The program is divided into two stages in order to maximize the physics output. In Phase-I, the relative phases of electric and magnetic form factors of the proton and transverse spin effects in $\overline{p}p$ hard elastic scatterings will be measured. In Phase-II a definite measurement of the last leading piece of the partonic description of the nucleon, the transversity distribution, will be possible. The storage of polarized antiprotons at HESR will open unique possibilities to test QCD in new domains. In this respect, the FAIR facility will have no competition.

References

1. PAX LoI, `http://www.fz-juelich.de/ikp/pax`.
2. M. Anselmino, V. Barone, A. Drago and N. Nikolaev, *Phys. Lett.* **B594** (2004) 97; A. Efremov, K. Goecke and P. Schweitzer, *Eur. Phys. J.* **C35** (2004) 207.
3. S. Brodsky, hep-ph/0411046 (2004). P. Zavada, hep-ph/0412206 (2004).
4. F. Rathmann *et al.*, *Phys. Rev. Lett.* **94** (2005) 014801; F. Rathmann and P. Lenisa, hep-ph/0412078 (2004); PAX TP, hep-ex/0505054.
5. J.C. Collins, *Nucl. Phys.* **B396** (1993) 161.
6. D. Sivers, *Phys. Rev.* **D41** (1990) 83; *Phys. Rev.* **D43** (1991) 261;
7. HERMES Collaboration, A. Airapetian *et al.*, *Phys. Rev. Lett.* **84** (2000) 4047; *Phys. Rev. Lett.* **90** (2003) 092002; *Phys. Rev.* **D64** (2001) 097101; K. Rith, *Prog. Part. Nucl. Phys.* **49** (2002) 245.
8. J.C. Collins, *Phys. Lett.* **B536** (2002) 43.
9. Jefferson Lab Hall A Collaboration, M.K. Jones *et al.*, *Phys. Rev. Lett.* **84** (2000) 1398; O. Gayou *et al.*, *Phys. Rev. Lett.* **88** (2002) 092301.
10. A.Z. Dubnickova, S. Dubnicka, and M.P. Rekalo, *Nuovo Cim.* **109** (1966) 241. S.J. Brodsky *et al.*; *Phys. Rev.* **D69** (2004) 054022.
11. V. Matveev *et al.*, *Lett. Nuovo Cim.* **7** (1972) 719; S. Brodsky and G. Farrar, *Phys. Rev. Lett.* **31** (1973) 1153; *Phys. Rev.* **D11** (1973) 1309; M. Diehl, T. Feldmann, R. Jakob and P. Kroll, *Phys. Lett.* **B460** (1999) 204.
12. D.G. Crabb *et al.*, *Phys. Rev. Lett.* **41** (1978) 1257.
13. F. Rathmann *et al.*, *Phys. Rev. Lett.* **71** (1993) 1379; K. Zapfe *et al.*, *Nucl. Instrum. Meth.* **A368** (1996) 293.
14. H.O. Meyer, *Phys. Rev.* **E50** (1994) 1485; C.J. Horowitz and H.O. Meyer, *Phys. Rev. Lett.* **72** (1994) 3981.
15. A.I. Milstein and V.M. Strakhovenko, physics/0504183; N.N. Nikolaev and F.F. Pavlov, contribution in this book.
16. P.G. Ratcliffe, hep-ph/0412157 (2004); H. Shimizu, hep-ph/0503270 (2005).
17. G. Bardin *et al.*, *Phys. Lett.* **B257** (1991) 514.
18. C.G. White *et al.*, *Phys. Rev.* **D49** (1994) 58.
19. BABAR coll., A. Aubert *et al.*, *Phys. Rev.* **D70** (2004) 091102.

SINGLE AND DOUBLE SPIN N–$\bar{N}$ INTERACTIONS AT GSI

M. Maggiora for the ASSIA Collaboration*

Dipartimento di Fisica "A. Avogadro"
Via Pietro Giuria 1, Torino, 10136, Italy
E-mail: marco.maggiora@to.infn.it

An HESR $\bar{p}^{\uparrow}p^{\uparrow}$ collider mode at GSI would provide new insights on the spin structure of the nucleon. Drell–Yan processes are a powerful tool to access chirally odd parton distribution functions like transversity $h_1(x)$, without their convolution with fragmentation functions, in reactions were all the quarks taking part can be valence quarks; spin asymmetries in hadron production, nucleonic form factors and open charm production are discussed as well.

1. Introduction

The new facility foreseen at GSI, and in particular the availability of antiproton beams, eventually polarised, offer new possibilities in the investigation of the nucleon structure. The first proposed scheme [1] focused on the extraction on a fixed proton target of an antiproton beam from the new superconducting synchrotron of rigidity 300 Tm (SIS300), while nowadays the GSI management seems more oriented to accept a scheme in which the

*V. Abazov, G. Alexeev, M. Alexeev, A. Amoroso, N. Angelov, S. Baginyan, F. Balestra, V.A. Baranov, Yu. Batusov, I. Belolaptikov, R. Bertini, A. Bianconi, R. Birsa, T. Blokhintseva, A. Bonyushkina, F. Bradamante, A. Bressan, M.P. Bussa, V. Butenko, M. Colantoni, M. Corradini, S. Dalla Torre, A. Demyanov, O. Denisov, V. Drozdov, J. Dupak, G. Erusalimtsev, L. Fava, A. Ferrero, L. Ferrero, M. Finger, Jr., M. Finger, V. Frolov, R. Garfagnini, M. Giorgi, O. Gorchakov, A. Grasso, V. Grebenyuk, V. Ivanov, A. Kalinin, V.A Kalinnikov, Yu. Kharzheev, Yu. Kisselev, N.V. Khomutov, A. Kirilov, E. Komissarov, A. Kotzinian, A.S. Korenchenko, V. Kovalenko, N.P. Kravchuk, N.A. Kuchinski, E. Lodi Rizzini, V. Lyashenko, V. Malyshev, A. Maggiora, M. Maggiora, A. Martin, Yu. Merekov, A.S Moiseenko, A. Olchevski, V. Panyushkin, D. Panzieri, G. Piragino, G.B. Pontecorvo, A. Popov, S. Porokhovoy, V. Pryanichnikov, M. Radici, M.P. Rekalo, A. Rozhdestvensky, N. Russakovich, P. Schiavon, O. Shevchenko, A. Shishkin, V.A. Sidorkin, N. Skachkov, M. Slunecka, S. Sosio, A. Srnka, V. Tchalyshev, F. Tessarotto, E. Tomasi, F. Tosello, E.P. Velicheva, L. Venturelli, L. Vertogradov, M. Virius, G. Zosi and N. Zurlo

HESR is modified in a $\bar{p} - p$ collider, eventually asymmetric, where one or both the beams can be polarised.

The knowledge of the nucleon structure requires the determination of the quark and gluon distribution functions, and of the quark fragmentation functions, at least at the first two leading twists, without integrating over the transverse momentum $\boldsymbol{\kappa}_\perp$ of the parton in the nucleon. If the quarks momenta are collinear to the nucleon momentum, or an integration on $\boldsymbol{\kappa}_\perp$ is performed, the quark structure of the nucleon is completely described at the leading twist by three distribution functions: the unpolarised distribution $f_1(x)$, the longitudinal ($g_1(x)$) and the transverse ($h_1(x)$) polarised distributions. If $\boldsymbol{\kappa}_\perp$ is no more neglected, at twist two and three eight parton distribution functions (PDF) are needed, some of them $\boldsymbol{\kappa}_\perp$-dependent and basically unknown. In case of hadron production, the fragmentation functions are needed as well; a complete review of the theoretical and experimental aspects relative to the parton distribution and fragmentation functions can be found in Ref. 2.

An excellent case to study is the Drell–Yan (DY) di-lepton production where protons and antiprotons annihilate in the initial state. Although affected by low cross-sections, these processes allow for the direct investigation of chirally odd PDF like transversity $h_1(x)$, without the suppression proper of the deep-inelastic scattering data, thanks to the non-perturbative vertices of the diagram (the two quark lines are uncorrelated in Fig. 1), and without their convolution with the unknown polarised quark fragmentation functions, like in the semi-inclusive deep-inelastic scattering. Besides transversity $h_1(x)$, we will later focus also on two $\boldsymbol{\kappa}_\perp$-dependent distributions: $f_{1T}^\perp(x, \boldsymbol{\kappa}_\perp{}^2)$ and $h_{1T}^\perp(x, \boldsymbol{\kappa}_\perp{}^2)$, respectively the distribution functions of an unpolarised quark in a transversely polarised hadron, and of a transversely polarised quark inside an unpolarised parent hadron.

New insights on the nucleon structure may come also from the investigation of hadron pairs productions, namely from exclusive ($p\bar{p} \to \Lambda\bar{\Lambda}$) or semi-inclusive ($p\bar{p} \to \Lambda\bar{\Lambda}X$) strange hadrons production, where the convolution of parton distribution and quark fragmentation functions can be accessed; Λ's and $\bar{\Lambda}$'s detection could allow also the investigation of open-charm production in antiproton–proton scattering $\bar{p}p \to \Lambda_c^+ X$.

Spin dependent measurements would allow also to disentangle the electric and magnetic part of the electromagnetic form factors in the exclusive di-lepton production from $p\bar{p}$ annihilation.

A complete experiment would require polarised protons and antiprotons, the key issue being the availability of a $\bar{p}$ beam, with an energy suitable to investigate the parton distribution functions in a wide range of the Bjorken kinematic variable x; the PAC committee has pointed toward a minimum luminosity of $10^{31}\,\mathrm{cm}^{-2}\,\mathrm{s}^{-1}$ at a center of mass energy $s \sim 200\,\mathrm{GeV}^2$,[a] but excellent physics, namely regarding transversity effects, can be performed also making use of unpolarised antiprotons and polarised protons.

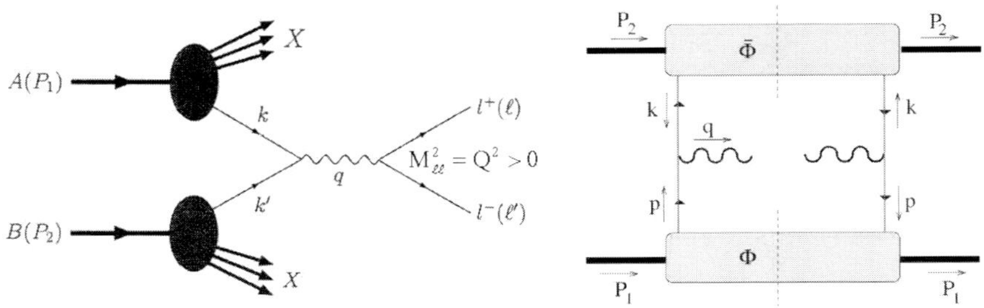

Fig. 1. Drell–Yan di-lepton production.

2. Drell–Yan di-lepton production

The possibility to directly access the parton distribution functions without their convolution with the quark fragmentation functions and without the suppression of the chirally odd amplitudes is not the only benefit of this kind of reactions; since an antiproton probe is involved, in $\bar{p}p$ DY diagrams all the quarks taking part to the annihilation vertex can be valence quarks.

We will focus on the muon pairs production $p\bar{p} \to \mu^+\mu^- X$ (Fig. 1), whose cross-section for a given M di-muon mass is given in the Collins–Soper frame[4] by:

$$\frac{d^2\sigma}{dM^2 dx_F} = \frac{4\pi\alpha^2}{9M^2 s}\frac{1}{(x_1 + x_2)}\sum_a e_a^2\left[f^a(x_1)f^{\bar{a}}(x_2) + f^{\bar{a}}(x_1)f^a(x_2)\right] \quad (1)$$

the summation being extended to the quark flavour a ($a = u, d, s$), $x_{1,2} = \frac{M^2}{2P_{1,2}q}$ the fractions of the longitudinal momenta of the incoming hadrons carried by the quark and anti-quarks taking part in the annihilation into the virtual photon, and the parameter $\tau = x_1 x_2 = \frac{M^2}{s}$.

The scaling properties and the kinematic behaviour of the $\mathrm{p\bar{p}} \to \mu^+\mu^- X$ reaction are the same as for the $pp \to \mu^+\mu^- X$; the DY cross-section[5,6]

[a]Recent studies show how even higher luminosity ($5 \cdot 10^{31}\,\mathrm{cm}^{-2}\,\mathrm{s}^{-1}$) and center of mass energy (up to $s \sim 900\,\mathrm{GeV}^2$) could be obtained by the mean of the HESR collider.[3]

scales as $d^2\sigma/d\sqrt{\tau}dx_F \propto 1/s$, and shows (Fig. 2) important resonance effects from J/ψ and Υ resonance families. Only data coming from the "safe" region (a selection of values of M ranging from 4 to 9 GeV/c^2, where the di-muon spectrum is essentially continuous) will be considered to extract the PDF, since no resonance effect has to be disentangled in the data analysis, and since, for the data arising from the region below the J/ψ resonance families, perturbative contributions can be important, and the formulæ that we present later on would have to be corrected by additional terms. Nevertheless, since there exists arguments in the favour of possibility of studying spin effects in the J/ψ region,[7] and we intend to investigate also the perturbative corrections in the kinematic region below the safe region, data in this kinematic region will be collected as well. The importance of these perturbative effects decreases with increasing s.[8]

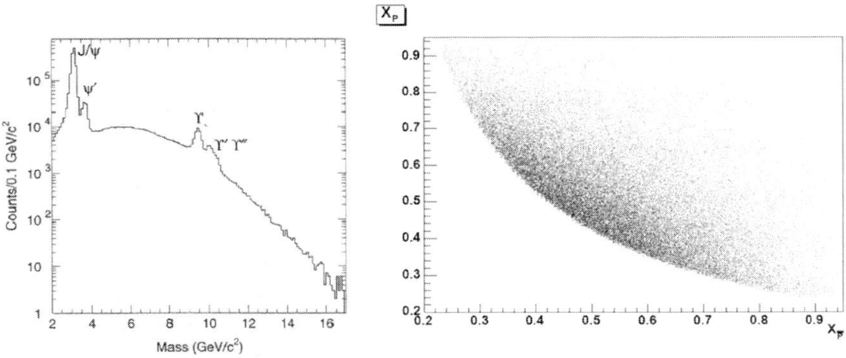

Fig. 2. Drell–Yan di-lepton production: a) cross-section; b) phase space (x1,x2) Montecarlo simulations from Bianconi and Radici.

To investigate the PDF in a wide (x_1, x_2) region means asking for a parameter $\tau \in [0, 1]$. For a larger statistics, data from the complete safe region should be collected: the upper limit of 9 GeV/c^2 for M defines the highest $\sqrt{s}$ needed to approach the region $\tau \sim 1$ (right upper corner in Fig. 3); the rejection of the events below the J/ψ peak, i.e. asking for $M \geq 4$ GeV/c^2, means to cut the allowed kinematic region, since a higher s means to access the very low x and τ region (left lower corner in Fig. 3).

The hyperbola of Fig. 3 show the τ region selected from the cut on the lowest side of the safe region for different values of s and thus of the beam(s) momentum(a). At fixed s, as shown in Ref. 9 most of the events fall near the lower cut hyperbole. The kinematic region that can be accessed making

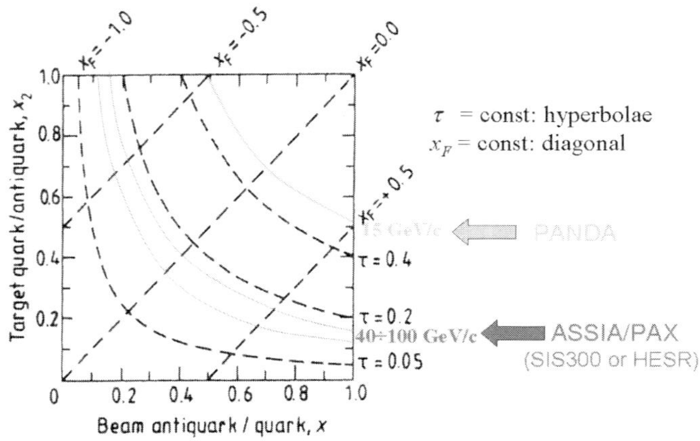

Fig. 3. Allowed (x_1, x_2) kinematic region in Drell–Yan processes: the regions above the hyperbola correspond to the cut on the di-lepton mass $M > 4\,\mathrm{GeV}/c^2$ for different energies of the beam: the lowest foreseen for PANDA at the HESR fixed target layout, the highest proposed for the new HESR collider layout.

use of antiprotons extracted from SIS 300 or in the HESR collider layout is wider then the region that could be explored by mean of antiprotons colliding on a fixed target in the HESR facility, if the beam energy would be the one foreseen for PANDA;[10] the outer left hyperbole corresponds to the PAC suggestion for the HESR collider mode at $s \sim 200\,\mathrm{GeV}/c^2$.

In the best conditions, i.e. if both polarised antiprotons and polarised protons were available, the following asymmetries could be observed:

$$A_{LL} = \frac{\sum_a e_a^2 g_1^a(x_1) g_1^{\bar{a}}(x_2)}{\sum_a e_a^2 f_1^a(x_1) f_1^{\bar{a}}(x_2)} \tag{2}$$

$$A_{TR} = \frac{\sin^2\theta\,\cos 2\phi}{1+\cos^2\theta}\,\frac{\sum_a e_a^2 h_1^a(x_1) h_1^{\bar{a}}(x_2)}{\sum_a e_a^2 f_1^a(x_1) f_1^{\bar{a}}(x_2)} \tag{3}$$

$$A_{LT} = \frac{2\sin 2\theta\,\cos\phi}{1+\cos^2\theta}\,\frac{M}{\sqrt{Q^2}}\,\frac{\sum_a e_a^2 (g_1^a(x_1) y g_T^{\bar{a}}(_2) - x h_L^a(x_1) h_1^{\bar{a}}(x_2))}{\sum_a e_a^2 f_1^a(x_1) f_1^{\bar{a}}(x_2)} \tag{4}$$

where the indices correspond to longitudinal and transverse polarisation of the target and of the beam respectively, and the polar (θ) and azimuthal (ϕ) angles are the ones defined in Ref. 4. The validity of these formulæ depends strongly on the assumptions that s and Q^2 are large enough: at the center of mass energy proposed by PAC ($\sqrt{s} \sim 14.5\,\mathrm{GeV}/c$) the higher order perturbative corrections would be strongly reduced[11] with respect to the $s \sim 30\,\text{GeV}/c^2$ for an HESR beam on a fixed target.

These asymmetries could be investigated at RHIC as well, but in the case of pp scattering, only $\bar{q}$ from the sea would contribute to the Drell–Yan diagram; moreover, due to the large $\sqrt{s} \sim 100\,\text{GeV}$, the data would be affected by a strongly reduced cross-section and by quite a small allowed kinematic range. The values of the asymmetries themselves would be reduced as well, due to the evolution of $h_1(x)$ on Q^2, much slower than that of the unpolarised distribution functions;[12] the numerator of the asymmetry ratio would grow more slowly than the denominator, thus leading to a suppression of the asymmetries for large values of $\sqrt{s}$. This would not be the case at GSI, where $\sqrt{s}$ would not be so large, a wide kinematic region could be accessed, and sizeable double spin asymmetries are expected.[7]

If we consider the completely unpolarised DY di-muon production, perturbative-QCD points to a cross-section independent of the azimuthal angle, once the acceptance is accounted for; experimental data[13] contradict this assumption, showing important azimuthal effects. Recently[14] it has been pointed out that initial state interaction in the unpolarised DY process could explain the observed asymmetries and be connected with the quark (anti-quark) T-odd distributions $h_{1q}^{\perp}$ and $h_{1\bar{q}}$. For the $\bar{p}p \to \mu^+\mu^- X$ process, the measurement of the $\cos 2\phi$ contribution to the angular distribution of the di-muon pair provides the product $h_1^{\perp}(x_2, \boldsymbol{\kappa}_{\perp}^2)\,\bar{h}_1^{\perp}(x_1, \boldsymbol{\kappa}_{\perp}^{'2})$. This asymmetry can be evaluated, also, by mean of the PANDA detector, where a polarised target cannot be installed because of the disturbance due to the magnetic field of the solenoid; however the maximal antiproton beam energy foreseen for PANDA at HESR considerably reduces the kinematic domain that can be accessed for the Bjorken x variable.

In the case of a transversely polarised protons, ϕ_{S_1} being the azimuthal angle of the proton spin in the Collins–Soper frame,[4] the asymmetry for the two target spin states:[15]

$$A_T = |\,\boldsymbol{S}_{\perp}\,|\;\frac{2\sin 2\theta\,\sin(\phi - \phi_{S_1})}{1+\cos^2\theta}\;\frac{M}{\sqrt{Q^2}} \times \\ \times\,\frac{\sum_a e_a^2[x(f_1^{a\perp}(x_1)f_1^{\bar{a}}(x_2) + y h_1^a(x_1)h_1^{\bar{a}\perp}(x_2))]}{\sum_a e_a^2 f_1^a(x_1)f_1^{\bar{a}}(x_2)} \tag{5}$$

is $\propto h_1(x_2, \boldsymbol{\kappa}_{\perp}^2)\,\bar{h}_1^{\perp}(x_1, \boldsymbol{\kappa}_{\perp}^{'2})$. The ideal approach would be to combine double spin measurements near the maximum value of the PDF with the investigation of single spin asymmetries (SSA) as a function of the Bjorken x to evaluate the x-dependence of the $h_1(x)$ function.[9] With unpolarised $\bar{p}$ and polarised p, the $\boldsymbol{\kappa}_\perp$-dependence of the quark distribution functions could be investigated. In particular, the measurement of the single-spin asymmetry, Eq. (5), in the absence of a polarised beam, is a unique tool to probe the $\boldsymbol{\kappa}_\perp$ effects. Recently, several papers have stressed the importance of measuring SSA in Drell–Yan processes;[16,17] these measurements allow the determination of new non perturbative spin properties of the proton, like the Sivers function, which describes the azimuthal distribution of quarks in a transversely polarised proton.[17]

Bianconi and Radici have shown that differing functional hypotheses of $h_1(x)$ can be much better resolved in the full x region, observing both the double[18] and the single[9] spin asymmetries, provided that an energy range larger than $s \sim 200\,\mathrm{GeV}/c^2$ were available, as suggested by the PAC committee; in this region the higher order perturbative corrections are expected to be low[11] and the cross-section is not yet strongly suppressed as in the higher energy range proper of RHIC. The study of $\bar{p}^\uparrow\, p \to \mu^+\, \mu^-\, X$ processes at GSI offers then unique possibilities.

3. Further physics

As described in detail in Ref. 19 several other items can be investigated at the new facility of GSI. The evaluation of spin observables in different hyperon production processes (Λ, $\bar{\Lambda}$ and $\Lambda\bar{\Lambda}$-pair production) can provide a better understanding of the spin dynamics in strong interactions, allowing for the selection among the different predictions of the proposed theoretical models.

The study of nucleon electromagnetic form factors is a powerful tool to investigate the nucleon structure, in particular as far as the time-like region is concerned, where data in the literature are poorer, and no spin effects have been investigated so far. Due to the cross-section energy behaviour, the spin observables for the reaction $\bar{p}p^\uparrow \to \mu^+\mu^-$ can be investigated only in an energy range up to $\sqrt{s} \sim 10\,\mathrm{GeV}/c$, but the different theoretical models available in literature provide for these quantities predictions very sensitive to the different underlying assumptions on the s-dependence of the form factors.

Another tool to investigate spin effects in hadron productions is the evaluation of the single-spin asymmetry:

$$A_N = \frac{d\sigma^\uparrow - d\sigma^\downarrow}{d\sigma^\uparrow + d\sigma^\downarrow} \tag{6}$$

for DY processes, that could add information on the spin properties of

QCD, by the mean also of the comparison with the data available in the literature for $\bar{p}\,p^{(\uparrow)} \to \pi\,X$ reactions.

Open charm production could be investigated as well by the mean of the detection, for example, of the $\bar{p}p \to \Lambda_c^+\,X$ production, where the $\Lambda_c^+ \to \Lambda\,\pi^+$ weak decay and the $\Lambda_c^+ \to \Lambda\,e^+\,\nu_c$ semi-leptonic decay can be used to infer the c polarisation.

4. Conclusions

An HESR p̄p collider mode at GSI, with a luminosity of $10^{31}\,\mathrm{cm}^{-2}\,\mathrm{s}^{-1}$ in the energy range $s \sim 200\,\mathrm{GeV}^2$, would provide excellent tools to investigate the nucleonic structure, even if only one of the two probes could be polarised.

The experimental asymmetries described above should not be suppressed as at higher energies, and higher order perturbative corrections proper of lower energies are expected to be negligible; these asymmetries should allow to resolve different functional behaviours of the PDF in a wide kinematic x-region.

References

1. ASSIA Collaboration, hep-ex/0507077.
2. V. Barone, A. Drago and P.G. Ratcliffe, *Phys. Rep.* **359** (2002) 1.
3. F. Bradamante *et al.*, *INFN/TC-05/14, physics/0511252.*
4. J. Collins and D.E. Soper, *Phys. Rev.* **D16** (1977) 2219.
5. P.L. McGaughey, J.M. Moss and J.C. Peng, *Ann. Rev. Nucl. Part. Sci.* **49** (1999) 217.
6. E.A. Hawker *et al.*, *Phys. Rev. Lett.* **80** (1998) 3715.
7. M. Anselmino *et al.*, *Phys. Rev.* **D71** (2005) 074006.
8. L.P. Gamberg, G.R. Goldstein and K.A. Oganessyan, *hep-ph/0411220.*
9. A. Bianconi and M. Radici, *Phys. Rev.* **D71** (2005) 074014.
10. PANDA Collaboration, Letter of Intent for 'Strong Interaction Studies with Antiprotons', `http://www.gsi.de/documents/DOC-2004-Jan-115-1.pdf`
11. H. Shimizu *et al.*, *Phys. Rev.* **D71** (2005) 114007.
12. V. Barone, T. Calarco and A. Drago, *Phys. Rev.* **D56** (1997) 527.
13. E. Anassontzis *et al.*, *Phys. Rev.* **D38** (1988) 1377.
14. D. Boer, S. Brodsky and D.S. Hwang, *Phys. Rev.* **D67** (2003) 054003-1; J.C. Collins, *Phys. Lett.* **B536** (2002) 43.
15. D. Boer, *Phys. Rev.* **D60** (1999) 014012.
16. N. Hammon, O. Teryaev and A. Schäfer, *Phys. Lett.* **B390** (1997) 409; D. Boer, P.J. Mulders and O. Teryaev, *Phys. Rev.* **D57** (1998) 3057; D. Boer and P.J. Mulders, *Nucl. Phys.* **B569** (2000) 505; D. Boer and J. Qiu, *Phys. Rev.* **D65** (2002) 034008.
17. M. Anselmino, U. D'Alesio and F. Murgia, *Phys. Rev.* **D67** (2003) 074010.
18. A. Bianconi and M. Radici, *hep-ph/0504261.*
19. M. Maggiora *et al.*, *Czech. J. Phys. Suppl.* **A55** (2005) A75; and refs. therein.

SPIN FILTERING IN STORAGE RINGS

N.N. Nikolaev[1,2] and F.F. Pavlov[1]

[1] *Institut für Kernphysik, Forschungszentrum Jülich, 52428 Jülich, Germany*
[2] *L.D. Landau Institute for Theoretical Physics, 142432 Chernogolovka, Russia*
E-mail: N.Nikolaev@fz-juelich.de
E-mail: F.Pavlov@fz-juelich.de

The spin filtering in storage rings is based on a multiple passage of a stored beam through a polarized internal gas target. Apart from the polarization by the spin-dependent transmission, a unique geometrical feature of interaction with the target in such a filtering process, pointed out by H.O. Meyer,[1] is a scattering of stored particles within the beam. A rotation of the spin in the scattering process affects the polarization buildup. We derive here a quantum-mechanical evolution equation for the spin-density matrix of a stored beam which incorporates the scattering within the beam. We show how the interplay of the transmission and scattering within the beam changes from polarized electrons to polarized protons in the atomic target. After discussions of the FILTEX results on the filtering of stored protons,[2] we comment on the strategy of spin filtering of antiprotons for the PAX experiment at GSI FAIR.[3]

1. Introduction

1.1. *Future QCD spin physics needs polarized antiprotons: PAX proposal*

The physics potential of experiments with high-energy stored polarized antiprotons is enormous. The list of fundamental issues includes the determination of transversity – the quark transverse polarization inside a transversely polarized proton – the last leading-twist missing piece of the QCD description of the partonic structure of the nucleon, which can only be investigated via double-polarized antiproton–proton Drell–Yan production. Without measurements of the transversity, the spin tomography of the proton would be ever incomplete. Other items of great importance for the perturbative QCD description of the proton include the phase of the time-like form factors of the proton and hard proton–antiproton scattering. Such an ambitious physics program with polarized antiproton–polarized proton collider has been proposed recently by the PAX Collaboration[3] for the new Facility for Antiproton and Ion Research (FAIR) at GSI in Darmstadt, Germany, aiming at luminosities of $10^{31}\,\mathrm{cm}^{-2}\,\mathrm{s}^{-1}$. An integral part of such a machine is a dedicated large-acceptance Antiproton Polarizer Ring (APR).

Here we recall, that for more than two decades, physicists have tried to produce beams of polarized antiprotons,[4] generally without success. Conventional methods like atomic beam sources (ABS), appropriate for the production of polarized protons and heavy ions cannot be applied, since antiprotons annihilate with matter. Polarized antiprotons have been produced from the decay in flight of $\bar{\Lambda}$ hyperons at Fermilab. The intensities achieved with antiproton polarizations $P > 0.35$ never exceeded $1.5 \cdot 10^5\,\mathrm{s}^{-1}$.[5] Scattering of antiprotons off a liquid hydrogen target could yield polarizations of $P \approx 0.2$, with beam intensities of up to $2 \cdot 10^3\,\mathrm{s}^{-1}$.[6] Unfortunately, both approaches do not allow efficient accumulation of antiprotons in a storage ring, which is the only practical way to enhance the luminosity. Spin splitting using the Stern–Gerlach separation of the given magnetic substates in a stored antiproton beam was proposed in 1985.[7] Although the theoretical understanding has much improved since then,[8] spin splitting using a stored beam has yet to be observed experimentally.

1.2. *FILTEX: proof of the spin-filtering principle*

At the core of the PAX proposal is spin filtering of stored antiprotons by multiple passage through a Polarized Internal hydrogen gas Target (PIT).[3,9] In contrast to the aforementioned methods, convincing proof of the spin-filtering principle has been produced by the FILTEX experiment at TSR-ring in Heidelberg.[2] It is a unique method to achieve the required high current of polarized antiprotons.

In the FILTEX experiment at TSR[2] the transverse polarization rate of $dP_B/dt = 0.0124 \pm 0.0006$ (only the statistical error is shown) per hour has been reached for 23 MeV stored protons interacting with an internal polarized atomic hydrogen target of areal density 6×10^{13} atoms/cm^2. The principal limitation on the observed polarization buildup was a very small acceptance of the TSR-ring. Extrapolations of the FILTEX result, in conjunction with the then available theoretical re-interpretation[1,10] of the FILTEX finding, suggested that in the custom-tailored large-acceptance Antiproton Polarizer Ring (APR) antiproton polarizations up to 35–40% are feasible.[9]

1.3. *Mechanisms of spin-filtering: transmission and scattering within the beam (pre-2005)*

Everyone is familiar with the polarization of the light transmitted through the plate of an optically active medium, which is usually the regime of weak absorption and predominantly real light-atom scattering amplitude. In the realm of particle physics, the absorption becomes the dominant feature of interaction. The transmitted beam becomes polarized by the polarization-dependent absorption, which is the standard mechanism, for instance, in neutron optics.[11] While the polarization of elastically scattered slow neutrons is a very important observable, the elastically scattered neutrons are never confused with the transmitted beam.

In his theoretical interpretation of the FILTEX result, H.O. Meyer made an important observation that the elastic scattering of stored particles within the beam is an intrinsic feature of the spin filtering in storage rings.[1] First, one takes a particle from the stored beam. Second, this particle is either absorbed (annihilation for antiprotons, meson production for sufficiently high energy protons and antiprotons) or scatters elastically on the polarized atom in the PIT. Third, if the scattering angle is smaller than the acceptance angle $\theta_{\rm acc}$ of the ring, the scattered particle ends up in the stored beam. Specifically, the polarization of the particle scattered within the beam would contribute to the polarization of a stored beam.

The FILTEX PIT used the hyperfine state of the hydrogen in which both the electron and proton were polarized. The familiar Breit Hamiltonian for the nonrelativistic *ep* interaction includes the hyperfine and tensor spin–spin interactions. Meyer and Horowitz [10] noticed that those spin–spin interactions give a sizeable cross section of the polarization transfer from polarized target electrons to scattered protons, which is comparable to that in the nuclear proton–proton scattering. (Incidentally, the transfer of the longitudinal polarization of accelerated electrons to scattered protons, suggested in 1957 by Akhiezer *et al.*,[12] is at the heart of the recent high precision measurements of the ratio of the charge and magnetic form factors of nucleons at Jlab and elsewhere, for the review see Ref. 13.) Furthermore, Meyer argued that the contribution from *pe* scattering is crucial for the quantitative agreement between the theoretical expectation for the polarization buildup of stored protons and the FILTEX result,[1] which prompted the idea to base the antiproton polarizer of the PAX on the spin filtering by polarized electrons in PIT.[9]

After the PAX proposal, the feasibility of the electron mechanism of spin filtering has become a major issue. Yu. Shatunov was, perhaps, the first to worry, and his discussions with A. Skrinsky prompted, eventually, A. Milstein and V. Strakhovenko of the Budker Institute to revisit the kinetics of spin filtering in storage rings.[14] Simultaneously and independently, similar conclusions on the self-cancellation of the polarized electron contribution to the spin filtering of (anti)protons were reached in Jülich by the present authors within a very different approach.

After this somewhat lengthy and Introduction, justified by the novelty of the subject, we review the basics of the quantum mechanical theory of spin filtering with full allowance for scattering within the beam.

2. Spin filtering in storage rings: transmission, scattering, kinematics and all that

The sky is blue because what we see is exclusively the elastically scattered light. The setting sun is reddish because we see exclusively the transmitted light. The sun changes its color because the transmission changes the frequency (wavelength) spectrum of the unscattered light. In the typical optical experiments, one never mixes the transmitted and scattered light. An unique feature of storage rings, noticed by Meyer, is a mixing of the transmitted and scattered beams.

Some kinematical features of the proton–atom scattering are noteworthy. First, the Coulomb fields of the proton and atomic electron screen each other beyond the Bohr radius a_B. To a good approximation, protons flying by an atom at impact parameters $> a_B$ do not interact with an atom. The cancellation of the proton and electron Coulomb fields holds at scattering angles (all numerical estimates are for $T_p = 23\,\mathrm{MeV}$)

$$\theta \gtrsim \theta_{\mathrm{min}} = \frac{\alpha_{\mathrm{em}} m_e}{\sqrt{2m_p T_p}} \approx 2 \cdot 10^{-2}\,\mathrm{mrad}, \tag{1}$$

at higher scattering angles one can approximate proton–atom interaction by an incoherent sum of quasielastic (E) scattering off protons and electrons,

$$d\sigma_E = d\sigma_{\mathrm{el}}^p + d\sigma_{\mathrm{el}}^e. \tag{2}$$

As Horowitz and Meyer emphasized, atomic electron is too light a target to deflect heavy protons, in *pe* scattering

$$\theta \le \theta_e = \frac{m_e}{m_p} \approx 5 \cdot 10^{-1}\,\mathrm{mrad}. \tag{3}$$

For 23 MeV protons in the TSR-ring, proton–proton elastic scattering is Coulomb interaction dominated for

$$\theta \lesssim \theta_{\mathrm{Coulomb}} \approx \sqrt{\frac{2\pi\alpha_{\mathrm{em}}}{m_p T_p \sigma^{pp}_{\mathrm{tot,nucl}}}} \approx 100\,\mathrm{mrad}. \tag{4}$$

Finally, the FILTEX ring acceptance angle equals

$$\theta_{\mathrm{acc}} = 4.4\,\mathrm{mrad}, \tag{5}$$

and we have a strong inequality

$$\theta_{\min} \ll \theta_e \ll \theta_{\mathrm{acc}} \ll \theta_{\mathrm{Coulomb}}. \tag{6}$$

The corollaries of this inequality are: (i) pe scattering is entirely within the stored beam, (ii) beam losses by single scattering are dominated by the Coulomb pp scattering.

At this point it is useful to recall the measurements of the pp total cross section in the transmission experiments with the liquid hydrogen target. With the electromagnetic pe interaction included, the proton–atom cross-section is gigantic:

$$\hat{\sigma}^{pe}_{\mathrm{tot}} = \hat{\sigma}^{e}_{\mathrm{el}}(>\theta_{\min}) \sim 4\pi\alpha^2_{\mathrm{em}} a^2_B \approx 2\cdot 10^4\,\mathrm{barn}. \tag{7}$$

How do we extract $\sigma^{pp}_{\mathrm{tot,nucl}} \sim 40\,\mathrm{mb}$ on top of such a background from pe scattering? Very simple: in view of (3) and its relativistic generalization, elastic scattering off electrons is entirely within the beam and does not cause any attenuation!

3. The in-medium evolution of the transmitted beam

In fully quantum-mechanical approach, the beam of stored antiprotons must be described by the spin-density matrix

$$\hat{\rho}(\boldsymbol{p}) = \tfrac{1}{2}[I_0(\boldsymbol{p}) + \boldsymbol{\sigma s}(\boldsymbol{p})], \tag{8}$$

where $I_0(\boldsymbol{p})$ is the density of particles with the transverse momentum $\boldsymbol{p}$ and $\boldsymbol{s}(\boldsymbol{p})$ is the corresponding spin density. As far as the pure transmission is concerned, it can be described by the polarization dependent refraction index for the hadronic wave, given by the Fermi–Akhiezer–Pomeranchuk–Lax formula:[11]

$$\hat{n} = 1 + \frac{1}{2p} N\hat{\mathsf{F}}(0). \tag{9}$$

The forward NN scattering amplitude $\hat{\mathsf{F}}(0)$ depends on the beam and target spins, and the polarized target acts as an optically active medium. It is convenient to use instead the Fermi Hamiltonian (with the distance z traversed in the medium playing the rôle of time)

$$\hat{H} = \tfrac{1}{2}N\hat{F}(0) = \tfrac{1}{2}N[\hat{R}(0) + i\hat{\sigma}_{\rm tot}], \tag{10}$$

where $\hat{R}(0)$ is the real part of the forward scattering amplitude and N is the volume density of atoms in the target. The anti-hermitian part of the Fermi Hamiltonian, $\propto \hat{\sigma}_{\rm tot}$, describes the absorption (attenuation) in the medium.

In terms of this Hamiltonian, the quantum-mechanical evolution equation for the spin-density matrix of the transmitted beam reads

$$\begin{aligned}\frac{d}{dz}\hat{\rho}(\boldsymbol{p}) &= i\Big(\hat{H}\hat{\rho}(\boldsymbol{p}) - \hat{\rho}(\boldsymbol{p})\hat{H}^{\dagger}\Big)\\ &= \underbrace{i\tfrac{1}{2}N\Big(\hat{R}\hat{\rho}(\boldsymbol{p}) - \hat{\rho}(\boldsymbol{p})\hat{R}\Big)}_{\text{(Pure refraction)}} - \underbrace{\tfrac{1}{2}N\Big(\hat{\sigma}_{\rm tot}\hat{\rho}(\boldsymbol{p}) + \hat{\rho}(\boldsymbol{p})\hat{\sigma}_{\rm tot}\Big)}_{\text{(Pure attenuation)}}.\end{aligned} \tag{11}$$

In the specific case of spin-$\frac{1}{2}$ protons interacting with the spin-$\frac{1}{2}$ protons (and electrons) the total cross section and real part of the forward scattering amplitude are parametrized as

$$\begin{aligned}\hat{\sigma}_{\rm tot} &= \sigma_0 + \underbrace{\sigma_1(\boldsymbol{\sigma}\cdot\boldsymbol{Q}) + \sigma_2(\boldsymbol{\sigma}\cdot\boldsymbol{k})(\boldsymbol{Q}\cdot\boldsymbol{k})}_{\text{(spin$-$sensitive loss)}},\\ \hat{R} &= R_0 + \underbrace{R_1(\boldsymbol{\sigma}\cdot\boldsymbol{Q}) + R_2(\boldsymbol{\sigma}\cdot\boldsymbol{k})(\boldsymbol{Q}\cdot\boldsymbol{k})}_{(\boldsymbol{\sigma}\cdot\text{Pseudomagnetic field})}.\end{aligned} \tag{12}$$

Then, upon some algebra, one finds the evolution equation for the beam polarization $\boldsymbol{P} = \boldsymbol{s}/I_0$

$$\begin{aligned}d\boldsymbol{P}/dz = &\underbrace{-N\sigma_1(\boldsymbol{Q} - (\boldsymbol{P}\cdot\boldsymbol{Q})\boldsymbol{P}) - N\sigma_2(\boldsymbol{Q}\boldsymbol{k})(\boldsymbol{k} - (\boldsymbol{P}\cdot\boldsymbol{k})\boldsymbol{P})}_{\text{(Polarization buildup by spin$-$sensitive loss)}}\\ &+ \underbrace{NR_1(\boldsymbol{P}\times\boldsymbol{Q}) + nR_2(\boldsymbol{Q}\boldsymbol{k})(\boldsymbol{P}\times\boldsymbol{k})}_{\text{(Spin precession in pseudomagnetic field)}},\end{aligned} \tag{13}$$

where we indicated the rôle of the anti-hermitian – attenuation – and hermitian – pseudomagnetic field – parts of the Fermi Hamiltonian. It is absolutely important that the cross sections $\sigma_{0,1,2}$ in the evolution equation for the transmitted beam describe all-angle scattering, in the proton–atom case that corresponds to $\theta \geq \theta_{\rm min}$.

Here we notice, that the precession effects are missed in the Milstein–Strakhovenko kinetic equation for the spin-state population numbers. The precession is the major observable in condensed matter studies with polarized neutrons.[15] Kinetic equation holds only if the spin-density matrix is diagonal one. In the case of the spin filtering in storage rings with pure transverse or longitudinal (supplemented by the Siberian snake for the compensation of the spin rotation) polarizations of PIT, the kinetic equation can be recovered, though, from the evolution of the density matrix upon the averaging over the precession. Hereafter we focus on the transverse polarization studied in the FILTEX experiment.

For the sake of completeness, we cite the full system of coupled evolution equations for the spin density matrix

$$\frac{d}{dz}\begin{pmatrix} I_0 \\ s \end{pmatrix} = -N\begin{pmatrix} \sigma_0(>\theta_{\rm min}) & Q\sigma_1(>\theta_{\rm min}) \\ Q\sigma_1(>\theta_{\rm min}) & \sigma_0(>\theta_{\rm acc}) \end{pmatrix}\cdot\begin{pmatrix} I_0 \\ s \end{pmatrix}, \qquad (14)$$

In has the eigen-solutions $\propto \exp(-\lambda_{1,2}Nz)$ with the eigenvalues $\lambda_{1,2} = \sigma_0 \pm Q\sigma_1$. Eq. (13) reduces to the Meyer's equation [1]

$$\frac{dP}{dz} = -N\sigma_1 Q\big(1 - \boldsymbol{P}^2\big). \qquad (15)$$

The polarization buildup follows the law $P(z) = -\tanh(Q\sigma_1 Nz)$.

4. Incorporation of the scattering within the beam into the evolution equation

For scattering angles of the interest, $\theta \gtrsim \theta_{\rm min}$, the differential cross section of the quasielastic proton–atom scattering equals

$$\begin{aligned}\frac{d\hat{\sigma}_E}{d^2\boldsymbol{q}} &= \frac{1}{(4\pi)^2}\hat{\mathsf{F}}(\boldsymbol{q})\hat{\rho}\hat{\mathsf{F}}^\dagger(\boldsymbol{q}) \\ &= \frac{1}{(4\pi)^2}\hat{\mathsf{F}}_e(\boldsymbol{q})\hat{\rho}\hat{\mathsf{F}}_e^\dagger(\boldsymbol{q}) + \frac{1}{(4\pi)^2}\hat{\mathsf{F}}_p(\boldsymbol{q})\hat{\rho}\hat{\mathsf{F}}_p^\dagger(\boldsymbol{q})\,. \end{aligned} \qquad (16)$$

The evolution equation for the spin-density matrix must be corrected for the lost-and-found protons, scattered quasielastically within the beam, $\theta \leq \theta_{\rm acc}$. The formal derivation from the multiple-scattering theory, in which the unitarity, i.e. the particle loss and recovery balance, is satisfied rigorously, is too lengthy to be reproduced here. The result is fairly transparent, though:

$$\frac{d}{dz}\hat{\rho} = i[\hat{H}, \hat{\rho}] = \underbrace{i\tfrac{1}{2}N\Big(\hat{R}\hat{\rho}(\boldsymbol{p}) - \hat{\rho}(\boldsymbol{p})\hat{R}\Big)}_{\text{Pure precession \& refraction}}$$
$$- \underbrace{\tfrac{1}{2}N\Big(\hat{\sigma}_{\rm tot}\hat{\rho}(\boldsymbol{p}) + \hat{\rho}(\boldsymbol{p})\hat{\sigma}_{\rm tot}\Big)}_{\text{Evolution by loss}}$$
$$+ \underbrace{N\int^{\Omega_{\rm acc}} \frac{d^2\boldsymbol{q}}{(4\pi)^2}\hat{\mathsf{F}}(\boldsymbol{q})\hat{\rho}(\boldsymbol{p}-\boldsymbol{q})\hat{\mathsf{F}}^{\dagger}(\boldsymbol{q})}_{\text{Lost \& found: scattering within the beam}} \,. \tag{17}$$

Notice the convolution of the transverse momentum distribution in the beam with the differential cross section of quasielastic scattering. This broadening of the momentum distribution is compensated for by the focusing and the beam cooling in a storage ring.

5. Needle-sharp scattering off electrons does not polarize the beam

The relevant parts of the nonrelativistic Breit ep interaction, found in all QED textbooks, are

$$U(\boldsymbol{q}) = \alpha_{\rm em}\left\{\frac{1}{\boldsymbol{q}^2} + \mu_p\frac{(\boldsymbol{\sigma}_p\boldsymbol{q})(\boldsymbol{\sigma}_e\boldsymbol{q}) - (\boldsymbol{\sigma}_p\,\boldsymbol{\sigma}_e)\boldsymbol{q}^2}{4m_pm_e\boldsymbol{q}^2}\right\}, \tag{18}$$

and give the contribution to the total proton–atom cross-section of the form (we suppress the condition $\theta > \theta_{\rm min}$)

$$\hat{\sigma}^e_{\rm tot} = \underbrace{\sigma^e_0}_{\text{Coulomb}} + \underbrace{\sigma^e_1(\boldsymbol{\sigma}_p\cdot\boldsymbol{Q}_e) + \sigma^e_2(\boldsymbol{\sigma}_p\cdot\boldsymbol{k})(\boldsymbol{Q}_e\cdot\boldsymbol{k})}_{\text{Coulomb}\times\text{(Hyperfine+Tensor)}}, \tag{19}$$

with $\sigma^e_2 = \sigma^e_1$.[10]

The pure electron target contribution to the transmission losses equals

$$\frac{1}{2}\frac{d}{dz}I_0(\boldsymbol{p})(1+\boldsymbol{\sigma}\cdot\boldsymbol{P}(\boldsymbol{p})) =$$
$$-\frac{1}{2}NI_0(\boldsymbol{p})\Big[\underbrace{\sigma^e_0 + \sigma^e_1\boldsymbol{P}\boldsymbol{Q}_e}_{\text{particle loss}} + \boldsymbol{\sigma}\underbrace{\Big(\sigma^e_0\boldsymbol{P} + \sigma^e_1\boldsymbol{Q}_e\Big)}_{\text{spin loss}}\Big]. \tag{20}$$

Here $\sigma^e_1 \approx -70\,\text{mb}$, which comes from the Coulomb-tensor and Coulomb-hyperfine interference,[10] is fairly large on the hadronic cross section scale.

Now note, that pe scattering is needle-sharp, $\theta \leq \theta_e \ll \theta_{\rm acc}$, and the lost-and-found contribution from the scattering within the beam can be evaluated as (we consider the transverse polarization)

$$\begin{aligned} &N \int \frac{d^2\boldsymbol{q}}{(4\pi)^2} \hat{\mathsf{F}}_e(\boldsymbol{q}) \hat{\rho}(\boldsymbol{p}-\boldsymbol{q}) \hat{\mathsf{F}}_e^{\dagger}(\boldsymbol{q}) \\ &\quad = \frac{1}{2} N I_0(\boldsymbol{p}) \int \frac{d^2\boldsymbol{q}}{(4\pi)^2} \hat{\mathsf{F}}_e(\boldsymbol{q}) \hat{\mathsf{F}}_e^{\dagger}(\boldsymbol{q}) + \frac{1}{2} N \boldsymbol{s}(\boldsymbol{p}) \int \frac{d^2\boldsymbol{q}}{(4\pi)^2} \hat{\mathsf{F}}_e(\boldsymbol{q}) \boldsymbol{\sigma} \hat{\mathsf{F}}_e^{\dagger}(\boldsymbol{q}) \\ &\quad = \underbrace{\frac{1}{2} N I_0(\boldsymbol{p}) \Big[\sigma_0^e + \sigma_1^e (\boldsymbol{P}\cdot\boldsymbol{Q})\Big]}_{\text{Lost \& found particle number}} + \underbrace{\frac{1}{2} N I_0(\boldsymbol{p}) \boldsymbol{\sigma} \Big[\sigma_0^e \boldsymbol{P} + \sigma_1^e \boldsymbol{Q}_e\Big]}_{\text{Lost \& found spin}}. \end{aligned} \tag{21}$$

One readily observes the exact cancellation of the transmission, Eq. (20), and scattering-within-the-beam, Eq. (21), electron target contributions to the evolution equation (17). The situation is entirely reminiscent of the cancellation of the effect of atomic electrons in the transmission measurements of the proton–proton total cross section. One concludes that polarized atomic electrons will not polarize stored (anti)protons.

6. Scattering within the beam in spin filtering by nuclear interaction

The angular divergence of the beam at the target position is much smaller than the ring acceptance $\theta_{\rm acc}$. Consequently, the contribution from the elastic pp scattering within the beam can be approximated by

$$\begin{aligned} &\int d^2\boldsymbol{p} \int^{\Omega_{\rm acc}} \frac{d^2\boldsymbol{q}}{(4\pi)^2} \hat{\mathsf{F}}_p(\boldsymbol{q}) \hat{\rho}(\boldsymbol{p}-\boldsymbol{q}) \hat{\mathsf{F}}_p^{\dagger}(\boldsymbol{q}) = \\ &\quad = \left[\int d^2\boldsymbol{p} I_0(\boldsymbol{p})\right] \cdot \int^{\Omega_{\rm acc}} \frac{d^2\boldsymbol{q}}{(4\pi)^2} \hat{\mathsf{F}}_p(\boldsymbol{q}) \frac{1}{2}\Big(1+\boldsymbol{\sigma}\boldsymbol{P}\Big) \hat{\rho}(\boldsymbol{q}) \hat{\mathsf{F}}_p^{\dagger}(\boldsymbol{q}) \\ &\quad = \hat{\sigma}^E(\leq \theta_{\rm acc}) \cdot \int d^2\boldsymbol{p} I_0(\boldsymbol{p}). \end{aligned} \tag{22}$$

The beam cooling amounts to averaging over azimuthal angles of scattered protons, upon this averaging

$$\begin{aligned} \hat{\sigma}^E(\leq \theta_{\rm acc}) = &\underbrace{\sigma_0^{\rm el}(\leq \theta_{\rm acc}) + \sigma_1^{\rm el}(\leq \theta_{\rm acc})(\boldsymbol{P}\cdot\boldsymbol{Q})}_{\text{Lost \& found particles}} \\ &+ \underbrace{\boldsymbol{\sigma}\cdot\left(\sigma_0^E(\leq \theta_{\rm acc})\boldsymbol{P}) + \sigma_1^E(\leq \theta_{\rm acc})\boldsymbol{Q})\right)}_{\text{Lost \& found spin}}. \end{aligned} \tag{23}$$

Now we decompose the pure transmission losses

$$
\begin{aligned}
\frac{d}{dz}\hat{\rho} = & \underbrace{-\frac{1}{2}N\Big(\hat{\sigma}_{\text{tot}}(>\theta_{\text{acc}})\hat{\rho}(\boldsymbol{p}) + \hat{\rho}(\boldsymbol{p})\hat{\sigma}_{\text{tot}}(>\theta_{\text{acc}})\Big)}_{\text{Unrecoverable transmission loss}} \\
& -\frac{1}{2}NI_0(\boldsymbol{p})\Big[\underbrace{\sigma_0^{\text{el}}(<\theta_{\text{acc}}) + \sigma_1^{\text{el}}(<\theta_{\text{acc}})\boldsymbol{PQ}}_{\text{Potentially recoverable particle loss}} \\
& + \boldsymbol{\sigma}\underbrace{\Big(\sigma_0^{\text{el}}(<\theta_{\text{acc}})\boldsymbol{P} + \sigma_1^{\text{el}}(<\theta_{\text{acc}})\boldsymbol{Q}\Big)}_{\text{Potentially recoverable spin loss}}\Big]
\end{aligned}
\tag{24}
$$

into the unrecoverable losses from scattering beyond the acceptance angle and the potentially recoverable losses from the scattering within the acceptance angle. Upon the substitution of (23) and (24) into the evolution equation (17), one finds the operator of mismatch between the potentially recoverable losses and the scattering within the beam of the form

$$
\begin{aligned}
\Delta\hat{\sigma} &= \frac{1}{4}\Big(\hat{\sigma}^{\text{el}}(<\theta_{\text{acc}})(1+\boldsymbol{\sigma P}) + (1+\boldsymbol{\sigma P})\hat{\sigma}^{\text{el}}(<\theta_{\text{acc}})\Big) - \hat{\sigma}^{E}(\le\theta_{\text{acc}}) \\
&= \boldsymbol{\sigma}\left(2\Delta\sigma_0\boldsymbol{P} + \Delta\sigma_1\boldsymbol{Q}\right).
\end{aligned}
\tag{25}
$$

The lost-and-found corrected coupled evolution equations take the form

$$
\frac{d}{dz}\begin{pmatrix} I_0 \\ s \end{pmatrix} = -N\begin{pmatrix} \sigma_0(>\theta_{\text{acc}}) & Q\sigma_1(>\theta_{\text{acc}}) \\ Q(\sigma_1(>\theta_{\text{acc}}) + \Delta\sigma_1) & \sigma_0(>\theta_{\text{acc}}) + 2\Delta\sigma_0 \end{pmatrix}\cdot\begin{pmatrix} I_0 \\ s \end{pmatrix}.
\tag{26}
$$

In the limit of vanishing mismatch, $\Delta\sigma_{0,1} = 0$, one would recover equations for pure transmission but with losses from scattering only beyond the acceptance angle. The corrections to the equation for the spin density do clearly originate from a difference between the spin of the particle taken away from the beam and the spin the same particle brings back into the beam after it was subjected to a small-angle elastic scattering. In terms of the standard observables as defined by Bystricky *et al.* (our θ is the scattering angle in the laboratory frame),[16]

$$
\begin{aligned}
\sigma_1^{\rm el}(>\theta_{\rm acc}) &= \tfrac{1}{2}\int_{\theta_{\rm acc}} d\Omega\Big(d\sigma/d\Omega\Big)\Big(A_{00nn}+A_{00ss}\Big),\\
\Delta\sigma_0 &= \tfrac{1}{2}[\sigma_0^{\rm el}(\leq\theta_{\rm acc})-\sigma_0^{E}(\leq\theta_{\rm acc})]\\
&= \tfrac{1}{2}\int_{\theta_{\rm min}}^{\theta_{\rm acc}} d\Omega\frac{d\sigma}{d\Omega}\Big(1-\tfrac{1}{2}D_{n0n0}-\tfrac{1}{2}D_{s'0s0}\cos(\theta)-\tfrac{1}{2}D_{k'0s0}\sin(\theta)\Big),\\
\Delta\sigma_1 &= \sigma_1^{\rm el}(\leq\theta_{\rm acc})-\sigma_1^{E}(\leq\theta_{\rm acc})\tfrac{1}{2}\int_{\theta_{\rm min}}^{\theta_{\rm acc}} d\Omega\frac{d\sigma}{d\Omega}\\
&\quad\times\Big(A_{00nn}+A_{00ss}-K_{n00n}-K_{s'00s}\cos(\theta)-K_{k'00s}\sin(\theta)\Big). \qquad (27)
\end{aligned}
$$

The difference between the spin of the particle taken away from the beam and put back after the small-angle elastic scattering corresponds to the spin-flip scattering, as Milstein and Strakhovenko correctly emphasized.[14] Here there is a complete agreement between the spin-density matrix and kinetic equation approaches.

7. Polarization buildup with the scattering within the beam

Coupled evolution equations with the scattering within the beam, Eq. (26), have the solutions $\propto \exp(-\lambda_{1,2}Nz)$ with the eigenvalues

$$
\begin{aligned}
\lambda_{1,2} &= \sigma_0+\Delta\sigma_0\pm\sigma_3,\\
\sigma_3 &= Q\sqrt{\sigma_1(\sigma_1+\Delta\sigma_1)+\Delta\sigma_0{}^2}. \qquad (28)
\end{aligned}
$$

The polarization buildup follows the law (see also Ref. 14)

$$
P(z) = -\frac{(\sigma_1+\Delta\sigma_1)\tanh(\sigma_3 Nz)}{\sigma_3+\Delta\sigma_0\tanh(\sigma_3 Nz)}. \qquad (29)
$$

The effective small-time polarization cross section equals

$$
\sigma_P \approx -Q(\sigma_1+\Delta\sigma_1). \qquad (30)
$$

8. Numerical estimates and the FILTEX result

We recall first the works by Meyer and Horowitz.[1,10] Meyer[1] initiated the whole issue of the scattering within the beam, correctly evaluated the principal double-spin dependent Coulomb–nuclear interference (CNI) effect, but an oversight has crept in when putting together the transmission and scattering-within-the-beam effects, which we shall correct below.

The FILTEX polarization rate as published in 1993, can be reinterpreted as $\sigma_P = 63 \pm 3\,(\text{stat.})\,\text{mb}$. The expectation from filtering by a pure nuclear elastic scattering at all scattering angles, $\theta > 0$, based on the pre-93 SAID database,[17] was

$$\sigma_1(\text{Nuclear}; \theta > 0) = 122\,\text{mb}. \tag{31}$$

The factor of two disagreement between σ_P and σ_1 called for an explanation, and Meyer made two important observations: (i) one only needs to include the filtering by scattering beyond the acceptance angle, (ii) the Coulomb–nuclear interference angle θ_{Coulomb} is large, $\theta_{\text{Coulomb}} \gg \theta_{\text{acc}}$, and one needs to correct for the Coulomb–nuclear interference (CNI) effects. Based on the pre-93 SAID database, he evaluated the CNI corrected

$$\sigma_1(\text{CNI}; \theta > \theta_{\text{acc}}) = 83\,\text{mb}. \tag{32}$$

The effect of pure nuclear elastic *pp* scattering within the acceptance angle would have been utterly negligible, this substantial departure from 122 mb of Eq. (31) is entirely due to the interference of the Coulomb and double-spin dependent nuclear amplitudes – there is a close analogy to the similar interference in *pe* scattering. As we shall argue below, for all the practical purposes Meyer's Eq. (32) is the final theoretical prediction for σ_P, but let the story unfold.

The estimate (32) was still about seven standard deviations from the above cited σ_P. Next Meyer noticed that protons scattered off electrons are polarized. They all go back into the beam. Based on the Horowitz–Meyer calculation of the polarization transfer from target electrons to scattered protons, that amounts to the correction to (32)

$$\delta\sigma_1^{ep} = -70\,\text{mb}. \tag{33}$$

Finally, adding the polarization brought into the beam by protons scattered elastically off protons within the acceptance angle,

$$\delta\sigma_1^{pp}(\text{CNI}; \theta_{\text{min}} < \theta < \theta_{\text{acc}}) = +52\,\text{mb}, \tag{34}$$

brings the theory to a perfect agreement with the experiment: $\sigma_1 = (83 - 70 + 52)\,\text{mb} = 65\,\text{mb}$.

Unfortunately, this agreement with σ_P must be regarded as an accidental one. In view of our discussion in Sec. 6, the starting point (32) corresponds to transmission effects already corrected for the scattering within the beam. As such, it correctly omits the transmission effects from the scattering off electrons. Then, correcting for (33) and (34) amounts to the double counting of the scattering within the beam. These corrections would have been legitimate only if one would have started with the sum of $\sigma_1(>\theta_{\rm min})$ for electron and proton targets rather than with (32).

In a more accurate treatment of the scattering within the beam, we encountered the mismatch cross-sections $\Delta\sigma_{0,1}$. They correspond to spin effects at extremely small scattering angles $\theta_{\rm min} < \theta < \theta_{\rm acc} \ll \theta_{\rm Coulomb}$. The elastic scattering that deep under the Coulomb peak can never be accessed in the direct scattering experiments, such observables can only be of the relevance to the storage rings. The existing SAID [17] and Nijmegen [18] databases have never been meant for the evaluation of the NN scattering amplitudes at so small angles. An important virtue of these databases is that they have a built-in procedure for the Coulomb–nuclear interference effects in all observables. If one would like to take an advantage of this feature, then one needs a careful extrapolation of these observables to the range of very extremely small angles of our interest. There are strong cancellations and it is prudent to extrapolate the whole integrands of $\Delta\sigma_{0,1}$ rather than the separate observables. Upon such an extrapolation, $\Delta\sigma_{0,1}$ are found to be negligible small, for the polarization cross section (30) of our interest we find $\Delta\sigma_1 \approx -6\cdot 10^{-3}\,$mb.

Milstein and Strakhovenko took a very different path:[14] they started with the nuclear scattering phases from the Nijmegen database,[18] added in all the Coulomb corrections following the Nijmegen prescriptions, and evaluated directly all the CNI effects. The numerical results for $\Delta\sigma_1$ from the two different evaluations are for all the practical purposes identical. The technical reason for negligible $\Delta\sigma_1$ in contrast to a very large difference between (31) and (32) is a vanishing interference between the hadronic spin-flip and dominant Coulomb amplitudes [14]

The principal conclusion is that the polarization buildup of stored protons is, for all the practical purposes, controlled by the transmission effects, described by Meyer's formula (32) for the CNI corrected nuclear proton–proton elastic scattering beyond the ring acceptance angle. Corrections to this formula for the spin-flip scattering prove to be negligible small. The polarization transfer from polarized electrons to scattered protons is a legitimate, and numerically substantial, effect, but it is exactly canceled by the electron contribution to the spin-dependent transmission effects.

The conversion of the FILTEX polarization rate, which by itself is the 20 standard deviation measurement, into the polarization cross section σ_P depends on the target polarization and the PIT areal density. The recent reanalysis [19] gave $\sigma_P = 72.5 \pm 5.8\,$mb, where both the statistical and systematical errors are included. The latest version of the SAID database,

SAID-SP05,[17] gives $\sigma_1(\text{CNI}; \theta > \theta_{\text{acc}}) = 85.6\,\text{mb}$, which is consistent with the FILTEX result within the quoted error bars. Following the direct evaluation of the CNI starting from the Nijmegen nuclear phase shifts, Milstein and Strakhovenko find for the same quantity 89 mb.[14]

9. Conclusions

We reported a quantum-mechanical evolution equation for the spin-density matrix of a stored beam interacting with the polarized internal target. The effects of the scattering within the beam are consistently included. An indispensable part of this description is a precession of the beam spin in the pseudomagnetic field of polarized atoms in PIT. In the specific application of our evolution equation to the spin filtering in the storage ring, the precession effects average out, and the spin-density matrix formalism and the kinetic equation formalism of Milstein and Strakhovenko become equivalent to each other.

Following Meyer, one must allow for the CNI contribution to the spin-dependent scattering within the beam, which has a very strong impact on the polarization cross section. There is a consensus between theorists from the Budker Institute and IP, Jülich on the self-cancellation of the transmission and scattering-within-the-beam contributions from polarized electrons to the spin filtering of (anti)protons. Both groups agree that corrections from spin-flip scattering within the beam to Eq. (32) for the polarization cross section are negligible small. There is only a slight disagreement between the reanalyzed FILTEX result, $\sigma_P = 72.5 \pm 5.8\,\text{mb}$ [19] and the theoretical expectations, $\sigma_P \approx 86\,\text{mb}$.

Regarding the future of the PAX suggestion,[3] the experimental basis for predicting the polarization buildup in a stored antiproton beam is practically non-existent. One must optimize the filtering process using the antiprotons available elsewhere(CERN, Fermilab). Several phenomenological models of antiproton–proton interaction have been developed to describe the experimental data from LEAR.[20–25] While the real part of the $p\bar{p}$ potential can be obtained from the meson-exchange nucleon–nucleon potentials by the G-parity transformation and is under reasonable control, the fully field-theoretic derivation of the anti-hermitian annihilation potential is as yet lacking. The double-spin $p\bar{p}$ observables necessary to constrain predictions for $\sigma_{1,2}$ are practically nonexistent (for the review see Ref. 26). Still, the expectations from the first generation models for double-spin dependence of $p\bar{p}$ interaction are encouraging, see Haidenbaur's review at the

Heimbach Workshop on Spin Filtering.[27] With filtering for two lifetimes of the beam, they suggest that in a dedicated large-acceptance polarizer storage ring, antiproton beam polarizations in the range of 15–25 % seem achievable, see Contalbrigo's talk at this Workshop.[28]

Acknowledgments

The reported study has been a part of the PAX scrutiny of the spin-filtering process. We are greatly indebted to F. Rathmann for prompting us to address this problem. We acknowledge discussions with M. Contalbrigo, N. Buttimore, J. Haidenbauer, P. Lenisa, Yu. Shatunov, B. Zakharov, and especially with H.O. Meyer, C. Horowitz, A. Milstein and V. Strakhovenko. Many thanks are due to M. Rentmeester, R. Arndt and I. Strakovsky for their friendly assistance with providing the custom-tailored outputs for the small-angle extrapolation purposes.

References

1. H.O. Meyer, *Phys. Rev.* **E50** (1994) 1485.
2. F. Rathmann *et al.*, *Phys. Rev. Lett.* **71** (1993) 1379.
3. Technical Proposal for *Antiproton–Proton Scattering Experiments with Polarization*, PAX Collaboration, spokespersons: P. Lenisa and F. Rathmann, available from arXiv:hep-ex/0505054 (2005).
4. Proc. of the *Workshop on Polarized Antiprotons*, Bodega Bay, CA, 1985, eds. A.D. Krisch, A.M.T. Lin, and O. Chamberlain, *AIP Conf. Proc.* **145** (AIP, New York, 1986).
5. D.P. Grosnick *et al.*, *Nucl. Instrum. Meth.* **A290** (1990) 269.
6. H. Spinka *et al.*, proc. of the *8th. Int. Symp. on Polarization Phenomena in Nuclear Physics*, Bloomington, Indiana, 1994, Eds. E.J. Stephenson and S.E. Vigdor, *AIP Conf. Proc.* **339** (AIP, Woodbury, NY, 1995), p. 713.
7. T.O. Niinikoski and R. Rossmanith, *Nucl. Instrum. Meth.* **A255** (1987) 460.
8. P. Cameron *et al.*, proc. of the *15th. Int. Spin Physics Symp.*, Upton, New York, 2002, Eds. Y.I. Makdisi, A.U. Luccio, and W.W. MacKay, *AIP Conf. Proc.* **675** (AIP, Melville, NY, 2003), p. 781.
9. F. Rathmann *et al.*, *Phys. Rev. Lett.* **94** (2005) 014801.
10. C.J. Horowitz and H.O. Meyer, *Phys. Rev. Lett.* **72** (1994) 3981.
11. I.I. Gurevich and L.V. Tarasov, *Low-Energy Neutron Physics*, North-Holland Publishing Company (1968).
12. A.I. Akhiezer, L.N. Rosenzweig and I.M. Shmushkevich, *Zh. Exp. Teor. Fiz.* **33** (1957) 765 [*Sov. Phys. JETP* **6** (1958) 588]; A.I. Akhiezer and M.P. Rekalo, *Sov. Phys. – Doklady* **13** (1968) 572.
13. V. Punjabi *et al.*, *Phys. Rev.* **C71** (2005) 055202 [*Erratum-ibid.* **C71** (2005) 069902].
14. A.I. Milstein and V.M. Strakhovenko, arXiv: physics/0504183.

15. S.V. Maleev, *Physics-Uspekhi* **45** (2002) 569.
16. J. Bystricky, F. Lehar and P. Winternitz, *J. Phys.* (France) **39** (1978) 1.
17. SAID Nucleon Nucleon scattering database, available from Center for Nuclear Studies, Department of Physics, George Washington University, USA, at website `http://gwdac.phys.gwu.edu/analysis/nn_analysis.html`.
18. Nijmegen Nucleon Nucleon Scattering Database, available from Radboud University Nijmegen, The Netherlands, at website `http://nn-online.org/`.
19. F. Rathmann, paper in preparation.
20. T. Hippchen, J. Haidenbauer, K. Holinde, V. Mull, *Phys. Rev.* **C44** (1991) 1323.
21. V. Mull, J. Haidenbauer, T. Hippchen, K. Holinde, *Phys. Rev.* **C44** (1991) 1337.
22. V. Mull, K. Holinde, *Phys. Rev.* **C51** (1995) 2360.
23. J. Haidenbauer, K. Holinde and A.W. Thomas, *Phys. Rev.* **C45** (1992) 952.
24. R. Timmermans, T.A. Rijken and J.J. de Swart, *Phys. Rev.* **C52** (1995) 1145.
25. M. Pignone, M. Lacombe, B. Loiseau and R. Vinh Mau, *Phys. Rev.* **C50** (1994) 2710.
26. E. Klempt, F. Bradamante, A. Martin, and J.-M. Richard, *Phys. Rep.* **368** (2002) 119.
27. F. Rathmann, Summary of the *Workshop on Spin Filtering in Storage Rings* (8th PAX Meeting), Heimbach, Germany September 2005; available from the PAX website at `http://www.fz-juelich.de/ikp/pax/`.
28. M. Contalbrigo, these proceedings.

THEORY LECTURES

SINGLE-SPIN ASYMMETRIES AND TRANSVERSITY IN QCD

S.J. Brodsky*

Stanford Linear Accelerator Center
Stanford University, Stanford, California 94309, USA

Initial- and final-state interactions from gluon exchange, normally neglected in the parton model, have a profound effect in QCD hard-scattering reactions, leading to leading-twist single-spin asymmetries, diffractive deep inelastic scattering, diffractive hard hadronic reactions, as well as nuclear shadowing and antishadowing–leading-twist physics not incorporated in the light-front wavefunctions of the target computed in isolation. The physics of such processes thus require the understanding of QCD at the amplitude level; in particular, the physics of spin requires an understanding of the phase structure of final-state and initial-state interactions, as well as the structure of the basic wavefunctions of hadrons themselves.

1. Introduction

Spin measurables, such as single-spin asymmetries in deep inelastic lepton scattering and transversity correlations in the Drell–Yan process $\overline{p}p \to \ell^+\ell^- X$, probe the structure of hadrons at a fundamental level. Initial- and final-state interactions from gluon-exchange, normally neglected in the parton model, have a profound effect in QCD hard-scattering reactions, leading to leading-twist single-spin asymmetries,[1] leading-twist diffractive deep inelastic scattering,[2] diffractive hard hadronic reactions, and nuclear shadowing and antishadowing–physics not incorporated in the light-front wavefunctions of the target computed in isolation. in particular, the single-spin asymmetry in semi-inclusive deep inelastic scattering depends on the phase difference of the final-state interactions in different partial waves and the same matrix elements which produce the anomalous magnetic moment of the target nucleon. The physics of spin thus requires an understanding of

*Work supported by the Department of Energy under contract number DE–AC02–76SF00515.

the phase structure of final-state and initial state interactions as well as knowledge of the basic wavefunctions of hadrons themselves. In this talk I will review the theory of single-spin asymmetries and the utility of light-front wavefunctions in representing the fundamental structure of hadrons in terms of quark and gluon degrees of freedom. I will also show how this formalism allows the computation of the transversity distribution arising from the quantum fluctuation of a physical electron.

2. Single-Spin Asymmetries from Final-State Interactions

Spin correlations provide a remarkably sensitive window to hadronic structure and basic mechanisms in QCD. Among the most interesting polarization effects are single-spin azimuthal asymmetries in semi-inclusive deep inelastic scattering, representing the correlation of the spin of the proton target and the virtual photon to hadron production plane: $\vec{S}_p \cdot \vec{q} \times \vec{p}_H$.[3] Such asymmetries are time-reversal odd, but they can arise in QCD through phase differences in different spin amplitudes.

The traditional explanation of pion electroproduction single-spin asymmetries in semi-inclusive deep inelastic scattering is that they are proportional to the transversity distribution of the quarks in the hadron h_1 [4–6] convoluted with the transverse momentum dependent fragmentation (Collins) function $H_1^\perp$, the distribution for a transversely polarized quark to fragment into an unpolarized hadron with non-zero transverse momentum.[7–11]

Dae Sung Hwang, Ivan Schmidt and I have showed that an alternative physical mechanism for the azimuthal asymmetries also exists.[1,12,13] The same QCD final-state interactions (gluon exchange) between the struck quark and the proton spectators which lead to diffractive events also can produce single-spin asymmetries (the Sivers effect) in semi-inclusive deep inelastic lepton scattering which survive in the Bjorken limit. See Fig. 1. In contrast to the SSAs arising from transversity and the Collins fragmentation function, the fragmentation of the quark into hadrons is not necessary; one predicts a correlation with the production plane of the quark jet itself $\vec{S}_p \cdot \vec{q} \times \vec{p}_q$.

The final-state interaction mechanism provides an appealing physical explanation within QCD of single-spin asymmetries. Remarkably, the same matrix element which determines the spin–orbit correlation $\vec{S} \cdot \vec{L}$ also produces the anomalous magnetic moment of the proton, the Pauli form factor, and the generalized parton distribution E which is measured in deeply virtual Compton scattering. Physically, the final-state interaction phase arises as the infrared-finite difference of QCD Coulomb phases for hadron wave

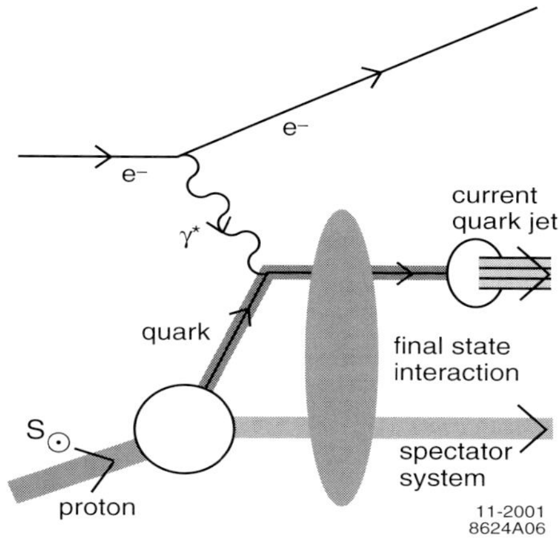

Fig. 1. A final-state interaction from gluon exchange in deep inelastic lepton scattering. The difference of the QCD Coulomb-like phases in different orbital states of the proton produces a single proton spin asymmetry.

functions with differing orbital angular momentum. An elegant discussion of the Sivers effect including its sign has been given by Burkardt.[14]

The final-state interaction effects can also be identified with the gauge link which is present in the gauge-invariant definition of parton distributions.[15] Even when the light-cone gauge is chosen, a transverse gauge link is required. Thus in any gauge the parton amplitudes need to be augmented by an additional eikonal factor incorporating the final-state interaction and its phase.[13,16] The net effect is that it is possible to define transverse momentum dependent parton distribution functions which contain the effect of the QCD final-state interactions.

A related analysis also predicts that the initial-state interactions from gluon exchange between the incoming quark and the target spectator system lead to leading-twist single-spin asymmetries in the Drell–Yan process $H_1 H_2^{\updownarrow} \to \ell^+\ell^- X$.[12,17] Initial-state interactions also lead to a $\cos 2\phi$ planar correlation in unpolarized Drell–Yan reactions.[18]

2.1. *Calculations of Single-Spin Asymmetries in QCD*

Hwang, Schmidt and I calculated[1] the single-spin Sivers asymmetry in semi-inclusive electroproduction $\gamma^* p^{\updownarrow} \to HX$ induced by final-state interactions in a model of a spin-1/2 proton of mass M with charged spin-1/2 and spin-0 constituents of mass m and λ, respectively, as in the QCD-motivated quark–diquark model of a nucleon. The basic electroproduction reaction is then $\gamma^* p \to q(qq)_0$. In fact, the asymmetry comes from the interference of two amplitudes which have different proton spin, but couple to the same final quark spin state, and therefore it involves the interference of tree and one-loop diagrams with a final-state interaction. In this simple model the azimuthal target single-spin asymmetry $A_{UT}^{\sin\phi}$ is given by

$$A_{UT}^{\sin\phi} = C_F \alpha_s(\mu^2) \, \frac{\Big(\Delta M + m\Big) r_\perp}{\Big[\Big(\Delta M + m\Big)^2 + \vec{r}_\perp^{\,2}\Big]} \times \Big[\vec{r}_\perp^{\,2} + \Delta(1-\Delta)\Big(-M^2 + \frac{m^2}{\Delta} + \frac{\lambda^2}{1-\Delta}\Big)\Big] \times \frac{1}{\vec{r}_\perp^{\,2}} \ln \frac{\vec{r}_\perp^{\,2} + \Delta(1-\Delta)(-M^2 + \frac{m^2}{\Delta} + \frac{\lambda^2}{1-\Delta})}{\Delta(1-\Delta)(-M^2 + \frac{m^2}{\Delta} + \frac{\lambda^2}{1-\Delta})} \,. \quad (1)$$

Here $r_\perp$ is the magnitude of the transverse momentum of the current quark jet relative to the virtual photon direction, and $\Delta = x_{\mathrm{Bj}}$ is the usual Bjorken variable. To obtain (1) from Eq. (21) of Ref. 1, we used the correspondence $|e_1 e_2|/4\pi \to C_F \alpha_s(\mu^2)$ and the fact that the sign of the charges e_1 and e_2 of the quark and diquark are opposite since they constitute a bound state. The result can be tested in jet production using an observable such as thrust to define the momentum $q + r$ of the struck quark.

Since the same matrix element controls the Pauli form factor, the contribution of each quark current to the SSA is proportional to the contribution $\kappa_{q/p}$ of that quark to the proton target's anomalous magnetic moment $\kappa_p = \sum_q e_q \kappa_{q/p}$.[1,14]

The HERMES collaboration has measured the SSA in pion electroproduction using transverse target polarization.[19] The Sivers and Collins effects can be separated using planar correlations; both contributions are observed to contribute, with values not in disagreement with theory expectations.[19,20] A recent comparison is given by Gamberg and Goldstein.[21] A related analysis also predicts that the initial-state interactions from gluon exchange between the incoming quark and the target spectator system lead to leading-twist single-spin asymmetries in the Drell–Yan process $H_1 H_2^{\updownarrow} \to \ell^+ \ell^- X$.[12,17] The SSA in the Drell–Yan process is the same as that obtained in SIDIS, with the appropriate identification of variables, but with the opposite Initial-state interactions also lead to a $\cos 2\phi$ planar correlation in unpolarized Drell–Yan reactions.[18] There is no Sivers effect in charged-current reactions since the W only couples to left-handed quarks.[22]

It should be emphasized that the Sivers effect occurs even for jet production; unlike transversity, hadronization is not required. There is no Sivers effect in charged current reactions since the W only couples to left-handed quarks.[22]

We can also consider the SSA of e^+e^- annihilation processes such as $e^+e^- \to \gamma^* \to \pi\Lambda^{\leftrightarrow}X$. The Λ reveals its polarization via its decay $\Lambda \to p\pi^-$. The spin of the Λ is normal to the decay plane. Thus we can look for a SSA through the T-odd correlation $\epsilon_{\mu\nu\rho\sigma} S^\mu_\Lambda p^\nu_\Lambda q^\rho_{\gamma^*} p^\sigma_\pi$. This is related by crossing to SIDIS on a Λ target.

Measurements from Jefferson Lab [23] also show significant beam single spin asymmetries in deep inelastic scattering. Afanasev and Carlson [24] have recently shown that this asymmetry is due to the interference of longitudinal and transverse photoabsorption amplitudes which have different phases induced by the final-state interaction between the struck quark and the target spectators. Their results are consistent with the experimentally observed magnitude of this effect. Thus similar FSI mechanisms involving quark orbital angular momentum appear to be responsible for both target and beam single-spin asymmetries.

3. Light-Front Wavefunctions and Spin

The light-front Fock expansion provides a convenient and rigorous wavefunction formalism for bound states in quantum field theory, analogous to the $\psi(\vec{p})$ momentum-space wavefunction description of nonrelativistic bound states of the Schrödinger theory, but allowing for particle creation and absorption. Unlike equal time theory (the instant form), boost transformations are kinematical in the front form, so that light-front wavefunctions are independent of the total four-momentum of the system. They are relativistic, and frame-independent, describing all particle number excitations n of the hadrons.

The light-front Fock expansion follows from the quantization of QCD at fixed light-front time $x^+ = x^0 + x^3$. The bound-state hadronic solutions $|\,\Psi_H\rangle$ are eigenstates of the light-front Heisenberg equation $H_{LF}\,|\,\Psi_H\rangle = M^2_H\,|\,\Psi_H\rangle$.[25] The spectrum of QCD is given by the eigenvalues M^2_H. The projection of each hadronic eigensolution on the free Fock basis: $\langle n\,|\,\Psi_H\rangle \equiv \psi_{n/H}(x_i, \vec{k}_{\perp i}, \lambda_i)$ then defines the LF Fock expansion in terms of the quark and transversely polarized gluon constituents in $A^+ = 0$ light-cone gauge. The light-front wavefunctions are functions of the constituent light-cone fractions $x_i = \frac{k_i^+}{P^+} = \frac{(k^0+k^z)_i}{P^+}$, relative transverse momenta $\vec{k}_{\perp i}$, and spin projections $S_i^z = \lambda_i$. The expansion has only transversely polarized gluons.

Recently Guy de Teramond and I have shown how one can use the AdS/CFT correspondence to obtain predictions for LFWFs of hadrons in a conformal approximation of QCD.[26]

The freedom to choose the light-like quantization four-vector provides an explicitly covariant formulation of light-front quantization and can be used to determine the analytic structure of light-front wave functions and define a kinematical definition of angular momentum. Angular momentum is consistently defined in the front form. The z component of angular momentum J^z is kinematical and conserved. The front form thus provides a consistent definition of relative orbital angular momentum and J^z conservation: the total spin projection $J^z = \sum_{i=1}^n S_i^z + \sum_i^{n-1} L_i^z$ is conserved in each Fock state. The cluster decomposition theorem and the vanishing of the "anomalous gravitomagnetic moment" $B(0)$ [27] are immediate properties of the LF Fock wavefunctions.[28]

Given the light-front wavefunctions $\psi_{n/H}(x_i, \vec{k}_{\perp i}, \lambda_i)$, one can compute a large range of hadron observables. For example, the valence and sea quark and gluon distributions which are measured in deep inelastic lepton scattering are defined from the squares of the LFWFS summed over all Fock states n. Form factors, exclusive weak transition amplitudes [29] such as $B \to \ell\nu\pi$. and the generalized parton distributions [30] measured in deeply virtual Compton scattering are (assuming the "handbag" approximation) overlaps of the initial and final LFWFS with $n = n'$ and $n = n' + 2$. The gauge-invariant distribution amplitude $\phi_H(x_i, Q)$ defined from the integral over the transverse momenta $\vec{k}_{\perp i}^2 \leq Q^2$ of the valence (smallest n) Fock state provides a fundamental measure of the hadron at the amplitude level;[31,32] they are the nonperturbative input to the factorized form of hard exclusive amplitudes and exclusive heavy hadron decays in perturbative QCD. The resulting distributions obey the DGLAP and ERBL evolution equations as a function of the maximal invariant mass, thus providing a physical factorization scheme.[33] In each case, the derived quantities satisfy the appropriate operator product expansions, sum rules, and evolution equations. However, at large x where the struck quark is far-off shell, DGLAP evolution is quenched,[34] so that the fall-off of the DIS cross sections in Q^2 satisfies inclusive-exclusive duality at fixed W^2.

4. Transversity of an Electron in QED

The Fock structure of the electron in QED provides an excellent example of the physics of spin distributions such as the transversity correlation and its relation to orbital angular momentum. The nonzero two-particle LFWFs of the electron in light-cone gauge $A^+ = 0$ are [35]

$$\begin{cases} \psi^{\uparrow}_{+\frac{1}{2}\,+1}(x,\vec{k}_\perp) = -\sqrt{2}\frac{(-k^1+ik^2)}{x(1-x)}\,\varphi\;, \\ \psi^{\uparrow}_{+\frac{1}{2}\,-1}(x,\vec{k}_\perp) = -\sqrt{2}\frac{(+k^1+ik^2)}{1-x}\,\varphi\;, \\ \psi^{\uparrow}_{-\frac{1}{2}\,+1}(x,\vec{k}_\perp) = -\sqrt{2}(M-\frac{m}{x})\,\varphi\;, \\ \psi^{\uparrow}_{-\frac{1}{2}\,-1}(x,\vec{k}_\perp) = 0\;, \end{cases} \tag{2}$$

where

$$\varphi = \varphi(x,\vec{k}_\perp) = \frac{e/\sqrt{1-x}}{M^2 - (\vec{k}_\perp^2 + m^2)/x - (\vec{k}_\perp^2 + \lambda^2)/(1-x)}\,. \tag{3}$$

$$\begin{cases} \psi^{\downarrow}_{+\frac{1}{2}\,+1}(x,\vec{k}_\perp) = 0\;, \\ \psi^{\downarrow}_{+\frac{1}{2}\,-1}(x,\vec{k}_\perp) = -\sqrt{2}(M-\frac{m}{x})\,\varphi\;, \\ \psi^{\downarrow}_{-\frac{1}{2}\,+1}(x,\vec{k}_\perp) = -\sqrt{2}\frac{(-k^1+ik^2)}{1-x}\,\varphi\;, \\ \psi^{\downarrow}_{-\frac{1}{2}\,-1}(x,\vec{k}_\perp) = -\sqrt{2}\frac{(+k^1+ik^2)}{x(1-x)}\,\varphi\;. \end{cases} \tag{4}$$

The photon has transverse polarization $S^z = \pm 1$. The pre-factors indicate the orbital angular momentum $L^z = 0, \pm 1$ of the two-parton state. Note that each state explicitly satisfies angular momentum conservation $J^z = \sum_{i=1,2} S_i^z + L^z$. The LF thus provides a rigorous definition of orbital angular momentum in relativistic quantum field theory.

We can generalize this form of the LFWFs to describe a quark spin-1 diquark model of the proton by giving the photon an effective mass λ and the fermionic constituent the mass m. A cut-off on the transverse momenta is introduced to regulate the distributions and provide a factorization scale. The model can be generalized further by introducing spectral weights in λ or M to mimic the fall-off of the LFWF of a composite system in QCD. The constituents can have arbitrary charges. Current matrix elements, structure functions, and generalized parton distributions are obtained from the overlap of the LFWFs. In the QED case, $m = M$ and $\lambda = 0$, one recovers the Schwinger anomalous magnetic moment $a_e = \frac{\alpha}{2\pi}$ and the standard formulae for the electron's Dirac and Pauli form factor. In addition one can verify that the anomalous gravitomagnetic moment for the electron scattering in external gravitational field vanishes identically for each Fock state, thus verifying a remarkable theorem based on the equivalence principle.

Dae Sung Hwang and I[36] have shown how the transverse polarization can be computed from the sum and difference of longitudinal LFWFs. The unpolarized, longitudinally polarized structure function and transversity distributions of the quark constituent in the quark–diquark model of the proton can then computed from the appropriate square of the LFWFs:

$$\begin{aligned}
&q(x,\Lambda^2)_{\text{spin}-1\ \text{diquark}}\\
&= \int \frac{\mathrm{d}^2\vec{k}_\perp \mathrm{d}x}{16\pi^3}\,\theta(\Lambda^2-\mathcal{M}^2)\,2\left[\frac{\vec{k}_\perp^2}{x^2(1-x)^2}+\frac{\vec{k}_\perp^2}{(1-x)^2}+\left(M-\frac{m}{x}\right)^2\right]|\varphi|^2\,,\\
&\Delta q(x,\Lambda^2)_{\text{spin}-1\ \text{diquark}}\\
&= \int \frac{\mathrm{d}^2\vec{k}_\perp \mathrm{d}x}{16\pi^3}\,\theta(\Lambda^2-\mathcal{M}^2)\,2\left[\frac{\vec{k}_\perp^2}{x^2(1-x)^2}+\frac{\vec{k}_\perp^2}{(1-x)^2}-\left(M-\frac{m}{x}\right)^2\right]|\varphi|^2\,,\\
&\delta q(x,\Lambda^2)_{\text{spin}-1\ \text{diquark}}\\
&= \int \frac{\mathrm{d}^2\vec{k}_\perp \mathrm{d}x}{16\pi^3}\,\theta(\Lambda^2-\mathcal{M}^2)\,4\left[\frac{\vec{k}_\perp^2}{x(1-x)^2}\right]|\varphi|^2\,.
\end{aligned} \tag{5}$$

The Soffer bound follows immediately from the LF Fock state construction. In this model $q(x)+\Delta q(x) = 2\delta q(x)$; *i.e.*, Soffer's inequality is satisfied with the equal sign.

5. Single-Spin Asymmetry and the Phase of Timelike Form Factors

As noted by Dubnickova, Dubnicka, and Rekalo,[37] and by Rock,[38] the existence of the T-odd single-spin asymmetry normal to the scattering plane in baryon pair production $e^-e^+ \to B\overline{B}$ requires a nonzero phase difference between the G_E and G_M form factors. The phase of the ratio of form factors G_E/G_M of spin-1/2 baryons in the timelike region can thus be determined from measurements of the polarization of one of the produced baryons. Carlson, Hiller, Hwang and I have shown that measurements of the proton polarization in $e^+e^- \to p\overline{p}$ strongly discriminate between the analytic forms of models which have been suggested to fit the proton G_E/G_M data in the spacelike region.[39]

There are three polarization observables, corresponding to polarizations in three directions, called longitudinal, sideways, and normal but often denoted z, x, and y, respectively. Longitudinal (z) when discussing the final state means parallel to the direction of the outgoing or in going baryon. Sideways (x) means perpendicular to the direction of the baryon but in the scattering plane. Normal (y) means normal to the scattering plane, in the direction of $\vec{k} \times \vec{p}$ where $\vec{k}$ is the electron momentum and $\vec{p}$ is the baryon momentum, with x, y, and z forming a right-handed coordinate system. The polarization P_y does not require polarization of a lepton and is[39]

$$A_N = P_y = \frac{\sin 2\theta \, \mathrm{Im} G_E^* G_M}{D\sqrt{\tau}} = \frac{(\tau - 1)\sin 2\theta \, \mathrm{Im} F_2^* F_1}{D\sqrt{\tau}} . \tag{6}$$

The expression for polarization P_y, Eq. (6), leads to results shown in Fig. 2. The polarizations are shown for four different fits to the spacelike data as referenced in the figure. The value of P_y should be the same for $e^+e^- \to p\bar{p}$ and $\bar{p}p \to \ell^+\ell^-$ up to an overall sign. The predicted polarizations are not small. Note that a purely polynomial fit to the spacelike data gives zero P_y. The normal polarization P_y is a single-spin asymmetry and requires a phase difference between G_E and G_M. It is an example of how time-reversal-odd observables can be nonzero if final-state or initial-state interactions give interfering amplitudes with different phases.

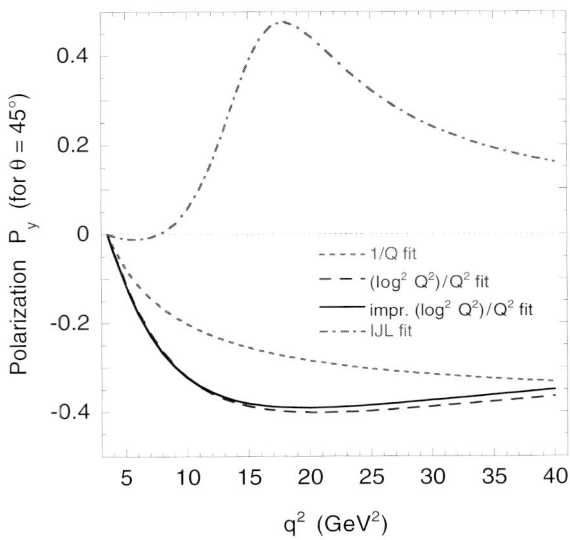

Fig. 2. Predicted polarization P_y in the timelike region for selected form factor fits described in the text. The plot is for $\theta = 45°$. The four curves are for an $F_2/F_1 \propto 1/Q$ fit; the $(\log^2 Q^2)/Q^2$ fit of Belitsky *et al.*;[40] an improved $(\log^2 Q^2)/Q^2$ fit; and a fit from Iachello *et al.*[41]

6. Exclusive Transversity

The most remarkable spin correlation ever observed in hadron physics is the normal spin—normal spin correlation asymmetry A_{NN} measured by A. Krisch and his collaborators [42] in large CM-angle pp elastic scattering. Both the projectile and target are polarized normal to the scattering plane. The ratio r_{NN} of cross sections where protons are both parallel to antiparallel becomes as large as 4:1. Since r_{NN} rises dramatically at the strangeness and charm thresholds $\sqrt{s} = 3, 5\,\mathrm{GeV}$, Guy de Teramond and I [43] have proposed that the these correlations are related to the onset of "octoquark" $uuduuds\bar{s}$, $uuduudc\bar{c}$ $J = L = S = 1$ resonances; such states only couple to the normal spin parallel channel. The resonance signal interferes with the hard QCD amplitude from quark-interchange, the dominant mechanism for hard scattering in conformal QCD, as predicted from AdS/CFT duality. This mechanism correctly predicts the value of r_{NN} away from the heavy quark thresholds. Furthermore, the resonance phenomenon can account for the dramatic quenching of color transparency observed by Bunce *et al.* in quasielastic pp scattering in nuclei.[44] The octoquark model can be tested at J-PARC and GSI by measuring the rate of charm production near threshold in pp collisions. In addition one can look for the corresponding octoquark states $uud\bar{u}\bar{u}\bar{d}c\bar{c}$ state in $\bar{p}p$ scattering at GSI.

7. Diffractive Deep Inelastic Scattering

A remarkable feature of deep inelastic lepton–proton scattering at HERA is that approximately 10% events are diffractive:[45,46] the target proton remains intact, and there is a large rapidity gap between the proton and the other hadrons in the final state. These diffractive deep inelastic scattering (DDIS) events can be understood most simply from the perspective of the color-dipole model: the $q\bar{q}$ Fock state of the high-energy virtual photon diffractively dissociates into a diffractive dijet system. The exchange of multiple gluons between the color dipole of the $q\bar{q}$ and the quarks of the target proton neutralizes the color separation and leads to the diffractive final state. The same multiple gluon exchange also controls diffractive vector meson electroproduction at large photon virtuality.[47] This observation presents a paradox: if one chooses the conventional parton model frame where the photon light-front momentum is negative $q^+ = q^0 + q^z < 0$, the virtual photon interacts with a quark constituent with light-cone momentum fraction $x = k^+/p^+ = x_{\mathrm{Bj}}$. Furthermore, the gauge link associated with the struck quark (the Wilson line) becomes unity in light-cone gauge

$A^+ = 0$. Thus the struck "current" quark apparently experiences no final-state interactions. Since the light-front wavefunctions $\psi_n(x_i, k_{\perp i})$ of a stable hadron are real, it appears impossible to generate the required imaginary phase associated with pomeron exchange, let alone large rapidity gaps.

This paradox was resolved by Paul Hoyer, Nils Marchal, Stephane Peigne, Francesco Sannino and myself.[2] Consider the case where the virtual photon interacts with a strange quark—the $s\bar{s}$ pair is assumed to be produced in the target by gluon splitting. In the case of Feynman gauge, the struck s quark continues to interact in the final state via gluon exchange as described by the Wilson line. The final-state interactions occur at a light-cone time $\Delta\tau \simeq 1/\nu$ shortly after the virtual photon interacts with the struck quark. When one integrates over the nearly-on-shell intermediate state, the amplitude acquires an imaginary part. Thus the rescattering of the quark produces a separated color-singlet $s\bar{s}$ and an imaginary phase. In the case of the light-cone gauge $A^+ = \eta \cdot A = 0$, one must also consider the final-state interactions of the (unstruck) $\bar{s}$ quark. The gluon propagator in light-cone gauge $d_{LC}^{\mu\nu}(k) = (i/k^2 + i\epsilon)\left[-g^{\mu\nu} + (\eta^\mu k^\nu + k^\mu \eta^\nu / \eta \cdot k)\right]$ is singular at $k^+ = \eta \cdot k = 0$. The momentum of the exchanged gluon k^+ is of $\mathcal{O}(1/\nu)$; thus rescattering contributes at leading twist even in light-cone gauge. The net result is gauge invariant and is identical to the color dipole model calculation. The calculation of the rescattering effects on DIS in Feynman and light-cone gauge through three loops is given in detail for an Abelian model in the references.[2] The result shows that the rescattering corrections reduce the magnitude of the DIS cross section in analogy to nuclear shadowing.

A new understanding of the role of final-state interactions in deep inelastic scattering has thus emerged. The multiple scattering of the struck parton via instantaneous interactions in the target generates dominantly imaginary diffractive amplitudes, giving rise to an effective "hard pomeron" exchange. The presence of a rapidity gap between the target and diffractive system requires that the target remnant emerges in a color-singlet state; this is made possible in any gauge by the soft rescattering. The resulting diffractive contributions leave the target intact and do not resolve its quark structure; thus there are contributions to the DIS structure functions which cannot be interpreted as parton probabilities;[2] the leading-twist contribution to DIS from rescattering of a quark in the target is a coherent effect which is not included in the light-front wave functions computed in isolation. One can augment the light-front wave functions with a gauge link corresponding to an external field created by the virtual photon $q\bar{q}$ pair current.[16,15] Such a gauge link is process dependent,[12] so the resulting augmented LFWFs are not universal.[2,16,48] We also note that the shadowing of nuclear structure functions is due to the destructive interference between multi-nucleon amplitudes involving diffractive DIS and on-shell intermediate states with a complex phase. In contrast, the wave function of a stable target is strictly real since it does not have on-energy-shell intermediate state configurations. The physics of rescattering and shadowing is thus not included in the nuclear light-front wave functions, and a probabilistic interpretation of the nuclear DIS cross section is precluded.

Rikard Enberg, Paul Hoyer, Gunnar Ingelman and I[49] have shown that the quark structure function of the effective hard pomeron has the same form as the quark contribution of the gluon structure function. The hard pomeron is not an intrinsic part of the proton; rather it must be considered as a dynamical effect of the lepton–proton interaction. Our QCD-based picture also applies to diffraction in hadron-initiated processes. The rescattering is different in virtual photon- and hadron-induced processes due to the different color environment, which accounts for the observed non-universality of diffractive parton distributions. This framework also provides a theoretical basis for the phenomenologically successful Soft Color Interaction (SCI) model[50] which includes rescattering effects and thus generates a variety of final states with rapidity gaps.

The phase structure of hadron matrix elements is thus an essential feature of hadron dynamics. Although the LFWFs are real for a stable hadron, they acquire phases from initial state and final state interactions. In addition, the violation of CP invariance leads to a specific phase structure of the LFWFs.[51] Dae Sung Hwang, Susan Gardner and I[51] have shown that this in turn leads to the electric dipole moment of the hadron and a general relation between the electric dipole moment and anomalous magnetic moment, Fock state by Fock state.

The rescattering of the struck parton in DIS generates dominantly imaginary diffractive amplitudes, giving rise to an effective "hard pomeron" exchange and a rapidity gap between the target and diffractive system, while leaving the target intact. This Bjorken-scaling physics, which is associated with the Wilson line connecting the currents in the virtual Compton amplitude survives even in light-cone gauge. Diffractive deep inelastic scattering in turn leads to nuclear shadowing at leading twist as a result of the destructive interference of multi-step processes within the nucleus. Shadowing and antishadowing thus arise because of the $\gamma^* A$ collision and the history of the $q\overline{q}$ dipole as it propagates through the nucleus. Thus there are contributions to the DIS structure functions which are not included in the light-front wave functions computed in isolation and which cannot be interpreted as parton probabilities.[2]

There are also leading-twist diffractive contributions $\gamma^* N_1 \rightarrow (q\overline{q})N_1$ arising from Reggeon exchanges in the t-channel.[52] For example, isospin–non-singlet $C = +$ Reggeons contribute to the difference of proton and neutron structure functions, giving the characteristic Kuti–Weisskopf $F_{2p} - F_{2n} \sim x^{1-\alpha_R(0)} \sim x^{0.5}$ behavior at small x. The x dependence of the structure functions reflects the Regge behavior $\nu^{\alpha_R(0)}$ of the virtual Compton amplitude at fixed Q^2 and $t = 0$. The phase of the diffractive amplitude is determined by analyticity and crossing to be proportional to $-1 + i$ for $\alpha_R = 0.5$, which together with the phase from the Glauber cut, leads to *constructive* interference of the diffractive and nondiffractive multi-step nuclear amplitudes. Furthermore, because of its x dependence, the nuclear structure function is enhanced precisely in the domain $0.1 < x < 0.2$ where antishadowing is empirically observed. The strength of the Reggeon amplitudes is fixed by the fits to the nucleon structure functions, so there is little model dependence. Ivan Schmidt, Jian-Jun Yang, and I [53] have applied this analysis to the shadowing and antishadowing of all of the electroweak structure functions. Quarks of different flavors will couple to different Reggeons; this leads to the remarkable prediction that nuclear antishadowing is not universal; it depends on the quantum numbers of the struck quark. This picture leads to substantially different antishadowing for charged and neutral current reactions, thus affecting the extraction of the weak-mixing angle θ_W. We find that part of the anomalous NuTeV result [54] for θ_W could be due to the non-universality of nuclear antishadowing for charged and neutral currents. Detailed measurements of the nuclear dependence of individual quark structure functions are thus needed to establish the distinctive phenomenology of shadowing and antishadowing and to make the NuTeV results definitive. Antishadowing can also depend on the target and beam polarization.

Acknowledgments

This talk is based on collaborations with Carl Carlson, Guy de Teramond, Rikard Enberg, Susan Gardner, Paul Hoyer, Dae Sung Hwang, Gunnar Ingelman, Hung Jung Lu, Ivan Schmidt and Jian-Jun Yang. The work was supported in part by the Department of Energy, contract No. DE-AC02-76SF00515.

References

1. S.J. Brodsky, D.S. Hwang and I. Schmidt, *Phys. Lett.* **B530** (2002) 99 [arXiv:hep-ph/0201296].
2. S.J. Brodsky, P. Hoyer, N. Marchal, S. Peigne and F. Sannino, *Phys. Rev.* **D65** (2002) 114025 [arXiv:hep-ph/0104291].
3. H. Avakian [CLAS Collaboration], *Presented at the Workshop on Testing QCD through Spin Observables in Nuclear Targets, Charlottesville, Virginia, 18–20 Apr 2002*
4. R.L. Jaffe, arXiv:hep-ph/9602236.
5. D. Boer, *Nucl. Phys. Proc. Suppl.* **105** (2002) 76 [arXiv:hep-ph/0109221].
6. D. Boer, *Nucl. Phys.* **A711** (2002) 21 [arXiv:hep-ph/0206235].
7. J.C. Collins, S.F. Heppelmann and G.A. Ladinsky, *Nucl. Phys.* **B420** (1994) 565 [arXiv:hep-ph/9305309].
8. V. Barone, A. Drago and P.G. Ratcliffe, *Phys. Rep.* **359** (2002) 1 [arXiv:hep-ph/0104283].
9. B.Q. Ma, I. Schmidt and J.J. Yang, *Phys. Rev.* **D66** (2002) 094001 [arXiv:hep-ph/0209114].
10. G.R. Goldstein and L. Gamberg, arXiv:hep-ph/0209085.
11. L.P. Gamberg, G.R. Goldstein and K.A. Oganessyan, *Phys. Rev.* **D67** (2003) 071504 [arXiv:hep-ph/0301018].
12. J.C. Collins, *Phys. Lett.* **B536** (2002) 43 [arXiv:hep-ph/0204004].
13. X. Ji and F. Yuan, *Phys. Lett.* **B543** (2002) 66 [arXiv:hep-ph/0206057].
14. M. Burkardt, *Nucl. Phys. Proc. Suppl.* **141** (2005) 86 [arXiv:hep-ph/0408009].
15. J.C. Collins and A. Metz, *Phys. Rev. Lett.* **93** (2004) 252001 [arXiv:hep-ph/0408249].
16. A.V. Belitsky, X. Ji and F. Yuan, *Nucl. Phys.* **B656** (2003) 165 [arXiv:hep-ph/0208038].
17. S.J. Brodsky, D.S. Hwang and I. Schmidt, *Nucl. Phys.* **B642** (2002) 344 [arXiv:hep-ph/0206259].
18. D. Boer, S.J. Brodsky and D.S. Hwang, *Phys. Rev.* **D67** (2003) 054003 [arXiv:hep-ph/0211110].
19. A. Airapetian *et al.* [HERMES Collaboration], *Phys. Rev. Lett.* **94** (2005) 012002 [arXiv:hep-ex/0408013].
20. H. Avakian and L. Elouadrhiri [CLAS Collaboration], *AIP Conf. Proc.* **698** (2004) 612.
21. L.P. Gamberg and G.R. Goldstein, arXiv:hep-ph/0509312.
22. S.J. Brodsky, D.S. Hwang and I. Schmidt, *Phys. Lett.* **B553** (2003) 223 [arXiv:hep-ph/0211212].
23. H. Avakian *et al.* [CLAS Collaboration], *Phys. Rev.* **D69** (2004) 112004 [arXiv:hep-ex/0301005].
24. A. Afanasev and C.E. Carlson, arXiv:hep-ph/0308163.
25. S.J. Brodsky, H.C. Pauli and S.S. Pinsky, *Phys. Rep.* **301** (1998) 299 [arXiv:hep-ph/9705477].
26. S.J. Brodsky and G.F. de Teramond, arXiv:hep-ph/0510240.

27. O.V. Teryaev, arXiv:hep-ph/9904376.
28. S.J. Brodsky, D.S. Hwang, B.Q. Ma and I. Schmidt, *Nucl. Phys.* **B593** (2001) 311 [arXiv:hep-th/0003082].
29. S.J. Brodsky and D.S. Hwang, *Nucl. Phys.* **B543** (1999) 239 [arXiv:hep-ph/9806358].
30. S.J. Brodsky, M. Diehl and D.S. Hwang, *Nucl. Phys.* **B596** (2001) 99 [arXiv:hep-ph/0009254].
31. G.P. Lepage and S.J. Brodsky, *Phys. Lett.* **B87** (1979) 359.
32. A.V. Efremov and A.V. Radyushkin, *Phys. Lett.* **B94** (1980) 245.
33. G.P. Lepage and S.J. Brodsky, *Phys. Rev.* **D22** (1980) 2157.
34. S.J. Brodsky and G.P. Lepage, SLAC-PUB-2294 *Presented at Workshop on Current Topics in High Energy Physics, Cal Tech., Pasadena, Calif., Feb 13–17, 1979*
35. S.J. Brodsky and S.D. Drell, *Phys. Rev.* **D22** (1980) 2236.
36. S.J. Brodsky and D.S. Hwang, to be published.
37. A.Z. Dubnickova, S. Dubnicka and M.P. Rekalo, *Nuovo Cim.* **A109** (1996) 241.
38. S. Rock, in proc. of *The e^+e^- Physics at Intermediate Energies Conference* ed. D. Bettoni, *eConf* **C010430** (2001) W14 [arXiv:hep-ex/0106084].
39. S.J. Brodsky, C.E. Carlson, J.R. Hiller and D.S. Hwang, *Phys. Rev.* **D69** (2004) 054022 [arXiv:hep-ph/0310277].
40. A.V. Belitsky, X. Ji and F. Yuan, *Phys. Rev. Lett.* **91** (2003) 092003 [arXiv:hep-ph/0212351].
41. F. Iachello, A.D. Jackson and A. Lande, *Phys. Lett.* **B43** (1973) 191.
42. A.D. Krisch, *Czech. J. Phys.* **50S1** (2000) 13.
43. S.J. Brodsky and G.F. de Teramond, *Phys. Rev. Lett.* **60** (1988) 1924.
44. J.L.S. Aclander *et al.*, *Phys. Rev.* **C70** (2004) 015208 [arXiv:nucl-ex/0405025].
45. C. Adloff *et al.* [H1 Collaboration], *Z. Phys.* **C76** (1997) 613 [arXiv:hep-ex/9708016].
46. J. Breitweg *et al.* [ZEUS Collaboration], *Eur. Phys. J.* **C6** (1999) 43 [arXiv:hep-ex/9807010].
47. S.J. Brodsky, L. Frankfurt, J.F. Gunion, A.H. Mueller and M. Strikman, *Phys. Rev.* **D50** (1994) 3134 [arXiv:hep-ph/9402283].
48. J.C. Collins, *Acta Phys. Polon.* **B34** (2003) 3103 [arXiv:hep-ph/0304122].
49. S.J. Brodsky, R. Enberg, P. Hoyer and G. Ingelman, *Phys. Rev.* **D71** (2005) 074020 [arXiv:hep-ph/0409119].
50. A. Edin, G. Ingelman and J. Rathsman, *Phys. Lett.* **B366** (1996) 371 [arXiv:hep-ph/9508386].
51. S.J. Brodsky, S.V. Gardner and D.S. Hwang, to be published.
52. S.J. Brodsky and H.J. Lu, *Phys. Rev. Lett.* **64** (1990) 1342.
53. S.J. Brodsky, I. Schmidt and J.J. Yang, *Phys. Rev.* **D70** (2004) 116003 [arXiv:hep-ph/0409279].
54. G.P. Zeller *et al.* [NuTeV Collaboration], *Phys. Rev. Lett.* **88** (2002) 091802 [*Erratum-ibid.* **90** (2003) 239902] [arXiv:hep-ex/0110059].

THE RELATIVISTIC HYDROGEN ATOM: A THEORETICAL LABORATORY FOR STRUCTURE FUNCTIONS

X. Artru* and K. Benhizia**

* *Institut de Physique Nucléaire de Lyon, IN2P3-CNRS and Université Claude Bernard, F-69622 Villeurbanne, France E-mail: x.artru@ipnl.in2p3.fr*

** *Laboratoire de Physique Mathématique et Physique Subatomique, Départ. de Physique, Faculté de Sciences, Université Mentouri, Constantine, Algeria E-mail: Beni.karima@laposte.net*

Thanks to the Dirac equation, the hydrogen-like atom at high Z offers a precise model of relativistic bound state, allowing to test properties of unpolarized and polarized structure functions analogous to the hadronic ones, in particular: Sivers effect, sum rules for the vector, axial, tensor charges and for the magnetic moment, positivity constraints, sea contributions and fracture functions.

1. Introduction

In this work we will study the hydrogen-like atom, treated by the Dirac equation, as a precise model of relativistic two-particle bound states when one of the constituent is very massive. We consider the case of large Z so that $Z\alpha \sim 1$ and relativistic effects are enhanced. What we neglect is

- the nuclear recoil
- the nuclear spin (but not necessarily the nuclear size)
- radiative corrections, e.g. the Lamb shift $\sim \alpha(Z\alpha)^4 m_e$.

In analogy with the quark distributions, we will study:

- the electron densities $q(k^+)$, $q(\mathbf{k}_T, k^+)$ or $q(\mathbf{b}, k^+)$ where $\mathbf{b} = (x, y)$ is the impact parameter;
- the corresponding polarized densities;
- the sum rules for the vector, axial and tensor charges and for the atom magnetic moment;
- the positivity constraints;
- the electron–positron sea and the fracture functions.

We use $k^+ = k^0 + k_z$ instead of the Björken scaling variable $k^+/P^+_{\rm atom}$ which would be of order $m_e/M_{\rm atom}$, hence very small. $|k^+|$ can run up to $M_{\rm atom}$, but typically $|k^- - m| \sim Z\alpha m$.

2. Deep-inelastic probes of the electron state

Deep-inelastic reactions on the atom are, for instance:

- Compton scattering: $\gamma_i + e^-(\text{bound}) \to \gamma_f + e^-_f(\text{free})$,
- Moeller or Bhabha scattering (replacing the γ by a $e^\pm$),
- annihilation: $e^+ + e^-(\text{bound}) \to \gamma + \gamma$,

the Mandelstam invariants $\hat{s}$, $\hat{t}$ and $\hat{u}$ being large compared to m_e^2. In Compton scattering for instance, taking the z axis along $\mathbf{Q} = \mathbf{k}^\gamma_f - \mathbf{k}^\gamma_i$, the particles γ_i, γ_f and e^-_f move almost in the $-z$ direction and we have

$$k^+ \simeq Q^+ \,, \tag{1}$$

$$\mathbf{k}_T \simeq \mathbf{k}^e_{fT} + \mathbf{k}^\gamma_{fT} - \mathbf{k}^\gamma_{iT} = -\mathbf{P}_{fT}(\text{nucleus})\,. \tag{2}$$

3. Joint $(\mathbf{b}, k^+)$ distribution

In the "infinite momentum frame" $P_z \gg M$, deep inelastic scattering measures the gauge-invariant *mechanical* longitudinal momentum

$$k_z = (P_z/M_{\rm atom})\, k^+_{\rm rest\ frame} = p_z - A_z \tag{3}$$

and not the canonical one $p_z = -i\partial/\partial z$. It is not possible to define a joint distribution $q(\mathbf{k}_T, k_z)$ in a gauge invariant way, since the three quantum operators $k_i = -i\partial_i - A_i(t,x,y,z)$ are not all commuting. On the other hand we can define unambiguously the joint distribution $q(\mathbf{b}, k_z)$. Note that $q(\mathbf{b}, k^+)$ can be measured in atom + atom collisions where two hard sub-collisions occur simultaneously: $\{e^-_1 + e^-_2$ and $N_1 + N_2\}$ or: $\{e^-_1 + N_2$ and $e^-_2 + N_1\}$.

Given the Dirac wave function $\Psi(\mathbf{r})$ at $t = 0$ in the atom frame, we have

$$dN_{e^-}/[d^2\mathbf{b}\; dk^+/(2\pi)] \equiv q(\mathbf{b}, k^+) = \Phi^\dagger(\mathbf{b}, k^+)\; \Phi(\mathbf{b}, k^+)\,, \tag{4}$$

$$\Phi(\mathbf{b}, k^+) = \int dz\;\exp\left\{-ik^+z + iE_n z - i\chi(\mathbf{b}, z)\right\} \Phi(\mathbf{r})\,, \tag{5}$$

$$\Phi(\mathbf{r}) = \begin{pmatrix} \Psi_1(\mathbf{r}) + \Psi_3(\mathbf{r}) \\ \Psi_2(\mathbf{r}) - \Psi_4(\mathbf{r}) \end{pmatrix}, \tag{6}$$

$$\chi(\mathbf{b}, z) = \int_{z_0}^{z} dz' \; V(x, y, z') = -Z\alpha \left[\sinh^{-1}\left(\frac{z}{b}\right) - \sinh^{-1}\left(\frac{z_0}{b}\right)\right] . \quad (7)$$

The two-component spinor Φ represents $(1 + \alpha_z)\Psi$. The "gauge link" $\exp\{-i\chi(\mathbf{b}, z)\}$ transforms Ψ in the Coulomb gauge to Ψ in the gauge $A^+ = 0$ ($A_z = 0$ in the infinite momentum frame). The choice of z_0 corresponds to a residual gauge freedom.

One can check that (4) is invariant under the gauge transformation $V(\mathbf{r}) \rightarrow V(\mathbf{r})+$ Constant, $E_n \rightarrow E_n+$ Constant. Such a transformation may result from the addition of electrons in outer shells; it does not change the mechanical 4-momentum of a K-shell electron.

4. Joint $(\mathbf{k}_T, k^+)$ distribution

The amplitude of the Compton reaction is given, modulo α matrices, by

$$\langle\Psi_f|e^{-i\mathbf{Q}\cdot\mathbf{r}}|\Psi_i\rangle \propto \int d^3\mathbf{r} \; e^{-i\mathbf{Q}\cdot\mathbf{r}-i\mathbf{k}_f\cdot\mathbf{r}-i\chi(\mathbf{b},z)} \; \Phi(\mathbf{r}) = \Phi(\mathbf{k}_T, k^+) \quad (8)$$

with $z_0 = -\infty$. $\exp\{-i\mathbf{k}_f \cdot \mathbf{r} - i\chi\}$ is the final wave function distorted by the Coulomb potential, in the eikonal approximation. Equivalently,

$$\Phi(\mathbf{k}_T, k^+) = \int d^3\mathbf{r} \; e^{-i\mathbf{k}_T\cdot\mathbf{b}-ik^+z-i\chi(\mathbf{r})} \; \Phi(\mathbf{r}) = \int d^2\mathbf{b} \; e^{-i\mathbf{k}_T\cdot\mathbf{b}} \; \Phi(\mathbf{b}, k^+) \, , \quad (9)$$

with the identification $k^+ = E_n + k_{f,z} + Q_z$. [*Check:* in a semi-classical approach, $k^0(\mathbf{r}) = E_n - V(\mathbf{r})$ and $k_{f,z} + Q_z = k_z(\mathbf{r}) - V(\mathbf{r})$ at the collision point.] For the annihilation reaction, the *incoming positron* wave is distorted, then $z_0 = +\infty$. Thus, the gauge link takes into account either an initial or a final state interaction.[1–3] The quantity

$$q(\mathbf{k}_T, k^+) = \Phi^\dagger(\mathbf{k}_T, k^+) \; \Phi(\mathbf{k}_T, k^+) \quad (10)$$

depends on the deep inelastic probe, contrary to $q(\mathbf{b}, k^+)$ and

$$q(k^+) = \int q(\mathbf{b}, k^+) \; d^2\mathbf{b} = \int q(\mathbf{k}_T, k^+) \; d^2\mathbf{k}_T/(2\pi)^2 \, . \quad (11)$$

5. Spin dependence of the electron density

Ignoring nuclear spin, the atom spin is $\mathbf{j} = \mathbf{L} + \mathbf{s}$. We will consider a $j = 1/2$ state and denote by $\bar{\mathbf{S}}^A = 2\langle\mathbf{j}\rangle$ and $\bar{\mathbf{S}}^e = 2\langle\mathbf{s}\rangle$ the atom and electron polarization vectors. $\mathbf{S}^A$ and $\mathbf{S}^e$ without bar indicate pure spin states. The *unpolarized* electron density in $(\mathbf{b}, k^+)$ space from a *polarized* atom is

$$q(\mathbf{b}, k^+; \mathbf{S}^A) = \Phi^\dagger(\mathbf{b}, k^+; \mathbf{S}^A)\, \Phi(\mathbf{b}, k^+; \mathbf{S}^A) \tag{12}$$

and the electron polarization is given by

$$\bar{\mathbf{S}}^e(\mathbf{b}, k^+; \mathbf{S}^A)\, q(\mathbf{b}, k^+; \mathbf{S}^A) = \Phi^\dagger(\mathbf{b}, k^+; \mathbf{S}^A)\, \vec{\sigma}\, \Phi(\mathbf{b}, k^+; \mathbf{S}^A)\,. \tag{13}$$

Taking into account the conservations of parity and angular momentum, the fully polarized density can be written as

$$q(\mathbf{b}, k^+, \mathbf{S}^e; \mathbf{S}^A) = q(b, k^+)\big[1 + C_{0n}\,(\mathbf{S}^A\cdot\hat{\mathbf{n}}) + C_{n0}\,(\mathbf{S}^e\cdot\hat{\mathbf{n}}) + C_{nn}\,(\mathbf{S}^e\cdot\hat{\mathbf{n}})(\mathbf{S}^A\cdot\hat{\mathbf{n}})$$

$$+\, C_{ll}\, S^e_z S^A_z + C_{l\pi}\, S^e_z(\mathbf{S}^A\cdot\hat{\pi}) + C_{\pi l}\,(\mathbf{S}^e\cdot\hat{\pi})\, S^A_z + C_{\pi\pi}\,(\mathbf{S}^e\cdot\hat{\pi})(\mathbf{S}^A\cdot\hat{\pi})\big], \tag{14}$$

where $\hat{\pi} = \mathbf{b}/b$ and $\hat{\mathbf{n}} = \hat{\mathbf{z}} \times \hat{\pi}$. The $C_{i,j}$'s are functions of b and k^+.

Similar equations work for $\mathbf{k}_T$ instead of $\mathbf{b}$. The link with Amsterdam notations [4] is, omitting kinematical factors,

$$q(k_T, k^+) = f_1 \qquad f_1\, C_{0n} = f_{1T}^\perp \tag{15}$$

$$f_1\, C_{ll} = g_1 \qquad f_1\, C_{n0} = -h_1^\perp \tag{16}$$

$$f_1\, C_{nn} = h_1 - h_{1T}^\perp \qquad f_1\, C_{l\pi} = g_{1T} \tag{17}$$

$$f_1\, C_{\pi\pi} = h_1 + h_{1T}^\perp \qquad f_1\, C_{\pi l} = h_{1L}^\perp\,, \tag{18}$$

The $\mathbf{b}$ – or $\mathbf{k}_T$ – integration washes out all correlations but C_{ll} and $C_{TT} = \frac{1}{2}\,[C_{nn} + C_{\pi\pi}]$, giving

$$q(k^+, \mathbf{S}^e; \bar{\mathbf{S}}^A) = q(k^+) + \Delta q(k^+)\, S^e_z \bar{S}^A_z + \delta q(k^+)\, \mathbf{S}^e_T \cdot \bar{\mathbf{S}}^A_T\,. \tag{19}$$

5.1. *Sum rules*

Integrating (19) over k^+, one obtains the vector, axial and tensor charges

$$q = \int_{-\infty}^{\infty} q(k^+)\, dk^+/(2\pi) = \int d^3\mathbf{r}\; \Psi^\dagger(\mathbf{r}; \mathbf{S}^A)\; \Psi(\mathbf{r};\ \text{any}\ \mathbf{S}^A)\,, \tag{20}$$

$$\Delta q = \int \Delta q(k^+)\, dk^+/(2\pi) = \int d^3\mathbf{r}\; \Psi^\dagger(\mathbf{r}; \mathbf{S}^A)\; \Sigma_z\; \Psi(\mathbf{r}; \mathbf{S}^A = \hat{\mathbf{z}})\,, \tag{21}$$

$$\delta q = \int \delta q(k^+)\, dk^+/(2\pi) = \int d^3\mathbf{r}\; \Psi^\dagger(\mathbf{r}; \mathbf{S}^A)\; \beta\; \Sigma_x\; \Psi(\mathbf{r}; \mathbf{S}^A = \hat{\mathbf{x}})\,. \tag{22}$$

For the hydrogen ground state,

$$q = 1\,, \qquad \Delta q = (1 - \xi^2/3)/(1 + \xi^2)\,, \qquad \delta q = (1 + \xi^2/3)/(1 + \xi^2)\,, \tag{23}$$

where $\xi = Z\alpha/(1 + \gamma)$, $\gamma = E/m_e = \sqrt{1 - (Z\alpha)^2}$.
Note the large "spin crisis" $\Delta q = 1/3$ for $Z\alpha = 1$.

5.2. *Results for the polarized densities in* $(\mathbf{b}, k^+)$

The 2-component wave functions of a $j_z = +1/2$ state write

$$\Phi(\mathbf{b}, k^+) = \begin{pmatrix} w \\ -iv\ e^{i\phi} \end{pmatrix}, \qquad \Phi(\mathbf{k}_T, k^+) = \begin{pmatrix} \tilde{w} \\ -\tilde{v}\ e^{i\phi} \end{pmatrix}. \tag{24}$$

Other orientations of $\mathbf{S}^A$ can be obtained by rotation in spinor space.

For the $(\mathbf{b}, k^+)$ representation we can ignore the second term of (7). Then $v(\mathbf{b}, k^+)$ and $w(\mathbf{b}, k^+)$ are real and given by

$$\begin{pmatrix} v \\ w \end{pmatrix} = \int_{-\infty}^{\infty} dz \begin{pmatrix} \xi b \\ r + i\xi z \end{pmatrix} e^{-ik^+ z + iEz - i\chi(\mathbf{b},z)}\ f(r)/r \tag{25}$$

where $f(r) \propto r^{\gamma-1}\ \exp(-mZ\alpha r)$ is the radial wave function. Then,

$$\begin{aligned} q(b, k^+) &= w^2 + v^2 \\ C_{nn}(b, k^+) &= 1 \\ C_{0n}(b, k^+) = C_{n0}(b, k^+) &= -2\ wu/(w^2 + v^2) \\ C_{ll}(b, k^+) = C_{\pi\pi}(b, k^+) &= (w^2 - v^2)/(w^2 + v^2) \\ C_{l\pi}(b, k^+) = C_{\pi l}(b, k^+) &= 0\,. \end{aligned} \tag{26}$$

5.2.1. *Sum rule the atom magnetic moment*

Consider a classical object at rest, of mass M, charge Q, spin $\mathbf{J}$ and time-averaged magnetic moment $\vec{\mu}$. Its center of mass $\mathbf{r}_G$ and the average center of charge $\langle \mathbf{r}_C \rangle$ coincide, say at $\mathbf{r} = 0$. Upon a boost of velocity $\mathbf{v}$, the center of energy $\mathbf{r}_G$ and $\langle \mathbf{r}_C \rangle$ are displaced laterally by

$$\mathbf{b}_G = \mathbf{v} \times \mathbf{J}/M\,, \quad \langle \mathbf{b}_C \rangle = \mathbf{v} \times \vec{\mu}/Q\,. \tag{27}$$

$\mathbf{b}_G$ and $\langle \mathbf{b}_C \rangle$ coincide if the gyromagnetic ratio has the Dirac value Q/M. For the hydrogen atom, $\mathbf{b}_G$ is negligible and the magnetic moment is almost fully anomalous. In the infinite momentum frame ($\mathbf{v} \simeq \hat{\mathbf{z}}$), we have an electric dipole moment [5]

$$-e\ \langle \mathbf{b} \rangle = \mu_A\ \hat{\mathbf{z}} \times \bar{\mathbf{S}}^A\,, \tag{28}$$

the transverse asymmetry of the $\mathbf{b}$ distribution coming from the C_{0n} term of (14). We recover the atom magnetic moment

$$\mu_A = -e\ (1 + 2\gamma)/(6m_e) \tag{29}$$

(the anomalous magnetic moment of the electron being omitted).

5.3. *Results for the polarized densities in* $(\mathbf{k}_T, k^+)$

For the $(\mathbf{k}_T, k^+)$ representation we should take $z_0 = \pm\infty$ but then (7) diverges. In practice we assume some screening of the Coulomb potential and take $|z_0|$ large but finite, which gives

$$\chi(\mathbf{b}, z) = -Z\alpha \left[\pm \ln(2|z_0|/b) + \sinh^{-1}(z/b) \right], \tag{30}$$

the upper sign corresponding to Compton scattering and the lower sign to annihilation. Modulo an overall phase,

$$\begin{pmatrix} \tilde{w} \\ \tilde{v} \end{pmatrix} = 2\pi \int_0^\infty b\, db\, b^{\mp iZ\alpha} \begin{pmatrix} J_0(k_T b)\; w(b, k^+) \\ J_1(k_T b)\; v(b, k^+) \end{pmatrix}, \tag{31}$$

$$\begin{aligned} q(k_T, k^+) &= |\tilde{w}|^2 + |\tilde{v}|^2 \\ C_{nn}(k_T, k^+) &= 1 \\ C_{0n}(k_T, k^+) &= C_{n0}(k_T, k^+) = 2\Im(\tilde{v}^* \tilde{w})/(|\tilde{w}|^2 + |\tilde{v}|^2) \\ C_{ll}(k_T, k^+) &= C_{\pi\pi}(k_T, k^+) = (|\tilde{w}|^2 - |\tilde{v}|^2)/(|\tilde{w}|^2 + |\tilde{v}|^2) \\ C_{l\pi}(k_T, k^+) &= -C_{\pi l}(k_T, k^+) = 2\Re(\tilde{v}^* \tilde{w})/(|\tilde{w}|^2 + |\tilde{v}|^2)\,. \end{aligned} \tag{32}$$

The factor $b^{-iZ\alpha}$ (Compton case) behaves like a converging cylindrical wave. Multiplying $\Phi(\mathbf{r})$, it mimics an additional momentum of the electron toward the z axis. In fact it takes into account the "focusing" of the final particle by the Coulomb field.[5] It also provides the relative phase between $\tilde{w}$ and $\tilde{v}$ which gives non-zero $C_{0n}(k_T, k^+)$ and $C_{n0}(k_T, k^+)$ (Sivers and Boer–Mulders–Tangerman effects). Similarly, $b^{+iZ\alpha}$ (annihilation case) takes into account the defocusing of the incoming positron.

5.4. *Positivity constraints*

The spin correlations between the electron and the atom can be encoded in a "grand density matrix" R,[6] which is the final density matrix of the crossed reaction *nucleus* $\to$ *atom*$(\mathbf{S}^A) + e^+(-\mathbf{S}^e)$.

Besides the trivial conditions $|C_{ij}| \leq 1$ the positivity of R gives

$$(1 \pm C_{nn})^2 \geq (C_{n0} \pm C_{0n})^2 + (C_{ll} \pm C_{\pi\pi})^2 + (C_{\pi l} \mp C_{l\pi})^2\,. \tag{33}$$

These two inequalities as well as $|C_{ll}| \leq 1$ are saturated by (26) and (32). In fact R is found to be of rank one. It means that the spin information of the atom is fully transferred to the electron, once the other degrees of freedom (k^+ and $\mathbf{b}$ or $\mathbf{k}_T$) have been fixed. If there is additional electrons or if we integrate over k^+, for instance, some information is lost and some positivity conditions get non-saturated. Conversely, the hypothesis that R is of rank one leaves only two possibilities:

$$C_{nn} = \pm 1\,, \quad C_{0n} = \pm C_{n0}\,, \quad C_{\pi\pi} = \pm C_{ll}\,, \quad C_{l\pi} = \mp C_{\pi l}\,, \qquad (34)$$

with (33) saturated. The hydrogen ground state chooses the upper sign. After integration over $\mathbf{b}$ or $\mathbf{k}_T$, we are left with the *Soffer inequality*,

$$2\ |\delta q(k^+)| \leq q(k^+) + \Delta q(k^+), \qquad (35)$$

which in our case is saturated, even after over k^+, see (23).

6. The electron–positron sea

The charge rule (20) involves positive contributions of both positive and negative values of k^+. So it seems that there is less than one electron (with $k^+ > 0$) in the atom. This paradox is solved by the introduction of the electron–positron sea.

Let $|n\rangle$ be an electron state in the Coulomb field. Negative n's are assigned to negative energy scattering states. Positive n's up to n_B label the bound states ($-m < E_n < +m$) and the remaining ones from n_B+1 to $+\infty$ are assigned to unbound states of positive energy, $E_n > +m$, considered as discrete. Let $|k, s\rangle$ be the plane wave of four-momentum k and spin s, solution the free Dirac equation. The destruction and creation operators in these two bases are related by

$$\alpha_{k,s} = \sum_n \langle k, s|n\rangle\ a_n\,, \qquad \alpha^\dagger_{k,s} = \sum_n a^\dagger_n\ \langle n|k, s\rangle\,. \qquad (36)$$

In the Dirac hole theory, the hydrogen-like atom is in the Fock state

$$|H_n\rangle = a^\dagger_n\ a^\dagger_{-1}\ a^\dagger_{-2} \cdots a^\dagger_{-\infty}\ |\text{QED-bare nucleus}\rangle. \qquad (37)$$

The number of electrons in the state $|k, s\rangle$ is

$$N^{\text{elec}}(\mathbf{k}, s) = \langle \alpha^\dagger_{k,s}\ \alpha_{k,s}\rangle = |\langle k, s|n\rangle|^2 + \sum_{n'<0} |\langle k, s|n'\rangle|^2. \qquad (38)$$

For a nucleus alone (but "QED-dressed") the first factor $a^\dagger_n$ of (37) is missing and the first term of (38) is absent. Therefore, passing to the continuum limit $\langle k, s|n\rangle \longrightarrow \Phi(\mathbf{k}_T, k^+)$,

$$N^{\text{elec}}_{\text{atom}} - N^{\text{elec}}_{\text{nucleus}} = \sum_{k,k^0>0} \sum_s |\langle k, s|n\rangle|^2 = \int_{k^+>0} \frac{dk^+}{2\pi}\ q(k^+)\,. \qquad (39)$$

One term n' of (38) corresponds, e.g., to Compton scattering on an electron of the Dirac sea, producing a fast electron plus the positron $|n'\rangle =$ vacant. It is the *fracture function* of reaction $\gamma_i + A \to \gamma_f + A + e^+ + X$.

Similarly the number of positrons is (with $k^0 > 0$):

$$N^{\text{posit}}(\mathbf{k}, s) = \langle \alpha_{-k,-s}\, \alpha^\dagger_{-k,-s} \rangle = \sum_{0<n'\neq n} |\langle -k, -s|n'\rangle|^2 \,. \tag{40}$$

For a nucleus alone the condition $n' \neq n$ is relaxed. By difference,

$$N^{\text{posit}}_{\text{nucleus}} - N^{\text{posit}}_{\text{atom}} = \int_{k^+<0} \frac{dk^+}{2\pi}\, q(k^+). \tag{41}$$

One term of Eq. (40) corresponds to the extraction of an electron of large negative energy, giving a fast positron and an electron in $|n'\rangle$. Eqs. (39) + (41) and (20) tell us that the atom and nuclear charges differ by e. But the charge of the electron–positron virtual cloud surrounding the nucleus is not zero (charge renormalization).

7. Conclusion

We have seen that the leading twist structure functions of the hydrogen-like atom at large Z has many properties that are supposed or verified for the hadronic ones, in particular: sum rules, longitudinal "spin crisis", Sivers effect, transverse electric dipole moment in the P_∞ frame, etc. It remains to evaluate these effects quantitatively. With this "theoretical laboratory" one may also investigate spin effects in fracture functions, non-leading twist structure functions, Isgur–Wise form factors, etc. The electron–positron sea may deserve further studies: to what extent is it polarized or asymmetrical in charge? Is the charge renormalization of the nucleus found in Sec. 6 the same as in standard QED? Is it finite and calculable for an extended nucleus?

Finally it may be interesting to do or re-analyse experiments on deep inelastic Compton scattering ("Compton profile measurements").

References

1. S.J. Brodsky, D.S. Hwang and I. Schmidt, *Phys. Lett.* **B530** (2002) 99.
2. A.V. Belitsky, X. Ji and F. Yuan, *Nucl. Phys.* **B656** (2003) 165.
3. J.C. Collins, *Phys. Lett.* **B536** (2002) 43.
4. A. Bacchetta, M. Boglione, A. Henneman and P.J. Mulders, *Phys. Rev. Lett.* **85** (2000) 712.
5. M. Burkardt, *Nucl. Phys.* **A735** (2004) 185.
6. X. Artru and J.-M. Richard, hep-ph/0401234.

GPD'S AND SSA'S

M. Burkardt

Department of Physics, New Mexico State University
Las Cruces, NM 88011, USA, E-mail: burkardt@nmsu.edu

Generalized parton distributions involving transverse polarization are transversely deformed. The deformation of chirally odd GPDs is related to a transversity decomposition of the quark angular momentum. Potential consequences for T-odd single-spin asymmetries (Sivers and Boer–Mulders effects) are discussed.

1. Introduction

Hadron form factors provide information about the Fourier transform of the charge distribution within the hadron. Generalized parton distributions (GPDs) provide a momentum decomposition of the form factor w.r.t. the average momentum fraction $x = \frac{1}{2}(x_i + x_f)$ of the active quark

$$\int dx H_q(x, \xi, t) = F_1^q(t) , \qquad \int dx E_q(x, \xi, t) = F_2^q(t) , \tag{1}$$

where $F_1^q(t)$ and $F_2^q(t)$ are the Dirac and Pauli form factors, respectively. x_i and x_f are the momentum fractions of the quark before and after the momentum transfer. The momentum direction of the active quark singles out a direction and it makes a difference whether the momentum transfer is along this momentum or in a different direction. GPDs thus not only depend on x and the invariant momentum transfer t but also on the longitudinal momentum transfer through the variable $2\xi = x_f - x_i$.

Since GPDs are the form factor of the same operator whose forward matrix elements yield the usual parton distribution functions (PDFs)

$$\int \frac{dx^-}{2\pi} e^{ix^- \bar{p}^+ x} \left\langle p' \left| \bar{q}\left(-\frac{x^-}{2}\right) \gamma^+ q\left(\frac{x^-}{2}\right) \right| p \right\rangle$$
$$= H(x, \xi, \Delta^2) \bar{u}(p') \gamma^+ u(p) + E(x, \xi, \Delta^2) \bar{u}(p') \frac{i\sigma^{+\nu}\Delta_\nu}{2M} u(p). \tag{2}$$

it is possible to develop a position space interpretation for GPDs.[1]

2. Position Space Interpretation for GPDs

Charge distributions in position space are usually measured in the center of mass frame, i.e. relative to the center of mass of the system. For impact parameter dependent PDFs, the analogous reference point is the $\perp$ center of momentum of all partons (quarks and gluons) $\mathbf{R}_\perp = \sum_{i=q,g} x_i \mathbf{r}_{\perp,i}$, where x_i is the momentum fraction carried by each parton and $\mathbf{r}_{\perp,i}$ is their $\perp$ position. One can form eigenstates of $\mathbf{R}_\perp$

$$\left|p^+, \mathbf{R}_\perp = \mathbf{0}_\perp, \lambda\right\rangle \equiv N \int d^2\mathbf{p}_\perp \left|p^+, \mathbf{p}_\perp, \lambda\right\rangle . \tag{3}$$

Impact parameter dependent PDFs are defined using the familiar light-cone correlation function in such transversely localized states [1,2]

$$q(x,\mathbf{b}_\perp) \equiv \int \frac{dx^-}{4\pi}\, e^{ixp^+x^-} \left\langle p^+, \mathbf{R}_\perp = \mathbf{0}_\perp \right| \bar{q}\big(-\frac{x^-}{2}, \mathbf{b}_\perp\big)\gamma^+ q\big(\frac{x^-}{2}, \mathbf{b}_\perp\big) \left|p^+, \mathbf{R}_\perp = \mathbf{0}_\perp\right\rangle \tag{4}$$

and with an additional γ_5 for the polarized distribution $\Delta q(x, \mathbf{b}_\perp)$. Impact parameter dependent PDFs are Fourier transforms of GPDs for $\xi = 0$ [1,3]

$$q(x, \mathbf{b}_\perp) = \int \frac{d^2\mathbf{\Delta}_\perp}{(2\pi)^2}\, e^{i\mathbf{\Delta}_\perp\cdot\mathbf{b}_\perp} H(x, 0, -\mathbf{\Delta}_\perp^2)\,, \tag{5}$$

$$\Delta q(x, \mathbf{b}_\perp) = \int \frac{d^2\mathbf{\Delta}_\perp}{(2\pi)^2}\, e^{i\mathbf{\Delta}_\perp\cdot\mathbf{b}_\perp} \tilde{H}(x, 0, -\mathbf{\Delta}_\perp^2)\,. \tag{6}$$

Due to a Galilean subgroup of $\perp$ boosts in the infinite momentum frame there are no relativistic corrections to Eq. (5). Furthermore, impact parameter dependent PDFs have a probabilistic interpretation very similar (and with the same limitations) as the usual PDFs.[4] For example, for $x > 0$ (quarks) one finds $q(x, \mathbf{b}_\perp) \geq |\Delta q(x, \mathbf{b}_\perp)| \geq 0$.

It is important to utilize theoretical constraints when parametrizing these functions to supplement experimental data. One such constraint arises directly from the fact that the reference point for impact parameter dependent PDFs is the $\perp$ center of momentum. For $x \to 1$ the active quark becomes the center of momentum and therefore $\mathbf{b}_\perp$ can never be large, and the $\perp$ width of $q(x, \mathbf{b}_\perp)$ should go to zero for $x \to 1$. For decreasing x the $\perp$ width is expected to increase gradually. Although the width in the valence region should still be relatively compact, its size should increase further once x is small enough for the pion cloud to contribute.[5] Therefore the t-dependence of GPDs should decrease with increasing x. This is consistent with recent lattice results, which showed that higher moments of GPDs have less t dependence than lower moments.[6]

3. Transversely Polarized Target

For a $\perp$ polarized target, impact parameter dependent PDFs are no longer axially symmetric. The deviation from axial symmetry is described by $E(x,0,t)$. For example, the unpolarized quark distribution $q_X(x,\mathbf{b}_\perp)$ for a target polarized in the $+\hat{x}$ direction reads [7]

$$q_X(x,\mathbf{b}_\perp) = q(x,\mathbf{b}_\perp) - \frac{1}{2M}\frac{\partial}{\partial b_y}\int \frac{d^2\mathbf{\Delta}_\perp}{(2\pi)^2} E(x,0,-\mathbf{\Delta}_\perp^2) e^{-i\mathbf{b}_\perp\cdot\mathbf{\Delta}_\perp}. \tag{7}$$

Here $q(x,\mathbf{b}_\perp)$ is the impact parameter dependent PDF in the unpolarized case (5). This distortion arises since the virtual photon in DIS couples more strongly to quarks that move towards it than quarks that move away from it (hence the γ^+ in the quark correlation function relevant for DIS).[7,8] If the orbital motion of the quarks and the spin of the target are correlated then quarks are more likely to move towards the virtual photon on one side of the target than the other and the distribution of quarks in impact parameter space appears deformed towards one side. The details of this deformation for each quark flavor are described by $E_q(x,0,t)$, which is not known yet. However, sign and overall scale can be estimated by considering the mean displacement of flavor q ($\perp$ flavor dipole moment)

$$d_y^q \equiv \int dx \int d^2\mathbf{b}_\perp q(x,\mathbf{b}_\perp) b_y = \frac{1}{2M}\int dx E_q(x,0,0) = \frac{\kappa_q^p}{2M}. \tag{8}$$

$\kappa_q = O(1-2)$ are the contributions from each quark flavor to the anomalous magnetic moment of the nucleon, i.e. $F_2(0) = \frac{2}{3}\kappa_u - \frac{1}{3}\kappa_d - \frac{1}{3}\kappa_s \ldots$, yielding $\left|d_y^q\right| = O(0.2\,fm)$, with opposite signs for u and d quarks (Fig. 1).

The $\perp$ distortion can also be linked to Ji's relation [9] between the 2nd moment of the GPDs H_q and E_q and the quark angular momentum

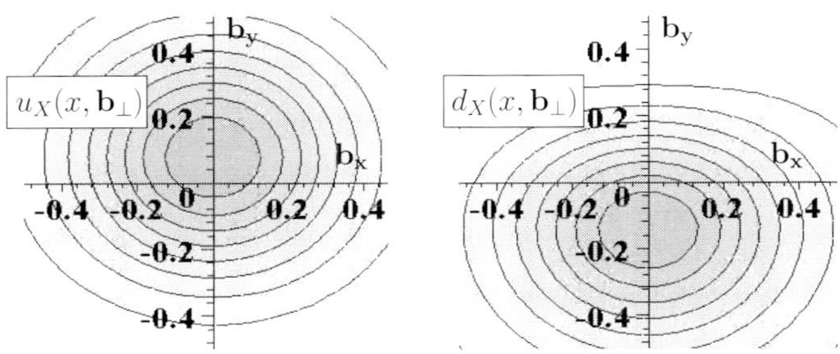

Fig. 1. Expected impact parameter dependent PDF for u and d quarks ($x_{\rm Bj} = 0.3$ is fixed) for a nucleon that is polarized in the x direction in the model from Ref. 7. For other values of x the distortion looks similar.

$$J_q^i = \varepsilon^{ijk} \int d^3x M_q^{0jk}. \tag{9}$$

Here $M_q^{0jk} = T_q^{0k}x^j - T_q^{0j}x^k$ and

$$T_q^{\mu\nu} = i\bar{q}\gamma^\mu \overset{\leftrightarrow}{D}{}^\nu q \tag{10}$$

(a symmetrization in μ and ν is implicit). Since the angular momentum is obtained by taking the weighted average of the position, where the weight factor is the momentum density, one would intuitively expect some connection between the transverse center of momentum for the quarks and their angular momentum. Indeed, as has been shown in Ref. 10, one can relate the $\perp$ shift of the center of momentum for a quarks with flavor q to the angular momentum carried by these quarks. Using Eq. (7) and taking into account an overall $\perp$ shift due to boosting to the infinite momentum frame one thus recovers the Ji relation [9]

$$\langle J_q^i \rangle = S^i \int dx\, x\left[H_q(x,0,0) + E_q(x,0,0)\right], \tag{11}$$

where S^i is the nucleon spin. In combination with measurements of the fraction of the quark spin contribution to the nucleon spin in polarized DIS, Eq. (11) is expected to provide novel information about the orbital angular momentum carried by the quarks.

The deformation of quark distributions in a $\perp$ polarized target also provides a very physical source for single-spin asymmetries (SSA) in semi-inclusive DIS. The Sivers function $f_{1T}^{\perp q}$, which parametrized the left–right asymmetry reads [11,12]

$$f_{q/p^\uparrow}(x, \mathbf{k}_\perp) = f_1^q(x, \mathbf{k}_\perp^2) - f_{1T}^{\perp q}(x, \mathbf{k}_\perp^2)\frac{(\hat{\mathbf{P}} \times \mathbf{k}_\perp)\cdot \mathbf{S}}{M}, \tag{12}$$

where $f_{q/p^\uparrow}(x, \mathbf{k}_\perp)$ represents the unintegrated parton density for quarks ejected from a $\perp$ polarized target. The phenomenology of these functions can be found for example in Ref. 13 and references therein. Although one may naively expect that these T-odd functions vanish, they survive the Bjorken limit due to final state interactions.[14–16]

For an (on average) attractive final state interaction, the position space deformation into the $+\hat{y}$ direction translates into a momentum space asymmetry for the ejected quark that prefers the $-\hat{y}$ direction and *vice versa* (Fig. 2) Since the sign of the position space distortion is governed by the sign of the anomalous magnetic moment contribution $\kappa_{q/P}$ from each quark flavor, this implies that the sign of the SSA is correlated to the sign of $\kappa_{q/P}$. Following the Trento convention,[12] this yields a negative Sivers function

$f_{1T}^{\perp u}$ in the proton, while $f_{1T}^{\perp d} > 0$.[17] For neutrons the signs are reversed. These predictions are consistent with recent HERMES data.[18]

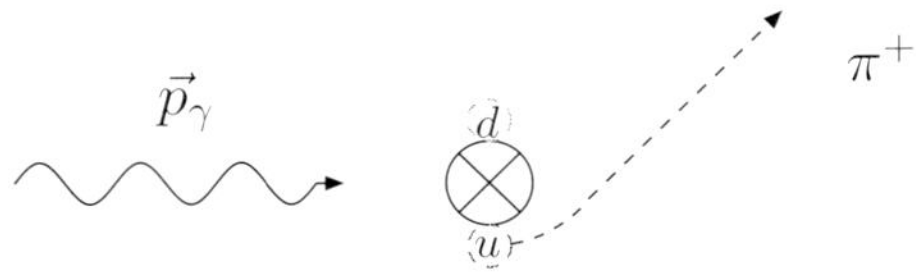

Fig. 2. The transverse distortion of the parton cloud for a proton that is polarized into the plane, in combination with attractive FSI, gives rise to a Sivers effect for u (d) quarks with a $\perp$ momentum that is on the average up (down).

4. Chirally Odd GPDs

The distribution of transversely polarized quarks in impact parameter space is described by the Fourier transform of chirally odd GPDs.[19] For an unpolarized target the distribution of quarks with transversity s^i reads

$$q_i(x,\mathbf{b}_\perp) = -\frac{s^i\varepsilon^{ij}}{2M}\frac{\partial}{\partial b_j}\int\frac{d^2\mathbf{\Delta}_\perp}{(2\pi)^2}\left[2\tilde{H}_T(x,0,-\mathbf{\Delta}_\perp^2)+E_T(x,0,-\mathbf{\Delta}_\perp^2)\right]e^{-i\mathbf{b}_\perp\cdot\mathbf{\Delta}_\perp} \tag{13}$$

While Eq. (7) describes the $\perp$ deformation of unpolarized quark distributions in a $\perp$ polarized nucleon, Eq. (13) demonstrates that a similar deformation is present in the distribution of $\perp$ polarized quarks in an unpolarized nucleon — except the latter deformation is described by the chirally odd GPDs $2\tilde{H}_T + E_T$. In Sec. 3, we linked the $\perp$ deformation of the unpolarized quark distributions in a $\perp$ polarized nucleon to the angular momentum carried by those quarks, yielding the Ji relation (11) which tells us how the quark angular momentum is correlated to the nucleon spin. Intuitively, we thus expect that there is some connection between chirally odd GPDs, which describe the $\perp$ deformation of $\perp$ polarized quark distributions in an unpolarized nucleon, and the correlation between the quark spin and the quark angular momentum in an unpolarized nucleon.

In order to investigate the correlation between polarization and angular momentum of the quarks, we decompose J_q^x into transversity components. The projector on $\perp$ spin (transversity) eigenstates $P_{\pm\hat{x}} \equiv \frac{1}{2}(1 \pm \gamma^x\gamma_5)$ commutes with γ^0, γ^y, and γ^z. Hence all components of the energy momentum tensor that appear in the definition of J_q^x do not mix between transversity (in the $\hat{x}$ direction) states, defined as $q_{\pm\hat{x}} = \frac{1}{2}(1 \pm \gamma^x\gamma_5)\, q$. It is thus possible to decompose J^x_q into transversity components

$$J^x_q = J^x_{q,+\hat{x}} + J^x_{q,-\hat{x}}. \tag{14}$$

Transversity projections of Eq. (9) yield the transversity components $J^x_{q,\pm\hat{x}}$

$$J^x_{q,\pm\hat{x}} = \frac{i}{2}\int d^3x \bar{q}_{\pm\hat{x}}\left[\gamma^0 \overleftrightarrow{D}^z + \gamma^z \overleftrightarrow{D}^0\right] q_{\pm\hat{x}}\, y \;-\; \text{“}y \leftrightarrow z\text{''} = \frac{1}{2}\left[J^x_q \pm \delta^x J^x_q\right], \tag{15}$$

where the transversity dependent piece reads

$$\delta^x J^x_q = \frac{1}{2}\int d^3x \bar{q}\left[\sigma^{0x} \overleftrightarrow{D}^z + \sigma^{zx} \overleftrightarrow{D}^0\right] q\, y \;-\; \text{“}y \leftrightarrow z\text{''}. \tag{16}$$

Taking the matrix elements of J^x_q yields the Ji relation (11). In order to examine the contribution from the chirally odd term (16), we consider the form factor of the transversity density with one derivative [19,20]

$$\begin{aligned}\langle p'|\, \bar{q}\sigma^{\lambda\mu}\gamma_5 i \overleftrightarrow{D}^\nu\, q\,|p\rangle &= \bar{u}\sigma^{\lambda\mu}\gamma_5 u\, \bar{p}^\nu A_{T20}(t) + \frac{\varepsilon^{\lambda\mu\alpha\beta}\Delta_\alpha \bar{p}_\beta \bar{p}^\nu}{M^2}\bar{u}u\, \tilde{A}_{T20}(t) \\ &+ \frac{\varepsilon^{\lambda\mu\alpha\beta}\Delta_\alpha \bar{p}^\nu}{2M}\bar{u}\gamma_\beta u\, B_{T20}(t) + \frac{\varepsilon^{\lambda\mu\alpha\beta}\bar{p}_\alpha \Delta^\nu}{M}\bar{u}\gamma_\beta u\, \tilde{B}_{T21}(t).\end{aligned} \tag{17}$$

Antisymmetrization in λ and μ and symmetrization in μ and ν is implied. The form factors in Eq. (17) are the 2nd moments of chirally odd GPDs

$$\begin{aligned} A_{T20}(t) &= \int_{-1}^{1} dx x H_T(x,\xi,t), & \tilde{A}_{T20}(t) &= \int_{-1}^{1} dx x \tilde{H}_T(x,\xi,t), \\ B_{T20}(t) &= \int_{-1}^{1} dx x E_T(x,\xi,t), & -2\xi \tilde{B}_{T21}(t) &= \int_{-1}^{1} dx x \tilde{E}_T(x,\xi,t). \end{aligned} \tag{18}$$

The chirally odd GPDs entering Eq. (18) are defined as non-forward matrix elements of light-like correlation functions of the tensor charge

$$\begin{aligned} p^+\int \frac{dz^-}{2\pi} e^{ixp^+z^-} \langle p'|\, \bar{q}\left(-\frac{z}{2}\right)\sigma^{+j}\gamma_5 q\left(\frac{z}{2}\right)|p\rangle &= H_T(x,\xi,t)\bar{u}\sigma^{+j}\gamma_5 u + \\ \tilde{H}_T(x,\xi,t)\varepsilon^{+j\alpha\beta}\bar{u}\frac{\Delta_\alpha p_\beta}{M^2}u + \varepsilon^{+j\alpha\beta}\Big[E_T(x,\xi,t)\bar{u}\frac{\Delta_\alpha\gamma_\beta}{2M}u &+ \tilde{E}_T(x,\xi,t)\bar{u}\frac{p_\alpha\gamma_\beta}{M}u\Big] \end{aligned} \tag{19}$$

Upon taking the expectation value of $\delta^x J^x$ in a delocalized wave packet (rest frame), the factor y (z) projects out terms linear in Δ^i in Eq. (17)

$$\langle \delta^x J^x\rangle = \frac{1}{2}\int dx\, x\left[H_T(x,0,0) + 2\tilde{H}_T(x,0,0) + E_T(x,0,0)\right]. \tag{20}$$

yielding a decomposition of the Ji relation into transversity components

$$\begin{aligned}\langle J^x_{q,\pm\hat{x}}\rangle &= \frac{S^x}{2}\int dx\, x\left[H(x,0,0) + E(x,0,0)\right] \\ &\pm \frac{1}{4}\int dx\, x\left[H_T(x,0,0) + 2\tilde{H}_T(x,0,0) + E_T(x,0,0)\right].\end{aligned} \tag{21}$$

Here S^x is the spin of the nucleon and for an unpolarized target, only the second term contributes. Although the derivation presented above was for one specific component, it is evident that rotational invariance implies analogous relations for $J^y_{q,\pm\hat{y}}$ and $J^z_{q,\pm\hat{z}}$. Similar relations can also be derived for a spinless target, such as a pion. The scale dependence of Eq. (20) is the same as for the second moment of the quark transversity $\int dx\, xH_T(x,0,0)$.

It is instructive to apply our new relations to a point-like spin-$\frac{1}{2}$ particle, where the "quark" spin is always equal to the "nucleon" spin $H(x,0,0) = H_T(x,0,0) = \delta(x-1)$ and $E = \tilde{H}_T = E_T = 0$. In this case $\langle \delta^x J^x \rangle = \frac{1}{2}$. For an unpolarized target there is a 50% probability that $S^x = \frac{1}{2}$ and a 50% probability that $S^x = -\frac{1}{2}$. When a quark has $s^x = \frac{1}{2}$, which occurs with 50% probability, the quark also has $J^x = \frac{1}{2}$, resulting in $\langle J^x(s^x = +1/2)\rangle = 0.5 \times \frac{1}{2} = \frac{1}{4}$, which is consistent with Eqs. (20) and (21). As a second example, when the same point-like "nucleon" has $S^x = +\frac{1}{2}$, all of its angular momentum is carried by "quarks" with $s^x = +\frac{1}{2}$, while none is carried by "quarks" with $s^x = -\frac{1}{2}$. This is again consistent with Eq. (21). A constituent quark model estimate for Eq. (20) can be found in Ref. 21.

In the general case, when the second moments of the involved GPDs are nontrivial, it is expected that Eqs. (20) and (21) will provide novel insights about the spin structure and spin–orbit correlation for quarks in the nucleon. While experimental results for $H_T(x,0,0)$ are expected soon, measuring the other two chirally odd GPDs which enter Eq. (20) will be more challenging. Therefore, initial applications of Eq. (20) will have to rely on lattice QCD simulations.[22]

5. Chirally odd GPDs and the Boer–Mulders Function

The Boer–Mulders function $h_1^{\perp q}$ [23] is similar to the Sivers function (12) except that the nucleon spin is replaced by the quark spin $\mathbf{s}$

$$f_{q^\uparrow/p}(x,\mathbf{k}_\perp) = \frac{1}{2}\left[f_1^q(x,\mathbf{k}_\perp^2) - h_1^{\perp q}(x,\mathbf{k}_\perp^2)\frac{(\hat{\mathbf{P}}\times\mathbf{k}_\perp)\cdot\mathbf{s}}{M}\right] \tag{22}$$

and describes the correlation between the $\perp$ momentum and the $\perp$ spin of the ejected quark in semi-inclusive DIS from an unpolarized target. In Sec. 3, a mechanism was suggested through which the FSI in semi-inclusive DIS translates a position space asymmetry in the target into a momentum asymmetry for the outgoing quark. Applying the same mechanism here yields again a negative correlation between the sign of the momentum asymmetry and the sign of the deformation in position space, i.e. we expect

$$\frac{h_1^{\perp q}}{2\tilde{H}_T + E_T} \sim \frac{f_{1T}^{\perp q}}{E} < 0. \tag{23}$$

Tests of this qualitative relationship require lattice determinations of $2\tilde{H}_T + E_T$, while $h_1^{\perp q}$ is accessible in polarized Drell–Yan experiments.[23]

Acknowledgments

This work was supported by the DOE (DE-FG03-95ER40965).

References

1. M. Burkardt, *Phys. Rev.* **D62** (2000) 071503; *Erratum ibid.* **D66** (2002) 119903.
2. D.E. Soper, *Phys. Rev.* **D15** (1977) 1141.
3. M. Diehl, *Eur. Phys. J.* **C25** (2002) 223; J.P. Ralston and B. Pire, *Phys. Rev.* **D66** (2002) 111501; M. Diehl, *Phys. Rep.* **388** (2003) 41; A.V Belitsky, X. Ji and F. Yuan, *Phys. Rev.* **D69** (2004) 074014.
4. M. Burkardt, hep-ph/0105324; P.V. Pobilitsa, *Phys. Rev.* **D70** (2004) 034004.
5. M. Strikman and C. Weiss, *Phys. Rev.* **D69** (2004) 054012.
6. Ph. Hägler *et al.*, *Phys. Rev. Lett.* **93** (2004) 112001.
7. M. Burkardt, *Int. J. Mod. Phys.* **A18** (2003) 173.
8. X. Ji, *Phys. Rev. Lett.* **91** (2003) 062001.
9. X. Ji, *Phys. Rev. Lett.* **78** (1997) 610.
10. M. Burkardt, to appear in *Phys. Rev.* **D**, hep-ph/0505189.
11. D.W. Sivers, *Phys. Rev.* **D43** (1991) 261.
12. A. Bacchetta *et al.*, *Phys. Rev.* **D70** (2004) 117504.
13. M. Anselmino, M. Boglione, and F. Murgia, *Phys. Rev.* **D60** (1999) 054027; V. Barone, A. Drago, and P.G. Ratcliffe, *Phys. Rep.* **359** (2002) 1; M. Anselmino, U. D'Alesio, and F. Murgia, *Phys. Rev.* **D67** (2003) 074010.
14. S.J. Brodsky, D.S. Hwang, and I. Schmidt, *Phys. Lett.* **B530** (2002) 99.
15. X. Ji and F. Yuan, *Phys. Lett.* **B543** (2002) 66; A. Belitsky, X. Ji, and F. Yuan, *Nucl. Phys.* **B656** (2003) 165.
16. J.C. Collins, *Phys. Lett.* **B536** (2002) 43; Acta Phys. Pol. **B34** (2003) 3103.
17. M. Burkardt, *Nucl. Phys.* **A735** (2004) 185; *Phys. Rev.* **D69** (2004) 074032; *Phys. Rev.* **D69** (2004) 057501.
18. HERMES collaboration, *Phys. Rev. Lett.* **94** (2005) 012002.
19. M. Diehl and Ph. Hägler, hep-ph/0504175.
20. P. Hägler, *Phys. Lett.* **B594** (2004) 164.
21. B. Pasquini *et al.*, hep-ph/0510376.
22. M. Gockeler *et al.*, hep-lat/0507001.
23. D. Boer and P.J. Mulders, *Phys. Rev.* **D57** (1998) 5780.

TIME REVERSAL ODD DISTRIBUTION FUNCTIONS IN CHIRAL MODELS

Alessandro Drago

Dipartimento di Fisica, Università di Ferrara
and INFN, Sezione di Ferrara
Via Saragat, 1 – 44100 Ferrara – Italy
E-mail: drago@fe.infn.it

The possibility of computing the so-called time-reversal odd distribution functions in chiral models is discussed. It is shown that, within the subclass of chiral models in which a local gauge invariance is present, T-odd distributions can be computed unambiguously. The phenomenological consequences of this result are shortly addressed.

1. Introduction

The possible existence of time-reversal odd distribution functions was postulated in many phenomenological analysis of single-spin asymmetries by Sivers first [1,2] and then by Anselmino, Boglione and Murgia [3,4] and by Boer, Mulders and Tangerman.[5–7] The theoretical status of these quantities remained uncertain till the explicit calculation of Brodsky, Hwang and Schmidt where it was demonstrated the possibility of computing T-odd quantities in QCD. The crucial ingredient, as pointed out by Collins [8] and by Belitsky, Ji and Yuan [9,10] is a link operator, dressing the struck quark. The link needs to be introduced in order to preserve the gauge invariance of the matrix element defining the distribution. The expansion of the link operator provides the crucial phase needed for the existence of quantities odd under "naive" time-reversal.

The possibility of computing T-odd distributions in chiral models remains unclear as long as no gauge invariance exists in the lagrangian, providing a clear guideline about how to write the interaction of the struck quark with the target spectator. This difficulty can be circumvented by considering chiral models in which a local symmetry is present.

2. Vector mesons as gauge fields

It was shown in the 80's that vector mesons can be introduced in chiral lagrangians as gauge bosons of a hidden local symmetry.[11–15] Traditionally, a dynamics is then attributed to the gauge bosons, although it is probably possible to compute T-odd distributions also in a scheme in which the gauge bosons can be eliminated solving field equations. In the usual scheme, a kinetic term is introduced, so that vector mesons play the role of massive gauge fields.

Once a local symmetry is introduced, the symmetry has to be preserved also in the matter sector of the lagrangian. Therefore, a covariant derivative and a link operator have to be defined, to guarantee the invariance of bilocal operators. Clearly, at this point the chiral lagrangian becomes equivalent to QCD concerning the possibility of computing T-odd distributions.[16]

3. Phenomenology

The possibility of computing T-odd distributions in chiral models opens the possibility of evaluating these quantities in a scheme based on the $1/N_c$ expansion. In particular, it has been shown that an interesting relation can be obtained for the Sivers function:

$$\Delta_0^T f_u(x, \mathbf{k}_\perp) = -\Delta_0^T f_d(x, \mathbf{k}_\perp) \,. \tag{1}$$

This relation can be derived from the symmetry of the so-called hedgehog ansatz in chiral models,[17–20] but it can also be obtained directly from the $1/N_c$ expansion, as shown by Pobylitsa.[21] In non-chiral models the predictions for the isospin dependence of the Sivers function are rather different from the one based on Eq. (1) and the Sivers function for the down quarks turns out to be much smaller than that for the up quarks.[22,23] Recent phenomenological analysis of semi-inclusive data are providing some support in favour of Eq. (1).[24] An explicit evaluation of various T-odd distributions using chiral models is now in progress.

References

1. D.W. Sivers, *Phys. Rev.* **D41** (1990) 83.
2. D.W. Sivers, *Phys. Rev.* **D43** (1991) 261.
3. M. Anselmino, M. Boglione, and F. Murgia, *Phys. Lett.* **B362** (1995) 164.
4. M. Anselmino and F. Murgia, *Phys. Lett.* **B442** (1998) 470.
5. P.J. Mulders and R.D. Tangerman, *Nucl. Phys.* **B461** (1996) 197.
6. D. Boer and P.J. Mulders, *Phys. Rev.* **D57** (1998) 5780.
7. D. Boer, *Phys. Rev.* **D60** (1999) 014012.

8. J.C. Collins, *Phys. Lett.* **B536** (2002) 43.
9. X.-d. Ji and F. Yuan, *Phys. Lett.* **B543** (2002) 66.
10. A.V. Belitsky, X. Ji, and F. Yuan, *Nucl. Phys.* **B656** (2003) 165.
11. M. Bando, T. Kugo, S. Uehara, K. Yamawaki, and T. Yanagida, *Phys. Rev. Lett.* **54** (1985) 1215.
12. M. Bando, T. Kugo, and K. Yamawaki, *Prog. Theor. Phys.* **73** (1985) 1541.
13. U.-G. Meissner, N. Kaiser, A. Wirzba, and W. Weise, *Phys. Rev. Lett.* **57** (1986) 1676.
14. M. Bando, T. Kugo, and K. Yamawaki, *Phys. Rep.* **164** (1988) 217.
15. U.-G. Meissner, *Phys. Rep.* **161** (1988) 213.
16. A. Drago, *Phys. Rev.* **D71** (2005) 057501.
17. M. Anselmino, A. Drago, and F. Murgia, (1997) Prepared for 12th. International Symposium on High-energy Spin Physics (SPIN 96), Amsterdam, Netherlands, 10–14 Sep 1996, p. 283.
18. M. Anselmino, A. Drago, and F. Murgia, (1997), hep-ph/9703303.
19. M. Anselmino, V. Barone, A. Drago, and F. Murgia, *Nucl. Phys. Proc. Suppl.* **105** (2002) 132.
20. M. Anselmino, V. Barone, A. Drago, and F. Murgia, (2002), hep-ph/0209073.
21. P.V. Pobylitsa, (2003), hep-ph/0301236.
22. A. Bacchetta, A. Schaefer, and J.-J. Yang, *Phys. Lett.* **B578** (2004) 109.
23. Z. Lu and B.-Q. Ma, *Nucl. Phys.* **A741** (2004) 200.
24. M. Anselmino *et al.*, *Phys. Rev.* **D71** (2005) 074006.

SOFFER BOUND AND TRANSVERSE SPIN DENSITIES FROM LATTICE QCD

M. Diehl[1], M. Göckeler[2], Ph. Hägler[3], R. Horsley[4], D. Pleiter[5], P.E.L. Rakow[6], A. Schäfer[2], G. Schierholz[1,5] and J.M. Zanotti[4]

[1] *Deutsches Elektron-Synchrotron DESY, 22603 Hamburg, Germany*
[2] *Institut für Theoretische Physik Universität Regensburg, 93040 Regensburg, Germany*
[3] *Institut für Theoretische Physik T39, Physik-Department der TU München, James-Franck-Straße, D-85747 Garching, Germany*
[4] *School of Physics, University of Edinburgh, Edinburgh EH9 3JZ, UK*
[5] *John von Neumann-Institut für Computing NIC / DESY 15738 Zeuthen, Germany*
[6] *Theoretical Physics Division, Dept. of Math. Sciences, University of Liverpool, Liverpool L69 3BX, UK*

QCDSF/UKQCD collaborations

Generalized transversity distributions encode essential information on the internal structure of hadrons related to transversely polarized quarks. Lattice QCD allows us to compute the lowest moments of these tensor generalized parton distributions. In this talk, we discuss a first lattice study of the Soffer bound and show preliminary results for transverse spin densities of quarks in the nucleon.

1. Introduction

Generalized parton distributions (GPDs)[1,2] are an ideal tool to study many fundamental facets of hadron structure in terms of quarks and gluons. One key point is the relation of GPDs to (orbital) angular momentum, which plays a central role for the nucleon spin sum rule.[3] Moreover, GPDs allow us to investigate the nontrivial interplay of longitudinal momentum and transverse coordinate space degrees of freedom.[4–6] In this contribution, we will focus our attention on the recently observed correlation of transverse quark spin and impact parameter which shows up in transverse spin densities of quarks in the nucleon.[7] It turns out that these correlations in the transverse plane are governed by quark helicity flip (or tensor) GPDs.[8] As in the case of the unpolarized and the polarized GPDs,[2] they are defined via the parametrization of an off-forward nucleon matrix element of a bilocal quark operator as follows

$$\langle P',\Lambda'|\int_{-\infty}^{\infty}\frac{d\lambda}{4\pi}e^{i\lambda x}\bar{q}\Big(-\frac{\lambda}{2}n\Big)\,n_\mu\sigma^{\mu\nu}\gamma_5\mathcal{U}q\Big(\frac{\lambda}{2}n\Big)\,|P,\Lambda\rangle=\overline{u}(P',\Lambda')n_\mu$$
$$\Big\{\sigma^{\mu\nu}\gamma_5\Big(H_T(x,\xi,t)-\frac{t}{2m^2}\widetilde{H}_T(x,\xi,t)\Big)+\frac{\epsilon^{\mu\nu\alpha\beta}\Delta_\alpha\gamma_\beta}{2m}\overline{E}_T(x,\xi,t)$$
$$+\frac{\Delta^{[\mu}\sigma^{\nu]\alpha}\gamma_5\Delta_\alpha}{2m^2}\widetilde{H}_T(x,\xi,t)+\frac{\epsilon^{\mu\nu\alpha\beta}\overline{P}_\alpha\gamma_\beta}{m}\widetilde{E}_T(x,\xi,t)\Big\}u(P,\Lambda)\,, \quad (1)$$

where $f^{[\mu\nu]}=f^{\mu\nu}-f^{\nu\mu}$, $\Delta=P'-P$ is the momentum transfer with $t=\Delta^2$, $\overline{P}=(P'+P)/2$, and $\xi=-n\cdot\Delta/2$ defines the longitudinal momentum transfer with the light-like vector n. The Wilson line ensuring gauge invariance of the bilocal operator is denoted by $\mathcal{U}$. Our parametrization in Eq. (1) in terms of the four independent tensor GPDs slightly differs from the literature[8,7] where a function E_T instead of $\overline{E}_T$ has been used. However, in Refs. 7, 9 it has been noted that E_T typically appears in linear combination with the tensor GPD $\widetilde{H}_T$. It is therefore reasonable to adopt a new notation and consider $\overline{E}_T=E_T+2\widetilde{H}_T$ and $\widetilde{H}_T$ as fundamental quantities.

One prominent feature of GPDs is that they reproduce the well known parton distributions in the forward limit, $\Delta=0$, and that their integral over the momentum fraction x leads directly to form factors. For this reason, the GPD $H_T(x,\xi,t)$ is called generalized transversity, since for vanishing momentum transfer it is equal to the transversity parton distribution, $H_T(x,0,0)=\delta q(x)=h_1(x)$ for $x>0$ and $H_T(x,0,0)=-\delta\bar{q}(-x)=-\bar{h}_1(-x)$ for $x<0$. On the other hand, integrating $H_T(x,\xi,t)$ over x gives the tensor form factor:

$$\int_{-1}^{1}dx\,H_T(x,\xi,t)=g_T(t). \quad (2)$$

Another feature of GPDs important for our investigations below is their interpretation as densities in the transverse plane for $\xi=0$.[4] To give an example, it has been shown that the impact parameter dependent quark distribution for the quark GPD H_q,

$$q(x,b_\perp)\equiv\int\frac{d^2\Delta_\perp}{(2\pi)^2}\,e^{-ib_\perp\cdot\Delta_\perp}H_q(x,\xi=0,t=-\Delta_\perp^2), \quad (3)$$

has the interpretation of a probability density for unpolarized quarks of flavor q with longitudinal momentum fraction x and transverse position $b_\perp=(b_x,b_y)$ relative to the center of momentum in a nucleon.

In order to facilitate the computation of the tensor GPDs in lattice QCD, we first transform the LHS of Eq. (1) to Mellin space by forming the integral $\int_{-1}^{1} dx x^{n-1} \cdots$. This results in nucleon matrix elements of towers of local tensor operators

$$\mathcal{O}_T^{\mu\nu\mu_1\ldots\mu_{n-1}}(0) = \bar{q}(0) i\sigma^{\mu\{\nu}\gamma_5 i\overleftrightarrow{D}^{\mu_1} \ldots i\overleftrightarrow{D}^{\mu_{n-1}\}} q(0)\,, \tag{4}$$

which are parametrized in terms of tensor generalized form factors (GFFs) A_{Tni}, $\overline{B}_{Tni}$, $\widetilde{A}_{Tni}$ and $\widetilde{B}_{Tni}$. Here, $\overleftrightarrow{D} = \frac{1}{2}(\overrightarrow{D} - \overleftarrow{D})$ and $\{\cdots\}$ indicates symmetrization of indices and subtraction of traces.[a] For $n = 1$, we have [8,10]

$$\begin{aligned}\langle P'\Lambda'|\, \bar{q}(0) i\sigma^{\mu\nu}\gamma_5 q(0)\, |P\Lambda\rangle &= \overline{u}(P',\Lambda')\Big\{\sigma^{\mu\nu}\gamma_5\Big(A_{T10}(t) - \frac{t}{2m^2}\widetilde{A}_{T10}(t)\Big) \\ &+ \frac{\epsilon^{\mu\nu\alpha\beta}\Delta_\alpha\gamma_\beta}{2m}\overline{B}_{T10}(t) + \frac{\Delta^{[\mu}\sigma^{\nu]\alpha}\gamma_5\Delta_{\alpha]}}{2m^2}\widetilde{A}_{T10}(t)\Big\} u(P,\Lambda)\,. \end{aligned} \tag{5}$$

The relation of the lowest moment of the tensor GPDs to the GFFs is simple and given by

$$\begin{aligned} H_T^{n=1}(\xi,t) &= A_{T10}(t) = g_T(t), & \widetilde{H}_T^{n=1}(\xi,t) &= \widetilde{A}_{T10}(t), \\ \overline{E}_T^{n=1}(\xi,t) &= \overline{B}_{T10}(t), & \widetilde{E}_T^{n=1}(\xi,t) &= \widetilde{B}_{T10}(t) = 0\,, \end{aligned} \tag{6}$$

where $H_T^n(\xi,t) \equiv \int_{-1}^{1} dx x^{n-1} H_T(x,\xi,t)$. The general parametrization in terms of GFFs and their relations to the moments of the GPDs for $n \geq 1$ can be found in Refs. 10, 11.

The calculation of moments of GPDs in lattice QCD follows standard methods, which have been described in detail in the literature.[12–14] In the following, we therefore give only an outline of the procedure we use to extract the GFFs. First, nucleon matrix elements in the form of two- and three-point functions are computed on an Euclidean space-time lattice. The typical suppression of the matrix elements by exponential factors $\exp(-\tau E)$ in the Euclidean time τ and the energy E is cancelled out by constructing an appropriate ratio $R(\tau)$ of three- to two-point functions, which is averaged over the plateau-region $R(\tau_{\text{plat.}}) \approx$ const. The averaged ratio is then renormalized and equated with the continuum parametrization of the corresponding nucleon matrix element, e.g. Eq. (5), for all contributing index (μ,ν) and momentum (P,P') combinations. This leads to an overdetermined set of linear equations which is solved to extract the GFFs. The statistical error on the GFFs is obtained from a jackknife analysis. Our

[a]The Mellin-moment index n used in this work differs from the n in Ref. 10 by one.

results have been non-perturbative renormalized[15] and transformed to the $\overline{\text{MS}}$ scheme at a scale of $4\,\text{GeV}^2$.

The lattice results to be discussed below have been obtained from simulations with $n_f = 2$ flavors of dynamical non-perturbatively $\mathcal{O}(a)$ improved Wilson fermions and Wilson glue. There are 12 datasets available consisting of four different couplings $\beta = 5.20, 5.25, 5.29, 5.40$ with three different $\kappa = \kappa_{\text{sea}}$ values per β. The pion masses of our calculation vary from 550 to 1000 MeV, and the lattice spacings and spatial volumes vary between 0.07–0.11 fm and $(1.4\text{–}2.0\,\text{fm})^3$ respectively. Our calculation does not include the computationally demanding disconnected contributions. We expect, however, that they are small for the tensor GFFs.[14] More details of the simulation can be found in Refs. 14, 16 and 17.

2. Lattice study of the Soffer bound

In this section, we investigate the Soffer bound[18]

$$|\delta q(x)| \leq \tfrac{1}{2}\big(q(x) + \Delta q(x)\big) \,, \tag{7}$$

which holds exactly only for quark and anti-quark distributions separately. For discussion of its validity, see e.g. Ref. 19 and section 3.12.3 of Ref. 2. For a lattice study of the Soffer bound, we take Mellin moments of Eq. (7) and consider the "Soffer-ratio"

$$S^n = \frac{2\left|\langle x^{n-1}\rangle_\delta\right|}{\langle x^{n-1}\rangle + \langle x^{n-1}\rangle_\Delta} \,, \qquad n = 1, 2, \ldots, \tag{8}$$

where $\langle x^{n-1}\rangle = \int_{-1}^{1} dx x^{n-1} q(x)$. At this point it is important to note that Mellin moments of distribution functions give always sums/differences of moments of quark and anti-quark distributions, e.g. $\langle x^{n-1}\rangle = \langle x^{n-1}\rangle_q + (-1)^n \langle x^{n-1}\rangle_{\bar{q}}$. Therefore, the ratio S^n in Eq. (8) is not necessarily smaller than one. Experience shows that contributions from anti-quarks are negligible in our calculation. In Fig. 1 we show our results for the ratio (8) versus the pion mass for up-quarks (similar results for down-quarks can be found in Ref. 14). The fact that the ratio is consistently below one for the lowest two moments of the up and the down quarks strongly suggests that the Soffer bound is satisfied in our lattice calculation. The lattice results show almost no dependence on the pion mass due to cancellations of the pion mass dependence of the individual distribution functions in the ratio (8). Linear chiral extrapolation in m_π^2 leads to the following predictions for the

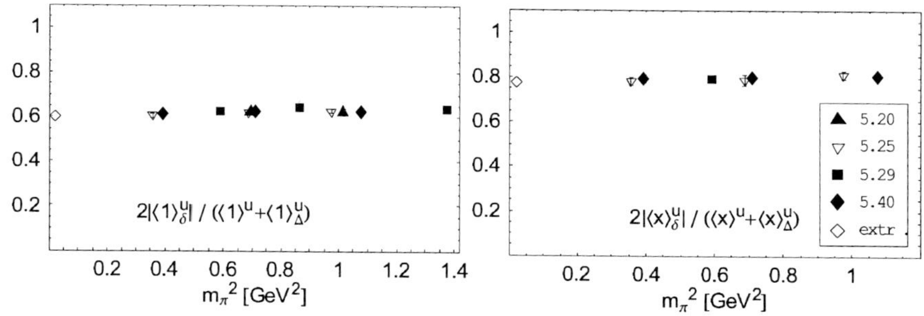

Fig. 1. Lattice results for the lowest two moments of the Soffer bound for up-quarks.

ratios at the physical pion mass

$$\begin{array}{lll} \text{up-quarks} & : & S^{n=1} = 0.60\pm.01,\ S^{n=2} = 0.78\pm.01\,, \\ \text{down-quarks} & : & S^{n=1} = 0.57\pm.02,\ S^{n=2} = 0.73\pm.05\,. \end{array} \tag{9}$$

3. Lattice results for the lowest moment of the transverse spin density

We now turn our attention to a discussion of our lattice results for the density of transversely polarized quarks in the nucleon. The lowest moment of the quark transverse spin density is given by [7]

$$\begin{aligned} \big\langle P^+, R_\perp = 0, S_\perp \big| \tfrac{1}{2}\,\overline{q}(b_\perp)\big[\gamma^+ - s^j_\perp i\sigma^{+j}\gamma_5\big] q(b_\perp) \big| P^+, R_\perp = 0, S_\perp \big\rangle \\ = \frac{1}{2}\bigg\{ A_{10}(b_\perp) + s^i_\perp S^i_\perp \bigg(A_{T10}(b_\perp) - \frac{1}{4m^2}\Delta_{b_\perp}\widetilde{A}_{T10}(b_\perp)\bigg) \\ + \frac{b^j_\perp \epsilon^{ji}}{m}\bigg(S^i_\perp B'_{10}(b_\perp) + s^i_\perp \overline{B}'_{T10}(b_\perp)\bigg) \\ + s^i_\perp (2b^i_\perp b^j_\perp - b^2_\perp \delta^{ij}) S^j_\perp \frac{1}{m^2}\widetilde{A}''_{T10}(b_\perp)\bigg\}\,. \end{aligned} \tag{10}$$

The transversity states

$$\big|P^+, R_\perp = 0, S_\perp\big\rangle = \frac{1}{\sqrt{2}}\Big(\big|P^+, R_\perp = 0, \Lambda = +\big\rangle + e^{i\chi}\,\big|P^+, R_\perp = 0, \Lambda = -\big\rangle\Big) \tag{11}$$

describe a nucleon with longitudinal momentum $P^+ = (P^0 + P^3)/\sqrt{2}$ which is localized in the transverse plane at $R_\perp = 0$ and has transverse spin $S_\perp = (\cos\chi, \sin\chi)$. The impact parameter dependent GFFs in Eq. (10) are just the Fourier-transforms of the momentum space GFFs at $\xi = 0$, as in

Eq. (3). The derivatives in Eq. (10) are defined by $f'(b_\perp) \equiv \partial_{b_\perp^2} f(b_\perp)$ and $\Delta_{b_\perp} f(b_\perp) \equiv 4\partial_{b_\perp^2} \left(b_\perp^2 \partial_{b_\perp^2}\right) f(b_\perp)$.

Since momenta are discretized on a finite lattice, we obtain the GFFs only for a limited number of different values of the momentum transfer squared t. We have in general 16 t-values available per dataset in a range of $0 \leq t < 4\,\mathrm{GeV}^2$. To facilitate the Fourier transformation to impact parameter space, we parametrize the GFFs using a p-pole ansatz

$$F(t) = \frac{F(0)}{\left(1 - t/m_p^2\right)^p}\,, \tag{12}$$

where the parameters $F(0)$, m_p and p for the individual GFFs are fixed by a fit to the lattice results. The ansatz in Eq. (12) is then Fourier transformed in order to get the GFFs as functions of the impact parameter $b_\perp$. Details of the p-pole parametrization and numerical results for the parameters will be given in a separate publication.[20] Here we only note that the values for the power p we are using for the different GFFs lead to a regular behavior of the transverse spin density in the limit $b_\perp \to 0$, as discussed in Ref. 7. In Figs. 2 and 3 we show preliminary results for the lowest moment of

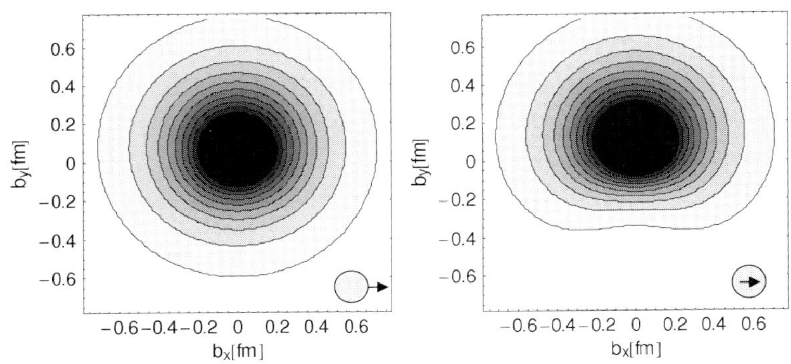

Fig. 2. Densities of up-quarks in the nucleon. The nucleon and quark spins are oriented in the transverse plane as indicated, where the inner arrow represent the quark and the outer arrow the nucleon spin. A missing arrow represents the unpolarized case.

transverse spin densities of quarks in the nucleon for up quarks. We note that the plots do not exactly show probability densities because the lowest moment corresponds to the difference of quark and anti-quark densities. The densities are however strictly positive for all $b_\perp$, indicating that the contributions from anti-quarks are small. On the LHS of Fig. 2, we show

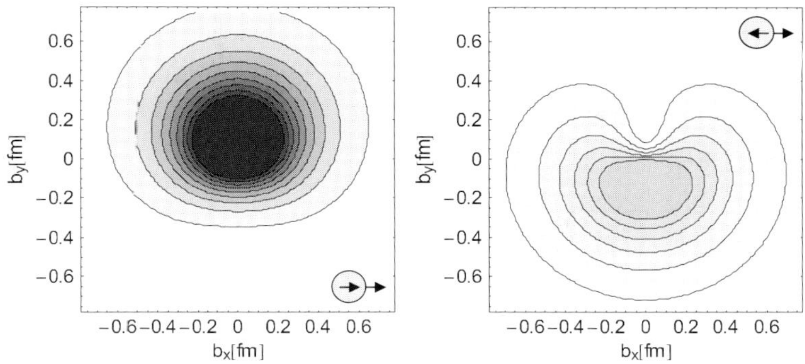

Fig. 3. Up-quark densities. Symbols are explained in the caption of Fig. 2.

the transversely distorted density of unpolarized quarks in a nucleon with spin in x-direction (coming from the dipole-term $\propto \epsilon^{ji} b_\perp^j S_\perp^i$ in Eq. (10)), which has already been discussed in Ref. 21. A new observation is that the GPD $\overline{E}_T$ also leads to a strong transverse distortion orthogonal to the transverse quark spin for an *unpolarized* nucleon (coming from the dipole-term $\propto \epsilon^{ji} b_\perp^j s_\perp^i$ in Eq. (10)) on the RHS of Fig. 2. It was argued in Ref. 9 that this shift in $+y$-direction[b] may correspond to a non-zero, negative Boer–Mulders function[22] $h_1^\perp < 0$ for up-quarks. The distortions due to transverse quark and nucleon spin add up for the density on the LHS in Fig. 3, while it goes in opposite $(-y)$-direction for quarks with spin opposite to the nucleon spin, as can be seen on the RHS of Fig. 3. Interestingly, there is practically no influence visible from the quadrupole-term $\propto s_\perp^i (2b_\perp^i b_\perp^j - b_\perp^2 \delta^{ij}) S_\perp^j$ in Eq. (10) for the up-quark densities.

4. Conclusions and outlook

Our lattice results for the transversity distribution suggests that the Soffer bound is saturated by $\approx 60 - 80\%$ for the lowest two x-moments. In addition, we have presented preliminary results for the lowest moment of the transverse spin density of quarks in the nucleon. The distortion of the density of transversely polarized quarks in an unpolarized nucleon is substantial and could give rise to a non-vanishing negative Boer–Mulders function for up-quarks through final state interactions as argued by Burkardt.[9]

---

[b] or equivalently a shift in $(-x)$-direction for quarks with spin in y-direction

We plan to extend our analysis of transverse spin densities in lattice QCD to the lowest two moments of up- and down-quarks and to investigate improved positivity bounds for GPDs which have been obtained in Ref. 7.

Acknowledgments

The numerical calculations have been performed on the Hitachi SR8000 at LRZ (Munich), on the Cray T3E at EPCC (Edinburgh),[23] and on the APEmille at NIC/DESY (Zeuthen). This work is supported by the DFG (Forschergruppe Gitter-Hadronen-Phänomenologie and Emmy-Noether-program), by the EU I3HP under contract number RII3-CT-2004-506078 and by the Helmholtz Association, contract number VH-NG-004.

References

1. D. Müller *et al.*, *Fortsch. Phys.* **42** (1994) 101 [hep-ph/9812448];
   X. Ji, *Phys. Rev.* **D55** (1997) 7114 [hep-ph/9609381];
   A.V. Radyushkin, *Phys. Rev.* **D56** (1997) 5524 [hep-ph/9704207].
2. M. Diehl, *Phys. Rep.* **388** (2003) 41 [hep-ph/0307382].
3. X. Ji, *Phys. Rev. Lett.* **78** (1997) 610 [hep-ph/9603249].
4. M. Burkardt, *Phys. Rev.* **D62** (2000) 071503 [*Erratum-ibid.* **D66** (2002) 119903] [hep-ph/0005108].
5. M. Diehl, *Eur. Phys. J.* **C25** (2002) 223 [*Erratum-ibid.* **C31** (2003) 277] [hep-ph/0205208].
6. Ph. Hägler *et al.*, *Phys. Rev. Lett.* **93** (2004) 112001 [hep-lat/0312014].
7. M. Diehl, Ph. Hägler, *Eur. Phys. J.* **C44** (2005) 87 [hep-ph/0504175].
8. M. Diehl, *Eur. Phys. J.* **C19** (2001) 485 [hep-ph/0101335].
9. M. Burkardt, hep-ph/0505189.
10. Ph. Hägler, *Phys. Lett.* **B594** (2004) 164 [hep-ph/0404138].
11. Z. Chen and X. Ji, *Phys. Rev.* **D71** (2005) 016003 [hep-ph/0404276].
12. M. Göckeler *et al.*, *Phys. Rev. Lett.* **92** (2004) 042002 [hep-ph/0304249].
13. Ph. Hägler *et al.*, *Phys. Rev.* **D68** (2003) 034505 [hep-lat/0304018].
14. M. Göckeler *et al.*, *Phys. Lett.* **B627** (2005) 113 [hep-lat/0507001].
15. G. Martinelli *et al.*, *Nucl. Phys.* **B445** (1995) 81 [hep-lat/9411010];
    M. Göckeler *et al.*, *Nucl. Phys.* **B544** (1999) 699 [hep-lat/9807044].
16. M. Göckeler *et al.*, *Few Body Syst.* **36** (2005) 111 [hep-lat/0410023].
17. M. Göckeler *et al.*, *Nucl. Phys.* **A755** (2005) 537 [hep-lat/0501029].
18. J. Soffer, *Phys. Rev. Lett.* **74** (1995) 1292 [hep-ph/9409254];
    D.W. Sivers, *Phys. Rev.* **D51** (1995) 4880.
19. G. Altarelli *et al.*, *Nucl. Phys.* **B534** (1998) 277 [hep-ph/9806345].
20. M. Göckeler *et al.*, *in preparation.*
21. M. Burkardt, *Int. J. Mod. Phys.* **A18** (2003) 173 [hep-ph/0207047].
22. D. Boer, P.J. Mulders, *Phys. Rev.* **D57** (1998) 5780 [hep-ph/9711485].
23. C.R. Allton *et al.*, *Phys. Rev.* **D65** (2002) 054502 [hep-lat/0107021].

SINGLE-SPIN ASYMMETRIES AND QIU–STERMAN EFFECT(S)*

A. Bacchetta

Theoretische Physik, Universität Regensburg
D-93040 Regensburg, Germany

Theory Group, Deutsches Elektronen-Synchroton DESY,
D-22603 Hamburg, Germany
E-mail: alessandro.bacchetta@desy.de

I discuss the relation between the Qiu–Sterman effects on one hand and the Collins, Sivers and Boer–Mulders effects on the other hand. It was suggested before that some of these effects are in fact the same, thus providing interesting connections between transverse-momentum dependent twist-2 functions and collinear twist-3 functions. Here I propose an alternative way to reach similar conclusions.

1. Introduction

Single-spin asymmetries have been observed in semi-inclusive deep inelastic scattering and proton–proton collisions.[1,2] Seemingly different mechanisms have been advocated to explain these effects. Qiu and Sterman proposed three possibilities, which I shall call chiral-even distribution, chiral-odd distribution and chiral-odd fragmentation Qiu–Sterman effects.[3] Earlier work in the same direction was carried out by Efremov and Teryaev.[4] On the other hand, the Sivers,[5] Boer–Mulders[6] and Collins[7] effects can give rise to the same asymmetries. It was argued by Boer, Mulders and Pijlman,[8] that these mechanisms are related, as all of them involve gluonic-pole matrix elements. This conclusion is apparently surprising, since the Sivers, Boer–Mulders and Collins effects can be described by *T-odd*, twist-2 distribution or fragmentation functions depending on intrinsic transverse momentum, while the effects discussed by Qiu–Sterman are T-odd, twist-3, and collinear. In my talk, I shall present an alternative derivation of the con-

*This work is supported by the Alexander von Humboldt foundation.

nection between the Sivers function and the Qiu–Sterman chiral-even distribution functions. A similar relation should hold also between the Boer–Mulders function and the Qiu–Sterman chiral-odd distribution function. On the other hand, I shall argue that the Collins function has a different origin compared to the Qiu–Sterman chiral-odd fragmentation function.

2. Semi-inclusive deep inelastic scattering

Single-spin asymmetries in deep inelastic scattering have been discussed in a large number of papers. I will now focus on the single-spin asymmetry that can be observed in the process $lp^{\uparrow} \to l\pi X$ when the target is transversely polarized and the transverse momentum of the final hadron is integrated over.[a] This particular asymmetry has been studied in a few references [9,6,10] in terms of T-odd distribution or fragmentation functions, while it has been studied in by Koike [11] in terms of Qiu–Sterman effects. No experimental measurement has been attempted so far, but it should be feasible at HERMES and COMPASS.

The general formula for the asymmetry up to subleading twist is [6]

$$A_{UT} = \frac{|\vec{S}_T|}{d\sigma_{UU}} \sum_q \frac{2\alpha^2 e_q^2}{sxy^2} V(y) \sin\phi_S \, \frac{M}{Q} \left[x\, f_T^q D_1^q - \frac{M_h}{M}\, h_1^q \frac{\tilde{H}^q}{z} \right], \qquad (1)$$

where $V(y) = 2\,(2-y)\,\sqrt{1-y}$.

The function f_T is a twist-3 distribution function and can be split in two parts, an interaction-dependent part, which can be related via equation of motions to quark–gluon–quark correlations, and a Wandzura–Wilczek part, which is related to a twist-2 distribution function, in this case the Sivers function:[12]

$$x\, f_T = x\, \tilde{f}_T^q - f_{1T}^{\perp(1)q}. \qquad (2)$$

Eq. (1) becomes

$$A_{UT} = \frac{|\vec{S}_T|}{d\sigma_{UU}} \sum_q \frac{2\alpha^2 e_q^2}{sxy^2} V(y) \sin\phi_S \, \frac{M}{Q} \left[\left(x\, \tilde{f}_T^q - f_{1T}^{\perp(1)q} \right) D_1^q - \frac{M_h}{M}\, h_1^q \frac{\tilde{H}^q}{z} \right], \qquad (3)$$

Eq. (3) contains two different kinds of twist-3 contributions. As mentioned before, the terms with a tilde are related to quark–gluon–quark correlations. They do not vanish even if transverse momentum is naively

[a] A similar case is when the final-state lepton is integrated over, but the transverse momentum of the hadron is detected.

neglected. They can be called *dynamical* twist-3 terms and should be related to the Qiu–Sterman contributions studied by Koike[11] (specifically to the chiral-even distribution G and chiral-odd fragmentation $\hat{E}$ described in his work). The term $f_{1T}^{\perp(1)}$ denotes the first moment (in transverse momentum space) of the Sivers function. This term would vanish if a collinear approximation was adopted from the beginning. Dynamically, it is a twist-2 term coming from the gauge-link contribution to quark–quark correlations, but it appears at twist 3 due to the fact that in this particular asymmetry (and in general whenever the transverse momentum of the outgoing hadron is not observed) off-collinear effects are kinematically suppressed. It can be therefore called a *kinematical* twist-3 term and it has been studied by Anselmino *et al.*[10] Note that in this asymmetry there is no contribution involving the Collins function due to off-collinear effects.

I come now to the main point of my talk. The A_{UT} asymmetry of Eq. (3) can be calculated also for totally inclusive deep-inelastic scattering (by replacing $D_1(z) \to \delta(1-z)$, $\tilde{H} \to 0$) and reduces to

$$A_{UT} = \frac{|\vec{S}_T|}{d\sigma_{UU}} \sum_q \frac{2\alpha^2 e_q^2}{sxy^2} V(y) \sin\phi_S \, \frac{M}{Q} \left(x\,\tilde{f}_T^q - f_{1T}^{\perp(1)q} \right). \tag{4}$$

However, in totally inclusive deep-inelastic scattering time-reversal invariance forbids the presence of such an asymmetry.[13,14] A relation is implied by this observation, namely

$$x\,\tilde{f}_T^q(x) - f_{1T}^{\perp(1)q}(x) = 0\,. \tag{5}$$

In principle, the relation holds only for the sum over all quark flavors, but repeating the above argument for a hypothetical photon that couples selectively to different flavors, one obtains the above relation.

Eq. (5) is the main result presented in this talk and provides a complementary way to state that there is a relation between the chiral-even Qiu–Sterman distribution function and the first moment of the Sivers function. Note that the vanishing of the function f_T, which is equivalent to Eq. (5), was already discussed by Goeke, Metz and Schlegel.[15]

An apprcpriate treatment of the T-odd distribution functions up to twist-3 should lead to the same result from a more formal point of view, making clear that $x\,\tilde{f}_T$ and $f_{1T}^{\perp(1)}$ are indeed the same object and both originate from gluonic-pole matrix elements.[8]

Note that the asymmetry in Eq. (3) – once the first term is dropped – turns out to be a good way to measure transversity, in particular in experiments which are sensitive to higher twist observables.[16] The function

$\tilde{H}$ was introduced for the first time by Jaffe and Ji,[9] who called it $\hat{e}_{\bar{1}}$. The absence of the Collins function in Eq. (3) suggests that it is intrinsically different from $\tilde{H}$. In fact, in the literature it was already observed that the Collins function is not related to gluonic poles.[17,18]

3. Drell–Yan

In analogy to semi-inclusive DIS, we can consider A_T asymmetries in Drell–Yan processes, $pp^\uparrow \to l\bar{l}X$, integrated over the transverse momentum of the lepton pair. This asymmetry has been discussed by Boer, Mulders and Teryaev,[12] but the conclusions reached by those authors are incomplete due to the fact that at that time the gauge link was not taken into account as a source of T-odd effects. The only contributions to the A_T asymmetry should be (assuming proton A to be transversely polarized)

$$A_T \propto \frac{|\vec{S}_T|}{d\sigma_{UU}} \sum_q e_q^2 \sin\phi_S \, \frac{M}{Q} \Big[x_A \Big((1-c)\, f_T^q + c\, \tilde{f}_T^q \Big)\, f_1^{\bar{q}} - h_1^q\, x_B \Big(c\, h^{\bar{q}} + (1-c)\, \tilde{h}^{\bar{q}} \Big) \Big], \qquad (6)$$

where the factor c depends on the frame of reference that is used to define the azimuthal angle ϕ_S and can assume values between 0 and 1.[12]

The first term of the asymmetry vanishes due to Eq. (5). Through a formal treatment of twist-3 distribution functions, it should be possible to prove that also the function h vanishes,[15] implying a relation between the Boer–Mulders function[6] and the Qiu–Sterman chiral-odd distribution function[19] similar to Eq. (5). The asymmetry reduces then to

$$A_T \propto \frac{|\vec{S}_T|}{d\sigma_{UU}} \sum_q e_q^2 \sin\phi_S \, \frac{M}{Q} \Big[c\, x_A \tilde{f}_T^q(x_A) f_1^{\bar{q}}(x_B) - (1-c)\, h_1^q(x_A) x_B\, \tilde{h}^{\bar{q}}(x_B) \Big]. \qquad (7)$$

Note that, when defined in the frame of reference where $c = 0$, this asymmetry gives the opportunity to measure the transversity distribution function in singly-polarized Drell–Yan, while in the frame where $c = 1$ gives an opportunity to study the Sivers function.

4. Proton–proton collisions

I now turn the attention to the A_N asymmetry in the process $pp^\uparrow \to \pi X$. The situation here is more involved than in deep inelastic scattering, due to the fact that partonic kinematics cannot be reconstructed completely, in particular in the transverse plane. Off-collinear kinematics at the partonic level has been analyzed in great detail by Anselmino *et al.*[20] It turns out that several T-odd distribution and fragmentation functions can contribute to the A_N asymmetry. These are again kinematical twist-3 contributions, in the sense that they involve twist-2 functions with a kinematical suppression due to off-collinear kinematics. Dynamical twist-3 effects in collinear kinematics are precisely those studied by Qiu and Sterman.[21]

As mentioned before, for distribution functions the two effects are identical and related to gluonic poles. The partonic cross sections to be used in both cases should not be normal partonic cross sections, but rather *gluonic-pole cross sections.* An example of the use of gluonic-pole cross sections with the Sivers and Boer–Mulders functions has been given for the process $pp^{\uparrow} \to \pi\pi X$.[22] Gluonic-pole cross sections are essentially equal to the standard partonic cross sections multiplied by overall color factors. Where do they come from and why they are not used in DIS and Drell–Yan? In fact, they are already used in DIS and Drell–Yan, but they go somewhat unnoticed! We know that T-odd functions arise from gluonic poles present in the gauge link. In deep inelastic scattering, the partonic process is $lq \to lq$. The gluons of the gauge link can attach *only* to the *outgoing* quark. The resulting gluonic-pole cross section, $l\widehat{gq} \to lq$, in this simple case corresponds to the normal partonic cross section. In Drell–Yan, the partonic process is $\bar{q}q \to l\bar{l}$, the gluon can attach *only* to the *incoming* antiquark and the resulting gluonic-pole cross section, $\bar{q}\widehat{gq} \to l\bar{l}$ is equal to *minus* the standard $\bar{q}q \to l\bar{l}$ cross section.

In the partonic processes involved in $pp \to \pi X$, colored partons are present both in the initial and the final state. The resulting gluonic-pole cross sections are then equal to the standard partonic cross section multiplied by nontrivial overall color factors, to be computed for each individual process (and each individual channel of the process). Note that gluonic-pole cross sections have been studied only for the exchange of a single gluon. It is not clear what happens when multiple gluon interactions are taken into account.

For fragmentation functions the situation is different. Since the Collins function is not related to gluonic poles, standard partonic cross sections can be used with it, as done by Anselmino *et al.*[23] On the contrary, gluonic-pole cross sections should be used with the chiral-odd fragmentation function $\tilde{H}$.

5. Conclusions

I discussed the Qiu–Sterman effects on one hand and the Sivers, Boer–Mulders and Collins functions on the other hand. I proposed a relation between the chiral-even Qiu–Sterman distribution function and the first moment of the Sivers function. A similar relation probably holds also between the Boer–Mulders function and the chiral-odd Qiu–Sterman distribution function. On the contrary, I argued that the Qiu–Sterman chiral-odd fragmentation function has a different origin compared to the Collins function.

Acknowledgments

I wish to thank D. Boer, M. Diehl, A. Metz, P. Mulders, F. Pijlman and O. Teryaev for fruitful discussions. I thank the organizers for the invitation to this pleasant and interesting workshop and the Dipartimento di Scienze e Tecnologie Avanzate, Università del Piemonte Orientale, for financial support.

References

1. HERMES, A. Airapetian *et al.*, *Phys. Rev. Lett.* **94** (2005) 012002 [hep-ex/0408013].
2. STAR, J. Adams *et al.*, *Phys. Rev. Lett.* **92** (2004) 171801 [hep-ex/0310058].
3. J. Qiu and G. Sterman, *Phys. Rev. Lett.* **67** (1991) 2264.
4. A.V. Efremov and O.V. Teryaev, *Phys. Lett.* **B150** (1985) 383.
5. D.W. Sivers, *Phys. Rev.* **D41** (1990) 83.
6. D. Boer and P.J. Mulders, *Phys. Rev.* **D57** (1998) 5780 [hep-ph/9711485].
7. J.C. Collins, *Nucl. Phys.* **B396** (1993) 161 [hep-ph/9208213].
8. D. Boer, P.J. Mulders and F. Pijlman, *Nucl. Phys.* **B667** (2003) 201 [hep-ph/0303034].
9. R.L. Jaffe and X. Ji, *Phys. Rev. Lett.* **71** (1993) 2547 [hep-ph/9307329].
10. M. Anselmino, M. Boglione, J. Hansson and F. Murgia, *Eur. Phys. J.* **C13** (2000) 519 [hep-ph/9906418].
11. Y. Koike, *AIP Conf. Proc.* **675** (2003) 449 [hep-ph/0210396].
12. D. Boer, P.J. Mulders and O.V. Teryaev, *Phys. Rev.* **D57** (1998) 3057 [hep-ph/9710223].
13. M. Anselmino, A. Efremov and E. Leader, *Phys. Rep.* **261** (1995) 1 [hep-ph/9501369].
14. R.L. Jaffe, *Comments Nucl. Part. Phys.* **19** (1990) 239.
15. K. Goeke, A. Metz and M. Schlegel, *Phys. Lett.* **B618** (2005) 90 [hep-ph/0504130].
16. P.J. Mulders and R.D. Tangerman, *Nucl. Phys.* **B461** (1996) 197 [hep-ph/9510301], *Erratum-ibid.* **B484** (1996) 538.

17. J.C. Collins and A. Metz, *Phys. Rev. Lett.* **93** (2004) 252001 [hep-ph/0408249].
18. D. Amrath, A. Bacchetta and A. Metz, *Phys. Rev.* **D71** (2005) 114018 [hep-ph/0504124].
19. Y. Kanazawa and Y. Koike, *Phys. Lett.* **B490** (2000) 99 [hep-ph/0007272].
20. M. Anselmino *et al.*, hep-ph/0509035.
21. J. Qiu and G. Sterman, *Phys. Rev.* **D59** (1999) 014004 [hep-ph/9806356].
22. A. Bacchetta, C.J. Bomhof, P.J. Mulders and F. Pijlman, *Phys. Rev.* **D72** (2005) 034030 [hep-ph/0505268].
23. M. Anselmino, M. Boglione, U. D'Alesio, E. Leader and F. Murgia, *Phys. Rev.* **D71** (2005) 014002 [hep-ph/0408356].

SIVERS FUNCTION: SIDIS DATA, FITS AND PREDICTIONS*

M. Anselmino[1], M. Boglione[1], U. D'Alesio[2], A. Kotzinian[3], F. Murgia[2], A. Prokudin[1]

[1]*Dipartimento di Fisica Teorica, Università di Torino and INFN, Sezione di Torino, Via P. Giuria 1, I-10125 Torino, Italy*

[2]*INFN, Sezione di Cagliari and Dipartimento di Fisica, Università di Cagliari, C.P. 170, I-09042 Monserrato (CA), Italy*

[3]*Dipartimento di Fisica Generale, Università di Torino and INFN, Sezione di Torino, Via P. Giuria 1, I-10125 Torino, Italy*

The most recent data on the weighted transverse single-spin asymmetry $A_{UT}^{\sin(\phi_h-\phi_S)}$ from HERMES and COMPASS collaborations are analysed within LO parton model; all transverse motions are taken into account. Extraction of the Sivers function for u and d quarks is performed. Based on the extracted Sivers functions, predictions for $A_{UT}^{\sin(\phi_h-\phi_S)}$ asymmetries at JLab are given; suggestions for further measurements at COMPASS, with a transversely polarized hydrogen target and selecting favourable kinematical ranges, are discussed. Predictions are also presented for Single Spin Asymmetries (SSA) in Drell–Yan processes at RHIC and GSI.

1. Introduction

In recent papers [1,2] we have discussed the role of intrinsic motions in inclusive and Semi-Inclusive Deep Inelastic Scattering (SIDIS) processes, both in unpolarized and polarized $\ell p \to \ell h X$ reactions. The LO QCD parton model computations have been compared with data on Cahn effect;[3] this allows an estimate of the average values of the transverse momenta of quarks inside a proton, $\boldsymbol{k}_\perp$, and of final hadrons inside the fragmenting quark jet, $\boldsymbol{p}_\perp$, with the best fit results: $\langle k_\perp^2 \rangle = 0.25$ (GeV/c)^2, $\langle p_\perp^2 \rangle = 0.20$ (GeV/c)^2.

More detail, both about the kinematical configurations and conventions [4] and the fitting procedure can be found in Ref. 1.

Equipped with such estimates, we have studied [1,2] the transverse single spin asymmetries $A_{UT}^{\sin(\phi_\pi-\phi_S)}$ observed by HERMES collaboration [5] and

*Talk presented by A. Prokudin

COMPASS collaboration;[6] that allowed extraction of the Sivers function [7]

$$\Delta^N f_{q/p^\uparrow}(x, k_\perp) = -\frac{2\,k_\perp}{m_p}\, f_{1T}^{\perp q}(x, k_\perp)\,, \tag{1}$$

defined by

$$f_{q/p^\uparrow}(x, \boldsymbol{k}_\perp) = f_{q/p}(x, k_\perp) + \frac{1}{2}\,\Delta^N f_{q/p^\uparrow}(x, k_\perp)\; \boldsymbol{S}\cdot(\hat{\boldsymbol{P}}\times\hat{\boldsymbol{k}}_\perp)\,, \tag{2}$$

where $f_{q/p}(x, k_\perp)$ is the unpolarized x and $k_\perp$ dependent parton distribution ($k_\perp = |\boldsymbol{k}_\perp|$); m_p, $\boldsymbol{P}$ and $\boldsymbol{S}$ are respectively the proton mass, momentum and transverse polarization vector ($\hat{\boldsymbol{P}}$ and $\hat{\boldsymbol{k}}_\perp$ denote unit vectors).

We consider here these whole new sets of HERMES [5] and COMPASS [6] data and perform a novel fit [2] of the Sivers functions. It turns out that the data well constrain the parameters, thus offering the first direct significant estimate of the Sivers functions – for u and d quarks – active in SIDIS processes. The sea quark contributions are found to be negligible, at least in the kinematical region of the available data. Finally, we exploit the QCD prediction [8] $f_{1T}^{\perp q}(x, k_\perp)|_{\rm D-Y} = -f_{1T}^{\perp q}(x, k_\perp)|_{\rm DIS}$ and compute a single-spin asymmetry, which can only originate from the Sivers mechanism,[9] for Drell–Yan processes at RHIC and GSI. The issue of QCD factorization of SIDIS and Drell–Yan processes was studied in Ref. 10.

2. Extracting the Sivers functions

Following Ref. 1, the inclusive ($\ell\, p \to \ell\, X$) unpolarized DIS cross section in non collinear LO parton model is given by

$$\frac{d^2\sigma^{\ell p\to\ell X}}{dx_{\scriptscriptstyle B}\, dQ^2} = \sum_q \int d^2\boldsymbol{k}_\perp\; f_q(x, k_\perp)\; \frac{d\hat{\sigma}^{\ell q\to\ell q}}{dQ^2}\; J(x_{\scriptscriptstyle B}, Q^2, k_\perp)\,, \tag{3}$$

and the semi-inclusive one ($\ell\, p \to \ell\, h\, X$) by

$$\frac{d^5\sigma^{\ell p\to\ell h X}}{dx_{\scriptscriptstyle B}\, dQ^2\, dz_h\, d^2\boldsymbol{P}_T} = \sum_q \int d^2\boldsymbol{k}_\perp\; f_q(x, k_\perp)\; \frac{d\hat{\sigma}^{\ell q\to\ell q}}{dQ^2}\; J\; \frac{z}{z_h}\; D_q^h(z, p_\perp)\,, \tag{4}$$

where

$$J = \frac{\hat{s}^2}{x_{\scriptscriptstyle B}^2 s^2}\,\frac{x_{\scriptscriptstyle B}}{x}\left(1 + \frac{x_{\scriptscriptstyle B}^2}{x^2}\frac{k_\perp^2}{Q^2}\right)^{-1}, \quad \frac{d\hat{\sigma}^{\ell q\to\ell q}}{dQ^2} = e_q^2\,\frac{2\pi\alpha^2}{\hat{s}^2}\,\frac{\hat{s}^2 + \hat{u}^2}{Q^4}\,\cdot \tag{5}$$

Q^2, $x_{\scriptscriptstyle B}$ and $y = Q^2/(x_{\scriptscriptstyle B}\, s)$ are the usual leptonic DIS variables and $z_h, \boldsymbol{P}_T$ the usual hadronic SIDIS ones, in the γ^*–p c.m. frame; x and z are light-cone momentum fractions, with (see Ref. 1 for exact relationships and further detail): $x = x_{\scriptscriptstyle B} + \mathcal{O}\left(\frac{k_\perp^2}{Q^2}\right)$, $z = z_h + \mathcal{O}\left(\frac{k_\perp^2}{Q^2}\right)$, $\boldsymbol{p}_\perp = \boldsymbol{P}_T - z_h\,\boldsymbol{k}_\perp + \mathcal{O}\left(\frac{k_\perp^2}{Q^2}\right)$.

The $\sin(\phi_h - \phi_S)$ weighted transverse single-spin asymmetry, measured by HERMES and COMPASS, which singles out the contribution of the Sivers function (1), is given by:

$$A_{UT}^{\sin(\phi_h-\phi_S)} = \left[\sum_q \int d\phi_S\, d\phi_h\, d^2\boldsymbol{k}_\perp\, \Delta^N f_{q/p^\uparrow}(x,k_\perp)\sin(\varphi-\phi_S)\cdot \frac{d\hat\sigma^{\ell q\to\ell q}}{dQ^2} J\, \frac{z}{z_h}\, D_q^h(z,p_\perp)\sin(\phi_h-\phi_S)\right] \Big/ \left[\sum_q \int d\phi_S\, d\phi_h\, \cdot d^2\boldsymbol{k}_\perp\, f_{q/p}(x,k_\perp)\, \frac{d\hat\sigma^{\ell q\to\ell q}}{dQ^2}\, J\, \frac{z}{z_h}\, D_q^h(z,p_\perp)\right], \quad (6)$$

where φ is the azimuthal angle of the quark transverse momentum, ϕ_h and ϕ_S are the azimuthal angles of produced hadron and polarization vector correspondingly. We shall use Eq. (6), in which we insert a parametrization for the Sivers functions, to fit the experimental data.

The $k_\perp$ integrated parton distribution and fragmentation functions $f_q(x)$ and $D_q^h(z)$ are taken from the literature, at the appropriate Q^2 values of the experimental data.[11,12]

We parametrize, for each light quark flavour $q = u, d$, the Sivers function in the following factorized form:

$$\Delta^N f_{q/p^\uparrow}(x,k_\perp) = 2\,\mathcal{N}_q(x)\, h(k_\perp)\, f_{q/p}(x,k_\perp)\,, \quad (7)$$

where

$$\mathcal{N}_q(x) = N_q\, x^{a_q}(1-x)^{b_q}\, \frac{(a_q+b_q)^{(a_q+b_q)}}{a_q^{a_q} b_q^{b_q}}\,, \quad h(k_\perp) = \frac{2k_\perp\, M_0}{k_\perp^2 + M_0^2}\,. \quad (8)$$

N_q, a_q, b_q and M_0 (GeV/c) are free parameters. $f_{q/p}(x,k_\perp)$ is the unpolarized distribution function. Since $h(k_\perp) \le 1$ and since we allow the constant parameter N_q to vary only inside the range $[-1,1]$ so that $|\mathcal{N}_q(x)| \le 1$ for any x, the positivity bound for the Sivers function is automatically fulfilled:

$$\frac{|\Delta^N f_{q/p^\uparrow}(x,k_\perp)|}{2 f_{q/p}(x,k_\perp)} \le 1\,. \quad (9)$$

We neglect the contributions of sea quark functions and consider only the contributions of $\Delta^N f_{u/p^\uparrow}$ and $\Delta^N f_{d/p^\uparrow}$, for a total of 7 free parameters:

$$N_u \quad a_u \quad b_u \quad N_d \quad a_d \quad b_d \quad M_0. \quad (10)$$

The results of our fits are shown in Figs. 1 and 2. In Fig. 1 we also show predictions, obtained using the extracted Sivers functions (see Table 1), for π^0 and K production; data on these asymmetries might be available soon from HERMES collaboration.

Table 1. Best fit values of the parameters of the Sivers functions.

$N_u = 0.32 \pm 0.11$	$N_d = -1.00 \pm 0.12$
$a_u = 0.29 \pm 0.35$	$a_d = 1.16 \pm 0.47$
$b_u = 0.53 \pm 3.58$	$b_d = 3.77 \pm 2.59$
$M_0^2 = 0.32 \pm 0.25\,(\mathrm{GeV}/c)^2$	$\chi^2/\mathrm{d.o.f.} = 1.06$

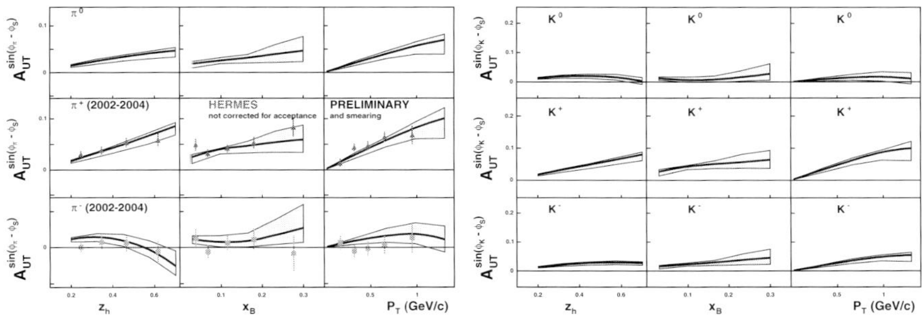

Fig. 1. HERMES data on $A_{UT}^{\sin(\phi_\pi-\phi_S)}$ for scattering off a transversely polarized proton target and charged pion production. The curves are the results of our fit. The shaded area spans a region corresponding to one-sigma deviation at 90% CL. Predictions for π^0 (upper-left panel) and kaon (right panels) asymmetries are also shown.

3. $A_{UT}^{\sin(\phi_h-\phi_S)}$ at COMPASS with polarized hydrogen target

By inspection of Eq. (6) it is easy to understand our numerical results for the u and d Sivers functions. In fact one can see that for scattering off a hydrogen target (HERMES), one has

$$\left(A_{UT}^{\sin(\phi_h-\phi_S)}\right)_{\text{hydrogen}} \sim 4\,\Delta^N f_{u/p^\uparrow}\,D_u^h + \Delta^N f_{d/p^\uparrow}\,D_d^h\,, \tag{11}$$

while, for scattering off a deuterium target (COMPASS),

$$\left(A_{UT}^{\sin(\phi_h-\phi_S)}\right)_{\text{deuterium}} \sim \left(\Delta^N f_{u/p^\uparrow} + \Delta^N f_{d/p^\uparrow}\right)\left(4\,D_u^h + D_d^h\right)\,. \tag{12}$$

Opposite u and d Sivers contributions suppress COMPASS asymmetries for any hadron h.

However, the COMPASS collaboration will soon be taking data with a transversely polarized hydrogen target. Adopting the same experimental cuts which were used for the deuterium target [1] the asymmetry is found to be around 5% (see Fig. 3). These expected values can be further increased by properly selecting the experimental data. For example, selecting events

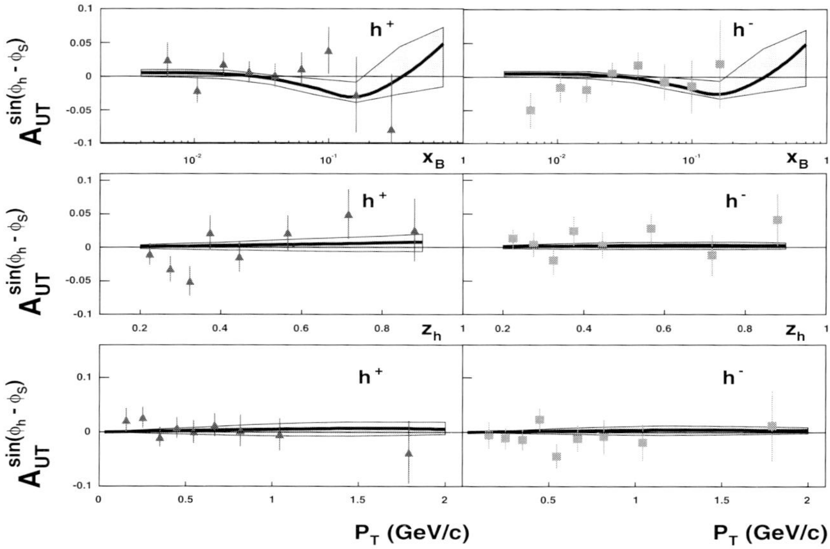

Fig. 2. COMPASS data on $A_{UT}^{\sin(\phi_h-\phi_S)}$ for scattering off a transversely polarized deuteron target and the production of positively (h^+) and negatively (h^-) charged hadrons. The curves are the results of our fit. The shaded area spans a region corresponding to one-sigma deviation at 90% CL.

with

$$0.4 \le z_h \le 1\,, \qquad 0.2 \le P_T \le 1\ \text{GeV}/c\,, \qquad 0.02 \le x_B \le 1\,, \qquad (13)$$

yields the predictions shown in the right panel of Fig. 3. The asymmetry for positively charged hadrons becomes larger, and one expects a clear observation of a sizeable azimuthal asymmetry also for the COMPASS experiment.

4. $A_{UT}^{\sin(\phi_h-\phi_S)}$ at JLab with polarized hydrogen target

Also JLab experiments are supposed to measure the SIDIS azimuthal asymmetry for the production of pions on a transversely polarised hydrogen target, at incident beam energies of 6 and 12 GeV. The kinematical region of this experiment is very interesting, as it will supply information on the behaviour of the Sivers functions in the large-x_B domain, up to $x_B \simeq 0.6$. Imposing the experimental cuts of JLab we obtain the predictions shown in Fig. 4. A large and healthy azimuthal asymmetry for π^+ production should be observed. Similar results have been obtained also in an approach based on a Monte Carlo event generator.[13]

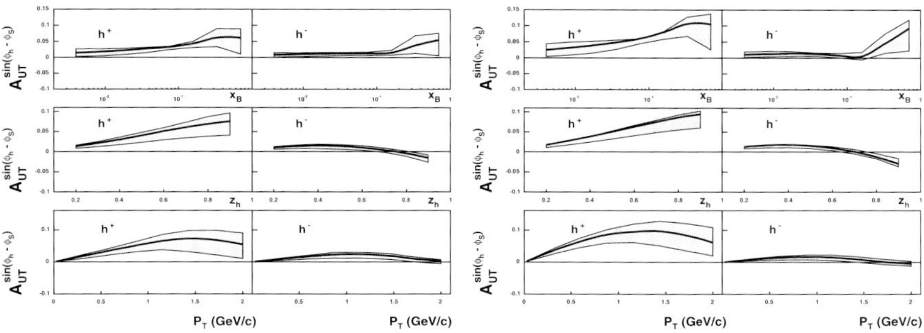

Fig. 3. Predictions for $A_{UT}^{\sin(\phi_h - \phi_S)}$ at COMPASS for scattering off a transversely polarized proton target and the production of positively (h^+) and negatively (h^-) charged hadrons. The plots in the left panel have been obtained by performing the integrations over the unobserved variables according to the standard COMPASS kinematical cuts; results with suggested new cuts, Eq. (13), are presented in the right panel.

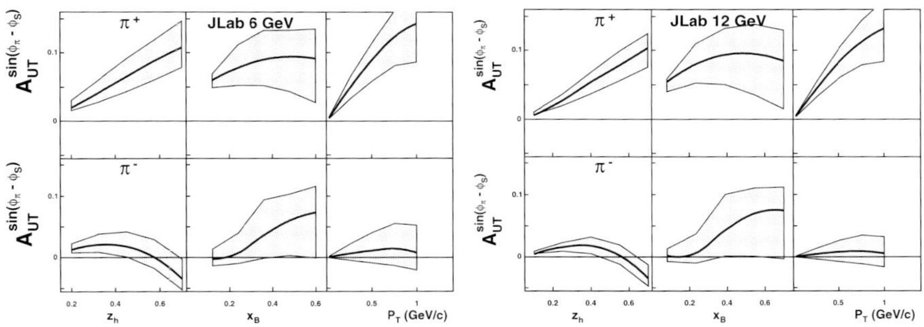

Fig. 4. Predictions for $A_{UT}^{\sin(\phi_\pi - \phi_S)}$ at JLab for the production of π^+ and π^- from scattering off a transversely polarized proton target.

5. Transverse SSA in Drell–Yan processes

Let us now consider the transverse single-spin asymmetry,

$$A_N = \frac{d\sigma^\uparrow - d\sigma^\downarrow}{d\sigma^\uparrow + d\sigma^\downarrow}, \tag{14}$$

for Drell–Yan processes, $p^\uparrow p \to \ell^+ \ell^- X$, $p^\uparrow \bar{p} \to \ell^+ \ell^- X$ and $\bar{p}^\uparrow p \to \ell^+ \ell^- X$, where $d\sigma$ stands for $d^4\sigma/dy\, dM^2\, d^2\boldsymbol{q}_T$ and y, M^2 and $\boldsymbol{q}_T$ are respectively the rapidity, the squared invariant mass and the transverse momentum of the lepton pair in the initial nucleon c.m. system.

In such a case the SSA (14) can only originate from the Sivers function and is given (selecting the region with $q_T^2 \ll M^2$, $q_T \simeq k_\perp$) by [9]

$$A_N = \Big[\sum_q e_q^2 \int d^2\boldsymbol{k}_{\perp q}\, d^2\boldsymbol{k}_{\perp \bar q}\, \delta^2(\boldsymbol{k}_{\perp q} + \boldsymbol{k}_{\perp \bar q} - \boldsymbol{q}_T)\, \Delta^N f_{q/p^\uparrow}(x_q, \boldsymbol{k}_{\perp q}) \cdot f_{\bar q/p}(x_{\bar q}, \boldsymbol{k}_{\perp \bar q})\Big] \Big/ \Big[2\sum_q e_q^2 \int d^2\boldsymbol{k}_{\perp q}\, d^2\boldsymbol{k}_{\perp \bar q}\, \delta^2(\boldsymbol{k}_{\perp q} + \boldsymbol{k}_{\perp \bar q} - \boldsymbol{q}_T) \cdot f_{q/p}(x_q, \boldsymbol{k}_{\perp q})\, f_{\bar q/p}(x_{\bar q}, \boldsymbol{k}_{\perp \bar q})\Big], \tag{15}$$

where $q = u, \bar u, d, \bar d, s, \bar s$ and $x_q = \frac{M}{\sqrt{s}}\, e^{y}$, $x_{\bar q} = \frac{M}{\sqrt{s}}\, e^{-y}$. Eq. (15) explicitly refers to $p^\uparrow\, p$ processes, with obvious modifications for $p^\uparrow\, \bar p$ and $\bar p^{\,\uparrow}\, p$ ones.

Inserting into Eq. (15) the Sivers functions extracted from our fit to SIDIS data and *reversed in sign*,[8] we obtain the predictions (shown in Fig. 5) for RHIC (left panel) and PAX (right panel) experiment [14] planned at the proposed asymmetric $p\,\bar p$ collider at GSI.

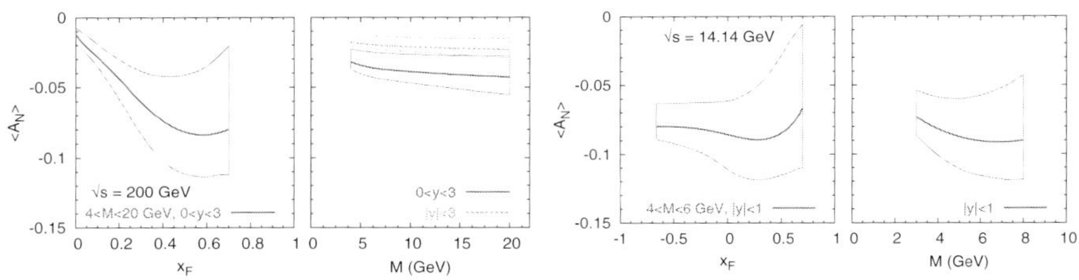

Fig. 5. Predictions for single spin asymmetries in Drell–Yan, $p^\uparrow\, p \to \ell^+\, \ell^-\, X$, processes at RHIC (left panel) and GSI (right panel), according to Eq. (15) of the text.

6. Comments and conclusions

The Sivers functions $\Delta^N f_{u/p^\uparrow}(x, k_\perp)$ and $\Delta^N f_{d/p^\uparrow}(x, k_\perp)$ have been extracted using recent HERMES [5] and COMPASS [6] collaborations data on $A_{UT}^{\sin(\phi_h - \phi_S)}$.

A sizeable h^+ asymmetry should be measured by COMPASS collaboration once they switch, as planned, to a transversely polarized hydrogen target.

Large values of $A_{UT}^{\sin(\phi_h - \phi_S)}$ are expected at JLab, both in the 6 and 12 GeV operational modes, for π^+ inclusive production.

We have then used basic QCD relations and computed the single spin asymmetries in Drell–Yan processes. We have used the same Sivers functions as extracted from SIDIS data, with opposite signs. The predicted A_N

could be measured at RHIC in pp collisions and, in the long range, at the proposed PAX experiment at GSI,[14] in $p\bar{p}$ interactions. It would provide a clear and stringent test of basic QCD properties.

References

1. M. Anselmino, M. Boglione, U. D'Alesio, A. Kotzinian, F. Murgia and A. Prokudin, *Phys. Rev.* **D71** (2005) 074006.
2. M. Anselmino, M. Boglione, U. D'Alesio, A. Kotzinian, F. Murgia and A. Prokudin, *Phys. Rev.* **D72** (2005) 094007.
3. R.N. Cahn, *Phys. Lett.* **B78** (1978) 269; *Phys. Rev.* **D40** (1989) 3107.
4. A. Bacchetta, U. D'Alesio, M. Diehl and C.A. Miller, *Phys. Rev.* **D70** (2004) 117504.
5. HERMES Collaboration, M. Diefenthaler, talk delivered at DIS 2005, Madison, Wisconsin (USA), April 27 – May 1, e-print archive: hep-ex/0507013.
6. COMPASS Collaboration, V.Yu. Alexakhin *et al.*, *Phys. Rev. Lett.* **94** (2005) 202002.
7. D. Sivers, *Phys. Rev.* **D41** (1990) 83; **D43** (1991) 261.
8. J.C. Collins, *Phys. Lett.* **B536** (2002) 43.
9. M. Anselmino, U. D'Alesio and F. Murgia, *Phys. Rev.* **D67** (2003) 074010.
10. X.D. Ji, J.P. Ma and F. Yuan, *Phys. Rev.* **D71** (2005) 034005; *Phys. Lett.* **B597** (2004) 299.
11. A.D. Martin, R.G. Roberts, W.J. Stirling and R.S. Thorne, *Phys. Lett.* **B531** (2002) 216.
12. S. Kretzer, *Phys. Rev.* **D62** (2000) 054001.
13. A. Kotzinian, e-print archive: hep-ph/0504081.
14. PAX Collaboration, e-print archive: hep-ex/0505054.

TWIST-3 EFFECTS IN SEMI-INCLUSIVE DEEP INELASTIC SCATTERING

M. Schlegel, K. Goeke and A. Metz

Ruhr-Universitaet Bochum, Institut fuer Theoretische Physik II,
Universtitaetsstr. 150, D-44780 Bochum, Germany
E-mail: marc.schlegel@tp2.rub.de

The general parametrization of the quark–quark correlation function for a spin-$\frac{1}{2}$ hadron is considered. Historically, in several steps additional terms of this parametrization were found. We briefly review this development and discuss how the Wilson line ensuring color gauge invariance of the correlator influences the parametrization. Recently, the complete structure of the quark–quark correlator was given and new time-reversal odd parton distributions were found at the twist-3 level.

1. Introduction

Hard scattering processes like semi-inclusive deep inelastic scattering (SIDIS) provide insight into the spin structure of the nucleon. On the theoretical side there has been a tremendous activity during the last years in order to investigate important aspects of these processes such as factorization, universality, time-reversal odd (T-odd) effects and the role of the gauge link.[1–14] For SIDIS key-ingredients are the two correlators that describe the distribution of partons inside the nucleon as well as the parton fragmentation. In a factorized picture they represent the non-perturbative part of the cross section and are given by ($\tilde{\xi} = (\xi^-, 0, \vec{\xi_T})$ and $\xi' = (0, \xi^+, \vec{\xi_T})$)

$$\Phi_{ij}(x, \vec{p}_T; S|n_-) = \int \frac{d\xi^- d^2\xi_T}{(2\pi)^3}\, e^{ip\cdot\tilde{\xi}}\, \langle P, S|\, \bar{\Psi}_j(0)\, \mathcal{L}^{[+]}[0, \tilde{\xi}|n_-]\, \Psi_i(\tilde{\xi})\, |P, S\rangle \tag{1}$$

and

$$\Delta_{ij}(z, \vec{k}_T; S_h|n_+) = \sum_X \int \frac{d\xi^+ d^2\xi_T}{(2\pi)^3}\, e^{ik\cdot\xi'} \times \langle 0|\, \mathcal{L}^{[-]}[0, \xi'|n_+]\, \Psi_i(\xi')\, |P_h, S_h; X\rangle\, \langle X; P_h, S_h|\, \bar{\Psi}_j(0)\, |0\rangle\,. \tag{2}$$

P denotes the 4-momentum of the target and S its covariant spin vector ($P^2 = M^2$, $S^2 = -1$, $P \cdot S = 0$). A corresponding notation (P_h, S_h) is used for the produced hadron. The plus-momentum of the quark in the target is given by $p^+ = xP^+$, whereas the minus-momentum of the fragmenting quark k^- is connected to P_h^- via $k^- = \frac{P_h^-}{z}$. The correlator $\Phi_{ij}(x, \vec{p}_T)$ in (1) describes how momentum and spin of a quark are distributed inside the nucleon. It can be parametrized by transverse momentum dependent (TMD) parton distribution functions (PDFs). On the other side, the correlator $\Delta_{ij}(z, \vec{k}_T)$ describes the fragmentation process of a quark into a hadron and is represented by TMD fragmentation functions (FFs). Note that in these correlators the gauge link $\mathcal{L}$, which is a path-ordered exponential of gluon fields, is included in order to guarantee color gauge invariance. In SIDIS, the path of the gauge link (Wilson line) cannot be chosen arbitrarily. It rather has to run along the lightcone up to infinity, then in transverse direction and back along the lightcone to the endpoint, i.e.,

$$\mathcal{L}^{[+]}[0, \tilde{\xi}|n_-] = [0; \infty^-, 0, \vec{0_T}] \times [\infty^-, 0, \vec{0_T}; \infty^-, 0, \vec{\xi_T}] \times [\infty^-, 0, \vec{\xi_T}; \tilde{\xi}], \quad (3)$$

where $[a; b]$ denotes a straight Wilson line connecting the points a and b.

To discuss the complete parametrization of the quark–quark correlator $\Phi_{ij}(x, \vec{p}_T)$ into TMD PDFs is the aim of this note which is based on Ref. 15. Recently, a lot of work has been devoted to the experimental investigation of TMD PDFs and FFs.[16–21] In the mentioned studies subleading (twist-3) effects contribute, and thus it is important to have a complete description of the quark–quark correlators including the twist-3 level. We intend to present such a description.

2. Parametrization

In order to express observables in terms of PDFs and FFs, the quark–quark correlators $\Phi_{ij}(x, \vec{p}_T)$ and $\Delta_{ij}(z, \vec{k}_T)$ have to be parametrized. This procedure was used and established in Ref. 22. Here we describe this method for $\Phi_{ij}(x, \vec{p}_T)$. It is convenient to introduce the more general, totally unintegrated quark–quark correlator $\Phi_{ij}(p; P, S|n_-)$ which depends on the 4-momentum p of the quark. It is defined by

$$\Phi_{ij}(p; P, S|n_-) \equiv \int \frac{d^4\xi}{(2\pi)^4} \, \mathrm{e}^{ip\cdot\xi} \, \langle P, S| \, \bar{\Psi}_j(0) \, \mathcal{L}[0, \xi|n_-] \, \Psi_i(\xi) \, |P, S\rangle \, . \quad (4)$$

In TMD SIDIS, this object a priori is unphysical since it does not enter the factorization formula for the cross section. The totally unintegrated correlator – and especially its Wilson line – is defined in such a way that the TMD correlator $\Phi_{ij}(x, \vec{p}_T; S)$ in (1) is reconstructed after integration upon p^-,

$$\Phi_{ij}(x, \vec{p}_T; S|n_-) = \int dp^- \;\; \Phi_{ij}(p; P, S|n_-)|_{p^+=xP^+}\,. \tag{5}$$

Instead of decomposing the TMD correlator $\Phi_{ij}(x, \vec{p}_T; S|n_-)$ directly into TMD parton distributions one parametrizes the fully unintegrated correlator (4). Such a parametrization was presented for the first time in Ref. 22. In that work $\Phi_{ij}(p; P, S|0)$ was decomposed into eight structures,

$$\begin{aligned}\Phi(p; P, S|0) = A_1 \mathbb{1} + A_2 \not{P} + A_3 \not{p} + A_4 \gamma_5 \not{S} + A_5 [\not{P}, \not{S}]\gamma_5 \\ + A_6 [\not{p}, \not{S}]\gamma_5 + A_7 (p \cdot S)\not{P}\gamma_5 + A_8 (p \cdot S)\not{p}\gamma_5.\end{aligned} \tag{6}$$

All structures are accompanied by unknown scalar functions $A_i(p^2, p \cdot P)$. In this parametrization no gauge link – and therefore no light-cone vector n_- – was taken into account. The correlator (4) obeys constraints due to hermiticity, parity and time reversal symmetry (with $\bar{A} = (A_0, \vec{A})$),

$$\Phi^\dagger(p; P, S|n_-) = \gamma_0\, \Phi(p; P, S|n_-)\, \gamma_0, \tag{7}$$

$$\Phi(p; P, S|n_-) = \gamma_0\, \Phi(\bar{p}; \bar{P}, -\bar{S}|\bar{n}_-)\, \gamma_0, \tag{8}$$

$$\Phi^*(p; P, S|0) = (-i\gamma_5 C)\, \Phi(\bar{p}; \bar{P}, -\bar{S}|0)\, (-i\gamma_5 C). \tag{9}$$

While the hermiticity constraint (7) and the parity constraint (8) are valid if the gauge link is included in the definition of (4), the time reversal constraint in (9) only holds for the correlator without link. Note, however, that this latter condition is fulfilled by the decomposition (6). The amplitudes A_i are connected unambiguously to the TMD parton distribution by a p^--integration over certain traces $\Phi^\Gamma \equiv \mathrm{Tr}(\Phi\Gamma)/2$ of the unintegrated correlator. For instance the PDF $f_1(x, \vec{p}_T)$ can be expressed in terms of the amplitudes A_2 and A_3,

$$\Phi^{[\gamma^+]}(x, \vec{p}_T; S|0) = 2P^+\!\int dp^- (A_2 + xA_3) \equiv f_1(x, \vec{p}_T). \tag{10}$$

The parametrization (6) was extended afterwards in Ref. 23. First of all, another structure $A_9(p \cdot S)[\not{P}, \not{p}]\gamma_5$ – obeying the time reversal constraint (9), therefore called a T-even structure – was found and added. Second, the time reversal constraint (9) does not hold for the fragmentation correlator $\Delta_{ij}(z, \vec{k_T})$, even if the gauge link is neglected. Consequently, structures which are not restricted by this condition have to be included in a decomposition of $\Delta_{ij}(k; P_h, S_h|0)$, similar to (6). The corresponding so-called T-odd structures in (6) are

$$\sigma^{\mu\nu} P_\mu p_\nu, \quad (p \cdot S) i\gamma_5, \quad \varepsilon_{\mu\nu\rho\sigma}\gamma^\mu P^\nu p^\rho S^\sigma. \tag{11}$$

Parton distributions which are related to these structures are called T-odd PDFs.

3. Gauge Link

The inclusion of the gauge link in the definition of the quark–quark correlators has various consequences. First of all, it affects observables as was shown by Brodsky, Hwang and Schmidt.[1] They calculated the single-spin asymmetry (SSA) A_{UT} for a transversely polarized target using a simple scalar diquark spectator model. It was demonstrated that a nonvanishing asymmetry is generated by the interference between the tree-level amplitude of the fragmentation process and the imaginary part of the one-loop amplitude, where the latter describes the gluon exchange between the struck quark and the target system, i.e. the quark is rescattered by the spectator diquark due to final state interactions. The model calculation of A_{UT} involving the rescattering effect was interpreted later on as a model calculation of the T-odd Sivers parton distribution [24] including its gauge link.[2] The investigation of Ref. 1 has therefore shown for the first time explicitly that T-odd parton distributions can be non-zero.

From this discovery one can draw the following conclusions concerning the parametrization (6). If time-reversal symmetry does not forbid T-odd parton distributions, it is essential to include T-odd structures (11) in the parametrization (6).

These T-odd structures are not the only consequence of the gauge link.[25] Because the Wilson line has to be chosen according to (3) the quark–quark correlator $\Phi_{ij}(x, \vec{p}_T; S|n_-)$ depends on the light-cone vector n_-. Therefore the fully unintegrated correlator (4) must also depend on this light-cone direction. Hence, the parametrization (6) contains additional structures depending on n_-. In Ref. 25 such structures were introduced and three of them written down explicitly,

$$\frac{M^2}{P \cdot n_-} \not{n}_-, \quad \frac{iM}{2(P \cdot n_-)} [\not{P}, \not{n}_-], \quad \frac{iM}{2(P \cdot n_-)} [\not{p}, \not{n}_-]. \tag{12}$$

Note that these terms do not contain the spin vector S.

A further step was initiated by the study of single spin asymmetries at the twist-3 level. Such subleading observables are the longitudinal SSAs A_{LU} and A_{UL}, the beam spin asymmetry and the asymmetry for a longitudinally polarized target. These asymmetries have been measured by the HERMES [17] and CLAS [18] Collaborations and have also been under phenomenological investigation.[26–29] They were also calculated [30,31] (cf. also

Ref. 32) in the diquark spectator model as was done for A_{UT}. Due to final state interactions such a calculation leads again to nonvanishing results, $A_{LU} \neq 0$ and $A_{UL} \neq 0$. On the basis of this outcome it was supposed that the parton model description of longitudinal asymmetries contains also T-odd, twist-3 parton densities.[30] This conjecture was confirmed later on in a revised parton model analysis.[33]

Interestingly enough, this leads back to the problem of the parametrization of the quark–quark correlator. It was argued in Ref. 33 that the parametrization (6) must contain in addition to the structures (12) a fourth spin independent structure

$$\frac{1}{(P \cdot n_-)} \varepsilon_{\mu\nu\rho\sigma} \gamma^\mu \gamma_5 P^\nu n_-^\rho p^\sigma , \tag{13}$$

which explicitly contains the light-cone vector n_-. However, in contrast to the structures (12), it gives rise to a new T-odd, twist-3 parton distribution $g^\perp$ which has never been discussed before in the literature. It is this new PDF which enters the parton model analysis of the beam spin asymmetry A_{LU}.

4. New Parton Distributions

In order to complete this decomposition one has to include structures which depend on both the light-cone vector n_- and the spin vector S. Such a complete decomposition has been presented in Ref. 15 for the first time. One finds 16 additional structures that are relevant once the target spin is involved. Altogether, the unintegrated correlator consists of 32 independent terms. We refrain from writing all structures and refer to Ref. 15.

It is important to note that once the PDFs are extracted from such a complete decomposition in the same manner as in (10) via (5) one finds new parton distributions, called $f_T^\perp$, $f_T^{\perp\prime}$ and $e_T^\perp$ in Ref. 15. We present the result of such an extraction of PDFs in terms of certain twist-2 and twist-3 traces of the TMD quark–quark correlator:

Twist-2 traces: (we use $\vec{p}_T \times \vec{S}_T = \epsilon_T^{ij} p_{Ti} S_{Tj}$, $\epsilon_T^{ij} = \epsilon^{-+ij}$)

$$\Phi^{[\gamma^+]}(x, \vec{p}_T) = f_1 - \frac{(\vec{p}_T \times \vec{S}_T)}{M} f_{1T}^\perp , \tag{14}$$

$$\Phi^{[\gamma^+\gamma_5]} = \lambda g_{1L} + \frac{\vec{p}_T \cdot \vec{S}_T}{M} g_{1T} , \tag{15}$$

$$\Phi^{[i\sigma^{i+}\gamma_5]}(x,\vec{p}_T) = S_T^i h_{1T} + \frac{p_T^i}{M}\left(\lambda h_{1L}^\perp + \frac{\vec{p}_T\cdot\vec{S}_T}{M} h_{1T}^\perp\right) - \frac{\varepsilon_T^{ij} p_{Tj}}{M} h_1^\perp . \tag{16}$$

<u>Twist-3 traces:</u>

$$\Phi^{[\mathbb{1}]}(x,\vec{p}_T) = \frac{M}{P^+}\left[e - \frac{(\vec{p}_T\times\vec{S}_T)}{M} e_T^\perp\right], \tag{17}$$

$$\begin{aligned}\Phi^{[\gamma^i]}(x,\vec{p}_T) = \frac{M}{P^+}\Bigg[&\frac{p_T^i}{M}\left(f^\perp - \frac{(\vec{p}_T\times\vec{S}_T)}{M} f_T^{\perp\prime}\right)\\ &+ \frac{\varepsilon_T^{ij} p_{Tj}}{M}\left(\lambda f_L^\perp + \frac{\vec{p}_T\cdot\vec{S}_T}{M} f_T^\perp\right)\Bigg],\end{aligned} \tag{18}$$

$$\Phi^{[\gamma^i\gamma_5]}(x,\vec{p}_T) = \frac{M}{P^+}\left[S_T^i g_T' + \frac{p_T^i}{M}\left(\lambda g_L^\perp + \frac{\vec{p}_T\cdot\vec{S}_T}{M} g_T^\perp\right) - \frac{\varepsilon_T^{ij} p_{Tj}}{M} g^\perp\right], \tag{19}$$

$$\Phi^{[i\sigma^{+-}\gamma_5]}(x,\vec{p}_T) = \frac{M}{P^+}\left[\lambda h_L + \frac{\vec{p}_T\cdot\vec{S}_T}{M} h_T\right], \tag{20}$$

$$\Phi^{[i\sigma^{ij}\gamma_5]}(x,\vec{p}_T) = \frac{M}{P^+}\left[\frac{S_T^i p_T^j - p_T^i S_T^j}{M} h_T^\perp - \varepsilon_T^{ij} h\right]. \tag{21}$$

The new PDFs $f_T^\perp$, $f_T^{\perp\prime}$ and $e_T^\perp$ are T-odd, twist-3 functions. They can appear in observables if the target is transversely polarized. We would like to add the following remarks:

(1) In total there are 32 transverse momentum dependent PDFs (including the twist-4 case) which agrees exactly with the number of independent amplitudes in the decomposition of the fully unintegrated correlator. This is a non-trivial result. Once the n_- dependent structures in the parametrization of (4) are neglected, the number of amplitudes is smaller than the number of transverse momentum dependent PDFs. This feature gives rise to the Lorentz invariance relations between parton distributions.[23,34] The n_- dependent structures violate these relations.[25] Such a breakdown has also been found in explicit model calculations.[35,36]

(2) With the exception of $e_T^\perp$, $f_T^\perp$, $f_T^{\perp\prime}$, $g^\perp$ all other twist-3 functions were already given in Ref. 23 (for the fragmentation case). Actually $\Phi^{[\gamma^i]}$ in

Ref. 23 contains a term of the type $\epsilon_T^{ij} S_{Tj} f_T$, which is absent in (18). To get maximal symmetry of the final result we have eliminated such a contribution by means of the identity

$$\vec{k}_T^2 \epsilon_T^{ij} S_{Tj} = -k_T^i \epsilon_T^{jk} k_{Tj} S_{Tk} + \epsilon_T^{ij} k_{Tj} \vec{k}_T \cdot \vec{S}_T . \tag{22}$$

(3) The new functions appear in transverse momentum dependent SIDIS and in the unintegrated Drell–Yan process at subleading twist. To be specific, in SIDIS $e_T^\perp$ enters the double polarized cross section σ_{LT} (multiplied with the Collins function $H_1^\perp$), while $f_T^\perp$ and $f_T^{\perp\prime}$ enter σ_{UT} (multiplied with the usual unpolarized fragmentation function D_1).

(4) The results (14)–(21) depend in principle on an arbitrary light-cone vector n_-. However, once these results enter a factorized description of an observable, n_- can no longer be chosen arbitrarily. It is rather given unambiguously in terms of the kinematics of the process.

Summary

In summary, we discussed how the general structure of the fully unintegrated quark–quark correlator was derived. This allows one to write down the most general form of the transverse momentum dependent correlator $\Phi_{ij}(x, \vec{p}_T; S|n_-)$ appearing in the description of various hard processes. New twist-3 T-odd parton densities were found. The gauge link which is contained in the definition of the correlator not only allows T-odd structures to exist but also generates new terms at subleading twist which do not appear in a parametrization where the gauge link is neglected.

Acknowledgements

The work of M.S. has been supported by the Graduiertenkolleg "Physik der Elementarteilchen an Beschleunigern und im Universum". The work has also been partially supported by the Verbundforschung (BMBF) and the Transregio/SFB Bochum–Bonn–Giessen. This research is part of the EU Integrated Infrastructure Initiative Hadron Physics Project under contract number RII3-CT-2004-506078.

References

1. S.J. Brodsky, D.S. Hwang, I. Schmidt, *Phys. Lett.* **B530** (2002) 99.
2. J.C. Collins, *Phys. Lett.* **B536** (2002) 43.
3. X. Ji, F. Yuan, *Phys. Lett.* **B543** (2002) 66.
4. A.V. Belitsky, X. Ji, F. Yuan, *Nucl. Phys.* **B656** (2003) 165.

5. A. Metz, *Phys. Lett.* **B549** (2002) 139.
6. M. Burkardt, *Phys. Rev.* **D66** (2002) 114005.
7. P.V. Pobylitsa, hep-ph/0301236.
8. D. Boer, P.J. Mulders, F. Pijlman, *Nucl. Phys.* **B667** (2003) 201.
9. J.C. Collins, *Acta Phys. Pol.* **B34** (2003) 3103.
10. M. Burkardt, *Phys. Rev.* **D69** (2004) 091501.
11. X. Ji, J.P. Ma, F. Yuan, *Phys. Rev.* **D71** (2005) 034005.
12. C.J. Bomhof, P.J. Mulders, F. Pijlman, *Phys. Lett.* **B596** (2004) 277.
13. J.C. Collins, A. Metz, *Phys. Rev. Lett.* **93** (2004) 252001.
14. X. Ji, J.P. Ma, F. Yuan, *JHEP* **0507** (2005) 020.
15. K. Goeke, A. Metz, M. Schlegel, *Phys. Lett.* **B618** (2005) 90.
16. HERMES Coll., A. Airapetian *et al.*, *Phys. Rev. Lett.* **84** (2000) 4047.
17. HERMES Coll., A. Airapetian *et al.*, *Phys. Rev.* **D64** (2001) 097101.
18. CLAS Coll., H. Avakian *et al.*, *Phys. Rev.* **D69** (2004) 112004.
19. STAR Coll., J. Adams *et al.*, *Phys. Rev. Lett.* **92** (2004) 171801.
20. HERMES Coll., A. Airapetian *et al.*, *Phys. Rev. Lett.* **94** (2005) 012002.
21. COMPASS Coll., V.Y. Alexakhin *et al.*, *Phys. Rev. Lett.* **94** (2005) 202002.
22. J.P. Ralston, D.E. Soper, *Nucl. Phys.* **B152** (1979) 109.
23. P.J. Mulders, R.D. Tangerman, *Nucl. Phys.* **B461** (1996) 197; **B484** (1997) 538(E).
24. D.W. Sivers, *Phys. Rev.* **D41** (1990) 83; D.W. Sivers, *Phys. Rev.* **D43** (1991) 261.
25. K. Goeke, A. Metz, P.V. Pobylitsa, M.V. Polyakov, *Phys. Lett.* **B567** (2003) 27.
26. E. De Sanctis, W.D. Nowak, K.A. Oganessyan, *Phys. Lett.* **B483** (2000) 69.
27. B.Q. Ma, I. Schmidt, J.J. Yang, *Phys. Rev.* **D63** (2001) 037501.
28. A.V. Efremov, K. Goeke, P. Schweitzer, *Phys. Lett.* **B522** (2001) 37; **B544** (2002) 389(E).
29. A.V. Efremov, K. Goeke, P. Schweitzer, *Phys. Rev.* **D67** (2003) 114014.
30. A. Metz, M. Schlegel, *Eur. Phys. J.* **A22** (2004) 489.
31. A. Afanasev, C.E. Carlson, hep-ph/0308163.
32. A. Metz, M. Schlegel, *Annalen Phys.* **13** (2004) 699.
33. A. Bacchetta, P.J. Mulders, F. Pijlman, *Phys. Lett.* **B595** (2004) 309.
34. D. Boer, P.J. Mulders, *Phys. Rev.* **D57** (1998) 5780.
35. R. Kundu, A. Metz, *Phys. Rev.* **D65** (2002) 014009.
36. M. Schlegel, A. Metz, hep-ph/0406289; M. Schlegel, K. Goeke, A. Metz, M.V. Polyakov, *Phys. Part. Nucl.* **35** (2004) 44.

QUARK AND GLUON SIVERS FUNCTIONS*

Ivan Schmidt

Departamento de Física,
Universidad Técnica Federico Santa María,
Casilla 110-V, Valparaíso, Chile
E-mail: ischmidt@fis.utfsm.cl

The physics of hadron single transverse spin asymmetries is discussed. Possible measurements of both the quark and gluon Sivers functions are proposed.

1. Introduction

In the usual QCD factorization formalism, a collinear approximation for the partonic intrinsic motion is used, and therefore the total inclusive hadronic cross section is written as a convolution of a hard elementary partonic cross section with distribution and fragmentation functions in which the transverse motion of the partons has been integrated. Nevertheless, the intrinsic quark and gluon transverse momenta are important, because for one thing they provide corrections to the collinear approximation, and moreover they are essential in order to explain single spin asymmetries (SSA) within a generalized transverse momentum dependent QCD factorization formalism. In this case the cross section for an inclusive process $AB \to CX$ is written as:

$$d\sigma = \sum_{abc} f_{a/A}(x_a, k_{\perp a}) \otimes f_{b/B}(x_b, k_{\perp b}) \otimes d\widehat{\sigma}^{ab\to c\ldots}(x_a, x_b, k_{\perp a}, k_{\perp b}) \otimes D_{C/c}(z, k_{\perp C}) \ . \quad (1)$$

In dealing with SSAs, the most important transverse momentum dependent functions are the Sivers distribution function $f_{1T}^{\perp} = f_{a/p\uparrow}(x_a, \vec{k}_{\perp a}) - f_{a/p\downarrow}(x_a, \vec{k}_{\perp a})$, which gives the probability distribution of finding unpolarized quarks inside a transversely polarized proton, and the Collins fragmentation function $H_1^{\perp}$, which gives the probability of unpolarized hadrons

*This work is supported by Fondecyt (Chile) under grant 1030355.

coming from the fragmentation of a transversely polarized quark. In several processes both functions can contribute.

2. Quark Sivers Function

The Sivers function is proportional to the T-odd correlation $\vec{S}_\perp \cdot (\vec{P} \times \vec{k}_\perp)$, and therefore contains an azimuthal asymmetry with respect to the direction of the hadron momentum $\vec{P}$. In 1993 Collins gave a proof about the vanishing of this function,[1] but which was shown later on by Collins himself to be incorrect.

In order to understand the physics that is present in the Sivers function let us consider a specific process: semi-inclusive deep inelastic lepton scattering $\ell p^\uparrow \to \ell' \pi X$. In the target rest frame, single-spin correlations correspond to the T-odd triple product $i\vec{S}_p \cdot \vec{p}_\pi \times \vec{q}$, where the phase i is required by time-reversal invariance. The differential cross section thus has an azimuthal asymmetry proportional to $|\vec{p}_\pi||\vec{q}| \sin\theta_{q\pi} \sin\phi$ where ϕ is the angle between the plane containing the photon and pion and the plane containing the photon and proton polarization vector $\vec{S}_p$. In a general frame, the azimuthal asymmetry has the invariant form $\frac{i}{M}\epsilon_{\mu\nu\sigma\tau}P^\mu S_p^\nu p_\pi^\sigma q^\tau$ where the polarization four-vector of the proton satisfies $S_p^2 = -1$ and $S_p \cdot P = 0$.

In general we can express the SSA as:

$$SSA = \frac{\langle\uparrow|\uparrow\rangle - \langle\downarrow|\downarrow\rangle}{\langle\uparrow|\uparrow\rangle + \langle\downarrow|\downarrow\rangle}, \tag{2}$$

where the transverse spin basis is related to the helicity basis by $|\uparrow / \downarrow\rangle = \frac{1}{\sqrt{2}}\left(|+\rangle \pm i\,|-\rangle\right)$, which means that the SSA can be written as:

$$\frac{2\mathrm{Im}\,\langle+|\,-\rangle}{\langle+|\,+\rangle + \langle-|\,-\rangle}\,. \tag{3}$$

Therefore in order to produce a correlation involving a transversely-polarized proton, there are two necessary conditions: (1) There must be two proton spin amplitudes $M[\gamma^* p(J_p^z) \to F]$ with $J_p^z = \pm\frac{1}{2}$ which couple to the same final-state $|F\rangle$; and (2) The two amplitudes must have different, complex phases. The correlation is proportional to $\mathrm{Im}(M[J_p^z = +\frac{1}{2}]^* M[J_p^z = -\frac{1}{2}])$.

As a result, we can reach the following conclusions: (1) The analysis of single-spin asymmetries requires an understanding of QCD at the amplitude level. (2) It also provides a handle on the proton angular momentum. Since we need the interference of two amplitudes which have different proton spin $J_p^z = \pm\frac{1}{2}$ but couple to the same final-state, the orbital angular momentum of the two proton wavefunctions must differ by $\Delta L^z = 1$. The anomalous magnetic moment for the proton is also proportional to the interference of amplitudes $M[\gamma^* p(J_p^z) \to F]$ with $J_p^z = \pm\frac{1}{2}$ which couple to the same final-state $|F\rangle$. (3) Since we need an Imaginary part, the SSA cannot come from tree level diagrams.

Final state interactions clearly fit into this picture. If a target is stable, its light-front wavefunction must be real. Thus the only source of a nonzero complex phase in leptoproduction in the light-front frame are final-state interactions. The rescattering corrections from final-state exchange of gauge particles produce Coulomb-like complex phases which, however, depend on the proton spin.

In Ref. 2 the single-spin asymmetry in semi-inclusive electroproduction $\gamma^* p \to HX$, induced by final-state interactions, was calculated in a model of a spin-1/2 proton of mass M with charged spin-1/2 and spin-0 constituents of mass m and λ, respectively, as in the QCD-motivated quark–diquark model of a nucleon. The basic electroproduction reaction is then $\gamma^* p \to q(qq)_0$.

There it was shown that the final-state interactions from gluon exchange between the outgoing quark and the target spectator system leads to single-spin asymmetries in deep inelastic lepton–proton scattering at leading twist in perturbative QCD; *i.e.*, the rescattering corrections are not power-law suppressed at large photon virtuality Q^2 at fixed $x_{\rm bj}$. The azimuthal single-spin asymmetry transverse to the photon-to-pion production plane decreases as $\alpha_s(r_\perp^2) x_{\rm bj} M r_\perp [\ln r_\perp^2]/r_\perp^2$ for large $r_\perp$, where $r_\perp$ is the magnitude of the momentum of the current quark jet relative to the virtual photon direction. The fall-off in $r_\perp^2$ instead of Q^2 compensates for the dimension of the $\bar{q}$–q–gluon correlation. The mass M of the physical proton mass appears here since it determines the ratio of the $L_z = 1$ and $L_z = 0$ matrix elements. This is the same type of physics that gives shadowing and antishadowing effects in both electromagnetic and weak deep inelastic scattering in nuclei.[3]

A related analysis also predicts that the initial-state interactions from gluon exchange between the incoming quark and the target spectator system lead to leading-twist single-spin asymmetries in the Drell–Yan process $H_1 H_2^{\updownarrow} \to \ell^+ \ell^- X$.[4,5] These final- and initial-state interactions can be identified as the path-ordered exponentials which are required by gauge invariance and which augment the basic light-front wavefunctions of hadrons.[4,6] Both pictures, final and initial state interactions and different gauge links, lead to the conclusion that the Sivers function is not really universal, but changes sign between SSAs in deep inelastic scattering and Drell–Yan processes.

3. How to obtain the Sivers function ?

There have been many theoretical and experimental analysis about ways in which to separate the Collins and Sivers effects. Probably the simplest is to use the SSAs which can be measured in weak interaction processes. For example, consider charged current neutrino semi-inclusive deep inelastic scattering, where a hadron (pion) is measured in the final state. In this case, the transversity distribution cannot contribute to the cross section since the produced quark from the weak interaction of the W boson is always left-handed. On the other hand, in the final-state interaction picture the SSA in charged and neutral current weak interactions will also be present, just as in the electromagnetic case. Thus these weak interaction processes will clearly distinguish the underlying physical mechanisms which produce target single-spin asymmetries.[7]

Let us see this in more detail. The quark distribution in the proton is described by a correlation matrix:

$$\Phi^{\alpha\beta}(x,\boldsymbol{p}_\perp) = \int \frac{d\xi^- d^2\boldsymbol{\xi}_\perp}{(2\pi)^3} e^{ip\cdot\xi} \left\langle P,S|\bar{\psi}^\beta(0)\psi^\alpha(\xi)|P,S\right\rangle |_{\xi^+=0} \,, \tag{4}$$

where $x = p^+/P^+$. The correlation matrix Φ is parametrized in terms of the transverse momentum dependent quark distribution functions:[8]

$$\Phi(x,\boldsymbol{p}_\perp) = \frac{1}{2}\left[f_1 \not{n} + f_{1T}^\perp \frac{\epsilon_{\mu\nu\rho\sigma}\gamma^\mu n^\nu p_\perp^\rho S_\perp^\sigma}{M} + g_{1s}\gamma_5 \not{n} \right. \tag{5}$$

$$\left. + h_{1T} i\gamma_5\sigma_{\mu\nu}n^\mu S_\perp^\nu + h_{1s}^\perp \frac{i\gamma_5\sigma_{\mu\nu}n^\mu p_\perp^\nu}{M} + h_1^\perp \frac{\sigma_{\mu\nu}p_\perp^\mu n^\nu}{M} \right],$$

where the distribution functions have arguments x and $\boldsymbol{p}_\perp$, and $n^\mu = (n^+, n^-, \boldsymbol{n}_\perp) = (0, 2, \boldsymbol{0}_\perp)$.

Similar expressions and parametrization can be obtained for the quark fragmentation correlation matrix $\Delta^{\alpha\beta}(z, \boldsymbol{k}_\perp)$.

3.1. *Electromagnetic case*

The hadronic tensor of the leptoproduction by the electromagnetic interaction in leading order in $1/Q$ is given by [8]

$$2MW^{\mu\nu}(q,P,P_h) = \int d^2\boldsymbol{p}_\perp\, d^2\boldsymbol{k}_\perp\, \delta^2(\boldsymbol{p}_\perp + \boldsymbol{q}_\perp - \boldsymbol{k}_\perp) \\ \times \tfrac{1}{4}\,\mathrm{Tr}\Big[\Phi(x_B,\boldsymbol{p}_\perp)\,\gamma^\mu\Delta(z_h,\boldsymbol{k}_\perp)\,\gamma^\nu\Big] \\ + \Big(\; q \leftrightarrow -q\;,\quad \mu \leftrightarrow \nu\;\Big), \tag{6}$$

where $x_B = Q^2/2P\cdot q$ and $z_h = P\cdot P_h/P\cdot q$. The momentum $\boldsymbol{q}_\perp$ is the transverse momentum of the exchanged photon in the frame where P and P_h do not have transverse momenta.

The single-spin asymmetry (SSA) in semi-inclusive deep inelastic scattering (SIDIS) $ep^\updownarrow \to e'\pi X$, which is given by the correlation $\vec{S}_p\cdot\vec{q}\times\vec{p}_\pi$, is obtained from (6). As mentioned before, for the electromagnetic interaction there are two mechanisms for this SSA: $h_1H_1^\perp$ and $f_{1T}^\perp D_1$ (Collins and Sivers effects), where h_1 is the transversity distribution and D_1 the unpolarized quark fragmentation function. .

We can also consider the SSA of e^+e^- annihilation processes such as $e^+e^- \to \gamma^* \to \pi\Lambda^\updownarrow X$. The Λ reveals its polarization via its decay $\Lambda \to p\pi^-$. The spin of the Λ is normal to the decay plane. Thus we can look for a SSA through the T-odd correlation $\epsilon_{\mu\nu\rho\sigma}S^\mu_\Lambda p^\nu_\Lambda q^\rho_{\gamma^*}p^\sigma_\pi$. This is related by crossing to SIDIS on a Λ target.

3.2. *Charged weak current case*

Charged currents: Let us consider the SSA in the charged current (CC) weak interaction process $\nu p^\updownarrow \to \ell\pi X$. For the CC weak interaction, the trace in (6) becomes

$$\mathrm{Tr}\Big[\Phi\gamma^\mu P_L\Delta\gamma^\nu P_L\Big] = \mathrm{Tr}\Big[\Phi P_R\gamma^\mu P_L\Delta P_R\gamma^\nu P_L\Big] = \mathrm{Tr}\Big[\Phi_{\mathrm{CC}}\gamma^\mu\Delta_{\mathrm{CC}}\gamma^\nu\Big]\,, \tag{7}$$

where $P_L = (1-\gamma_5)/2$, $P_R = (1+\gamma_5)/2$, and

$$\Phi_{\mathrm{CC}} \equiv P_L\Phi P_R\,, \qquad \Delta_{\mathrm{CC}} \equiv P_L\Delta P_R\,. \tag{8}$$

The result is that Φ_{CC} does not contain the chiral-odd distribution functions which are present in (5), and Δ_{CC} does not contain the corresponding chiral-odd fragmentation functions. The charged current only couples to a single quark chirality, and thus it is not sensitive to the transversity distribution. Thus SSAs can only arise in charged current weak interaction SIDIS from the Sivers FSI mechanism $f_{1T}^{\perp}D_1$ in leading order in $1/Q$; in contrast, both the Collins $h_1 H_1^{\perp}$ and Sivers $f_{1T}^{\perp}D_1$ mechanisms contribute to SSAs for the electromagnetic and neutral current (NC) weak interactions.

In an similar way, we can also consider the SSAs of the processes $\pi p^{\updownarrow}$ (or $pp^{\updownarrow}$) $\to WX \to \ell\nu X$. If y denotes the W rapidity, it can be shown that the region $y \sim -1$ is very sensitive to the antiquark Sivers functions, whereas the region $y \sim +1$ is sensitive to the quark Sivers functions. Furthermore, it turns out that the SSA for W^+ gives information about the Sivers u quark distribution in the region $y \to 1$ and about the Sivers $\bar{d}$ in the region $y \to -1$. Something similar happens for the W^- SSA (interchanging u and d). Therefore the measurement of the SSAs $A_N^{W^\pm}$ is a practical way to separate the u and d quarks Sivers functions and their corresponding antiquark distributions $\bar{u}$ and $\bar{d}$.[9]

Neutral currents: Let us now consider the SSA in the neutral current weak interaction process $\nu p^{\updownarrow} \to \nu\pi X$. For the NC weak interaction, the interaction vertex of Z-f-f is given by $(-ie/\sin\theta_{\rm W}\cos\theta_{\rm W})(c_L P_L + c_R P_R)$ with the weak isospin-dependent coefficients $c_{L,R} = I_{\rm W}^3 - Q\sin^2\theta_{\rm W}$. Explicit values of $c_{L,R}$ are given by $c_L = \frac{1}{2} - \frac{2}{3}\sin^2\theta_{\rm W}$, $c_R = -\frac{2}{3}\sin^2\theta_{\rm W}$ for u, c, t quarks, and $c_L = -\frac{1}{2} + \frac{1}{3}\sin^2\theta_{\rm W}$, $c_R = \frac{1}{3}\sin^2\theta_{\rm W}$ for d, s, b quarks.

The trace in (6) becomes

$$a\ \mathrm{Tr}\Big[\Phi\gamma^\mu(c_L P_L + c_R P_R)\Delta\gamma^\nu(c_L P_L + c_R P_R)\Big] \tag{9}$$

$$= a\ \mathrm{Tr}\Big[\Phi(c_L P_R + c_R P_L)\gamma^\mu\Delta\gamma^\nu(c_L P_L + c_R P_R)\Big] = a\ \mathrm{Tr}\Big[\Phi_{\rm NC}\gamma^\mu\Delta\gamma^\nu\Big]\,,$$

where $a = 1/\sin^2\theta_{\rm W}\cos^2\theta_{\rm W}$ and

$$\Phi_{\rm NC} \equiv (c_L P_L + c_R P_R)\,\Phi\,(c_L P_R + c_R P_L)\,. \tag{10}$$

In this case, for the Sivers effect we find that the SSA is given by that of the electromagnetic case with $f_{1T}^{\perp}D_1$ replaced by

$$a\ \frac{c_L^2 + c_R^2}{2}\ f_{1T}^{\perp}D_1\,. \tag{11}$$

However, f_1 is also weighted by the same factor $a\,(c_L^2 + c_R^2)/2$. Therefore, the SSA from the final-state interaction mechanism in the NC weak interaction is the same as that in the electromagnetic interaction. This can be confirmed in the simple quark–diquark model.

For the $h_1 H_1^{\perp}$ mechanism, we find that the SSA is given by that of the electromagnetic case with $(h_1 H_1^{\perp})/(f_1 D_1)$ replaced by

$$\frac{2c_L c_R}{c_L^2 + c_R^2}\ \frac{h_1 H_1^{\perp}}{f_1 D_1}\,. \tag{12}$$

That is, the SSAs are modified by the quark weak isospin-dependent factor $2c_L c_R/(c_L^2 + c_R^2)$ in comparison with the electromagnetic case. The same factor appears in the linear $\cos\theta$ forward-backward asymmetry in the $e^+e^- \to Z \to q\bar{q}$ reaction.

The SSA of the Drell–Yan processes at the Z^0, such as $\pi p^\updownarrow$ (or $pp^\updownarrow$) $\to ZX \to \ell^+\ell^- X$, can arise from the $h_1 h_1^\perp$ and $f_{1T}^\perp D_1$ mechanisms. We can also consider the SSA of the e^+e^- annihilation processes such as $e^+e^- \to Z \to \pi\Lambda^\updownarrow X$, which can arise from the $H_1 H_1^\perp$ and $D_{1T}^\perp D_1$ mechanisms.[10] The SSAs of these processes have the same situation as those of the above SIDIS case. The initial/final-state interaction mechanisms have the same formulas as the electromagnetic case, whereas the Collins mechanisms are weighted by the quark weak isospin-dependent factor $2c_L c_R/(c_L^2 + c_R^2)$ present in (12).

4. Gluon Sivers function

The gluon Sivers function was mentioned for the first time in Ref. 9, and recently it was also considered in jet correlations[11] and in D meson production[12] in $p^\uparrow p$ collisions.

The direct photon production in pp collisions can provide a clear test of short-distance dynamics as predicted by perturbative QCD, because the photon originates in the hard scattering subprocess and does not fragment, which immediately means that the Collins effect is *not* present.[13] This process is very sensitive to the gluon structure function, since it is dominated by the quark–gluon Compton subprocess in a large photon transverse momentum range. Prompt-photon production, $pp(p\bar{p}) \to \gamma X$, has been a useful tool for the determination of the unpolarized gluon density and it is considered one of the most reliable reactions for extracting information on the polarization of the gluon in the nucleon.[14]

It turns out that both Sivers functions for quarks and gluons are involved in the SSA for direct photon production $A^\gamma(s, x_F)$, and therefore it is necessary to identify a kinematic region where the gluon Sivers function dominates. The cross section contains two terms, the first one involves a product of the quark Sivers function at the light-cone momentum variable x_b and the transverse momentum dependent unpolarized gluon distribution at light-cone momentum x_a, while the second involves a product of the gluon Sivers function at x_b times the transverse momentum dependent unpolarized quark distribution at x_a. Thus it is necessary to determine the range of integration over x_a and to study the relative magnitude of x_a and x_b. As an example, taking $\sqrt{s} = 200\,\mathrm{GeV}$ and $p_T = 20\,\mathrm{GeV}$, we find that the minimum value of x_a, $x_{\min} \approx x_F$ in the region $x_F > 0.3$. On the other hand, we can also see that when x_a is integrated over the range $[x_{\min}, 1]$, the main contribution comes from the low x_b values. Therefore, when we look at the large x_F region, where x_a is large but x_b is small, the asymmetry can be approximately expressed as

$$A^{\gamma}(s, x_F) = \frac{\langle \Delta_N G \rangle}{\langle G \rangle} , \tag{13}$$

where $\langle \Delta_N G \rangle$ and $\langle G \rangle$ mean the corresponding values over an appropriate integrating range.

Acknowledgments

The work presented here was done in collaboration with Stanley Brodsky, Dae Sung Hwang, and Jacques Soffer.

References

1. J.C. Collins, *Nucl. Phys.* **B396** (1993) 161.
2. S.J. Brodsky, D.S. Hwang and I. Schmidt, *Phys. Lett.* **B530** (2002) 99.
3. S.J. Brodsky, I. Schmidt and J.-J. Yang, *Phys. Rev.* **D70** (2004) 116003.
4. J.C. Collins, *Phys. Lett.* **B536** (2002) 43.
5. S.J. Brodsky, D.S. Hwang and I. Schmidt, *Nucl. Phys.* **B642** (2002) 344.
6. X. Ji and F. Yuan, *Phys. Lett.* **B543** (2002) 66.
7. S.J. Brodsky, D.S. Hwang and I. Schmidt, *Phys. Lett.* **B553** (2002) 223.
8. D. Boer and P.J. Mulders, *Phys. Rev.* **D57** (1998) 5780
9. I. Schmidt and J. Soffer, *Phys. Lett.* **B563** (2003) 179.
10. D. Boer, R. Jakob and P.J. Mulders, *Phys. Lett.* **B424** (1998) 143.
11. D. Boer, W. Vogelsang, *Phys. Rev.* **D69** (2004) 094025.
12. M. Anselmino *et al.*, *Phys. Rev.* **D70** (2004) 074025.
13. I. Schmidt, J. Soffer and J.-J. Yang, *Phys. Lett.* **B612** (2005) 258.
14. G. Bunce, N. Saito, J. Soffer, W. Vogelsang, *Ann. Rev. Nucl. Part. Sci.* **50** (2000) 525.

SIVERS EFFECT IN SEMI-INCLUSIVE DEEPLY INELASTIC SCATTERING AND DRELL–YAN

J.C. Collins[1,3], A.V. Efremov[2], K. Goeke[3],
M. Grosse Perdekamp[4,5], S. Menzel[3], B. Meredith[4],
A. Metz[3], P. Schweitzer[3]

[1] *Penn State University, 104 Davey Lab, University Park PA 16802, U.S.A.*
[2] *Joint Institute for Nuclear Research, Dubna, 141980 Russia*
[3] *Institut für Theoretische Physik II, Ruhr-Universität Bochum, Germany*
[4] *University of Illinois, Department of Physics, Urbana, IL 61801, U.S.A.*
[5] *RIKEN BNL Research Center, Upton, New York 11973, U.S.A.*

The Sivers function is extracted from HERMES data on single spin asymmetries in semi-inclusive deeply inelastic scattering. The result is used for making predictions for the Sivers effect in the Drell–Yan process.

1. Introduction

The recent HERMES and COMPASS data [1–4] on transverse target single spin asymmetries (SSA) in semi-inclusive deeply inelastic scattering (SIDIS) can be described [5] — on the basis of generalized factorization theorems [6–8] — in terms of the Sivers [9–12] and Collins [13] effects. These effects may also contribute to SSA in hadron-hadron-collisions [14] and longitudinal SSA in SIDIS,[15–18] but the status of factorization for these SSA is not clear. All that is clear is that the longitudinal target SSA in SIDIS are dominated by subleading-twist effects [19] and more difficult to interpret.[20,21]

The Collins and Sivers effects were subject to intensive phenomenological studies in hadron-hadron-collisions [22–25] and in SIDIS.[26–34]. In this note we will concentrate on the Sivers effect and review our recent work.[30,33,34] Studies of the equally interesting Collins effect are reported elsewhere.[32,35]

The Sivers function belongs to the class of so-called "naively time-reversal-odd" distributions, which have been predicted to obey an unusual "universality property", namely to appear with opposite sign in SIDIS and in the Drell–Yan (DY) process.[11] We show that an experimental test of this prediction, which is among the most important issues for the future spin physics, is feasible in the running or planned experiments at RHIC, COMPASS and GSI.

2. Sivers function from preliminary $P_{h\perp}$-weighted data [1]

In SIDIS the Sivers effect gives rise to a transverse target SSA with a specific angular distribution of the produced hadrons $\propto \sin(\phi - \phi_S)$, where ϕ (ϕ_S) is the azimuthal angle of the produced hadron (target polarization vector) with respect to the axis defined by the hard virtual photon.[5]

Weighting the events entering the spin asymmetry with $\sin(\phi-\phi_S)P_{h\perp}$ ($P_{h\perp}$ = transverse momentum of the produced hadron) yields the Sivers SSA given [5] (if we neglect soft factors [6–8]) by

$$A_{UT}^{\sin(\phi-\phi_S)P_{h\perp}/M_N} = (-2)\,\frac{\sum_a e_a^2\, x f_{1T}^{\perp(1)a}(x)\, z D_1^{a/\pi}(z)}{\sum_a e_a^2\, x f_1^a(x)\, D_1^{a/\pi}(z)}\,, \tag{1}$$

where the transverse moment of the Sivers function is defined as

$$f_{1T}^{\perp(1)a}(x) \equiv \int \mathrm{d}^2\mathbf{p}_T\, \frac{\mathbf{p}_T^2}{2M_N^2}\, f_{1T}^{\perp a}(x,\mathbf{p}_T^2)\,. \tag{2}$$

Preliminary data analyzed in this way are available.[1] Neglecting $f_{1T}^{\perp\bar{q}}$ and taking the ansatz motivated by predictions from the large-N_c limit [36]

$$f_{1T}^{\perp(1)u}(x) \overset{\text{large } N_c}{=} -f_{1T}^{\perp(1)d}(x) \overset{\text{ansatz}}{=} A\,x^b(1-x)^5\,, \tag{3}$$

the Sivers function was extracted from these data.[30] The fit result (χ^2 per degree of freedom $\equiv \chi^2_{\rm dof} \sim 0.3$) refers to a scale of about $2.5\,\mathrm{GeV}^2$ and is shown in Fig. 1. For f_1^a and D_1^a the parametrizations [37,38] were used.

3. Transverse parton momenta and the Gaussian ansatz

However, the currently available *published* data [2,3] were analyzed without a $P_{h\perp}$-weight, and can only be interpreted by resorting to some *model* for the distribution of the transverse parton momenta in the "unintegrated" distribution and fragmentation functions.[39] Here (for other models see Refs. 31, 32) we assume the distributions of transverse parton momenta to be Gaussian:

$$\begin{aligned} f_1^a(x,\mathbf{p}_T^2) &\equiv f_1^a(x)\frac{\exp(-\mathbf{p}_T^2/p_{\rm unp}^2)}{\pi\, p_{\rm unp}^2}\,,\\ f_{1T}^{\perp a}(x,\mathbf{p}_T^2) &\equiv f_{1T}^{\perp a}(x)\,\frac{\exp(-\mathbf{p}_T^2/p_{\rm Siv}^2)}{\pi\, p_{\rm Siv}^2}\,,\\ D_1^a(z,\mathbf{K}_T^2) &\equiv D_1^a(z)\frac{\exp(-\mathbf{K}_T^2/K_{D_1}^2)}{\pi\, K_{D_1}^2} \end{aligned} \tag{4}$$

and take the Gaussian widths to be flavour and x- or z-independent. This model describes well the distributions of low (with respect to the relevant hard scale) transverse hadron momenta in various hard reactions [24] — which is the case at HERMES,[2] where $\langle P_{h\perp}\rangle \sim 0.4\,\text{GeV} \ll \langle Q^2\rangle^{1/2} \sim 1.5\,\text{GeV}$.

In order to test the ansatz (4) for f_1^a and D_1^a in SIDIS we consider the HERMES data [17] on the average transverse momentum of the produced hadrons given by $\langle P_{h\perp}(z)\rangle = \frac{\sqrt{\pi}}{2}\sqrt{z^2 p_{\text{unp}}^2 + K_{D_1}^2}$ in the Gaussian ansatz.[33]

With the (fitted) parameters $p_{\text{unp}}^2 = 0.33\,\text{GeV}^2$ and $K_{D_1}^2 = 0.16\,\text{GeV}^2$ the Gaussian ansatz provides a good description of the data [17] — Fig. 1. For comparison we also show the description one obtains using the parameters obtained from an analysis [31] of EMC data [40] on the Cahn effect,[41] which is equally satisfactory. The good agreement observed in Fig. 1 indicates that the mechanisms generating the Cahn effect in the EMC data [40] and transverse hadron momenta at HERMES [17] could be compatible.[31,33]

4. Sivers function from final (published) HERMES data [2]

In the Gaussian model the expression for the Sivers SSA weighted without a power of transverse hadron momentum is given by [28]

$$A_{UT}^{\sin(\phi-\phi_S)} = a_{\text{Gauss}}^{\text{SIDIS}}\, A_{UT}^{\sin(\phi-\phi_S)P_{h\perp}/M_N}\,; \quad a_{\text{Gauss}}^{\text{SIDIS}} = \frac{\sqrt{\pi}}{2}\,\frac{M_N}{\sqrt{p_{\text{Siv}}^2 + K_{D_1}^2/z^2}}\,. \tag{5}$$

Positivity [42] constrains p_{Siv}^2 to be in the range [33] $0 < p_{\text{Siv}}^2 < 0.33\,\text{GeV}^2$. Though vague, this information is *sufficient* for the extraction of the *transverse moment* of the Sivers function.

Using the same assumptions (large-N_c, neglect of $\bar{q}$, etc.) as in Sec. 2 the fit ($\chi_{\text{dof}}^2 \sim 0.3$) to the final data [2] was obtained shown in Fig. 1. The fit [33] to the final data [2] obtained *assuming the Gaussian model* is compatible — see Fig. 1 — with the *"model-independent"* fit [30] to the preliminary data.[1] This observation is a valuable test of the Gaussian ansatz for the Sivers function.

5. Sivers effect from deuteron at COMPASS [3]

At COMPASS the deuteron Sivers effect was found consistent with zero within error bars.[3] Notice that the Sivers SSA from deuteron is sensitive solely to $(f_{1T}^{\perp u} + f_{1T}^{\perp d})$ which is subleading in the large-N_c limit. Thus, the deuteron Sivers SSA is $\sim \mathcal{O}(N_c^{-1})$, while the proton Sivers SSA is $\sim \mathcal{O}(N_c^0)$.

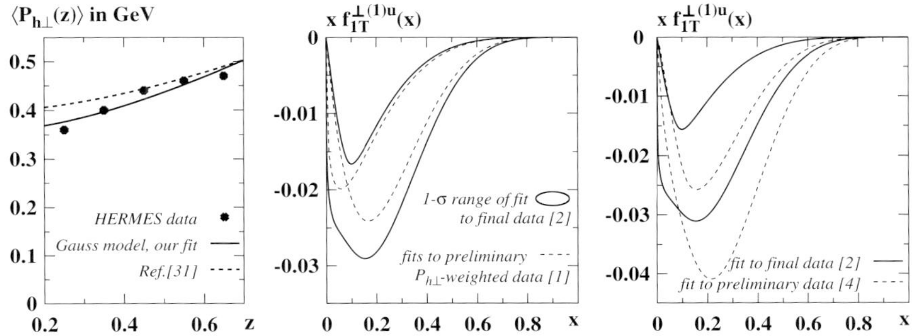

Fig. 1. Left: $\langle P_{h\perp}(z)\rangle$ of pions produced at HERMES [17] vs. z. The dashed (dotted) curve follows from the Gaussian ansatz with the parameters from our approach [33] (from a study of the Cahn effect [31]). Middle: the two alternative fits [30] of $xf_{1T}^{\perp(1)u}(x)$ to the $P_{h\perp}$-weighted *preliminary* data [1] and the 1-σ region of the fit [33] to *final* (non-$P_{h\perp}$-weighted) data.[2] Right: the 1-σ regions of the fits to the final data [2] and to the preliminary data.[4]

This suppression naturally explains [30,33] the compatibility of the HERMES and COMPASS results,[2,3] within errors. The COMPASS data [3] confirm the utility of the constraint (3) *at the present stage.*

6. Sivers function from most recent preliminary data [4]

Increasing precision of data will, sooner or later, require to relax the strict large-N_c constraint (3). The *preliminary* HERMES data released recently [4] are considerably more precise compared to the final (published) data.[2]

Could these data already constrain $1/N_c$-corrections? In order to answer this question, we repeat here the procedure of Ref. 33 described in Sec. 4 with the preliminary HERMES data.[4] The resulting fit is compatible with the fit obtained from the published data [2] — Fig. 1.

The $\chi^2_{\rm dof} \sim 2$ of this fit is larger than previously, see Sec. 4, which indicates that the description of the *preliminary* data [4] could be improved, e.g., by considering $1/N_c$ corrections. However, it could be equally sufficient to introduce more parameters in the large-N_c ansatz (3). Thus, our large-N_c ansatz is still useful to describe the most recent *preliminary* data.[4]

7. Sivers effect in the Drell–Yan process

On the basis of the first study [30] of the preliminary $P_{h\perp}$-weighted data [1] it was found that the Sivers effect can give rise to SSA in DY large enough to be measured in the planned COMPASS [43] and PAX [44,45] experiments. This conclusion is now solidified[33,34] by the study of the published HERMES data.[2] Thus, the predicted[11] sign reversal of the *quark* Sivers function in SIDIS and DY can be tested at COMPASS (PAX) in $p^\uparrow \pi^-$ ($p^\uparrow \bar{p}$) collisions.

At RHIC in DY from $p^\uparrow p$-collisions, however, *antiquark* Sivers distributions are of importance, which are not constrained by the present SIDIS data.[33] In order to see, what one can learn from RHIC about the Sivers SSA, let us assume the $\bar{q}$-Sivers distributions are given by

$$f_{1T}^{\perp(1)\bar{q}}(x) = f_{1T}^{\perp(1)q}(x) \times \begin{cases} 0.25 = \text{const} & \text{(model I)} \\ \frac{(f_1^{\bar{u}}+f_1^{\bar{d}})(x)}{(f_1^u+f_1^d)(x)} & \text{(model II)}, \end{cases} \tag{6}$$

with $f_{1T}^{\perp(1)q}(x)$ from the fit to the published HERMES data[2] — Sec. 4. The *models* I, II are consistent[33] with theoretical constraints, *and* with SIDIS data.[1–3] This makes them well suited to visualize possible effects of $\bar{q}$-Sivers distributions at RHIC.

The Sivers SSA in DY is defined similarly to that in SIDIS. Here $(\phi - \phi_S)$ is the azimuthal angle between the virtual photon and the polarization vector (the polarized proton moves into the positive z-direction). We again neglect soft factors and assume the Gaussian model, so that the SSA is

$$A_{UT}^{\sin(\phi-\phi_S)} = 2\, \frac{a_{\text{Gauss}}^{\text{DY}} \sum_a e_a^2\, x_1 f_{1T\,\text{DY}}^{\perp(1)a}(x_1)\, x_2 f_1^{\bar{a}}(x_2)}{\sum_a e_a^2\, x_1 f_1^a(x_1)\, x_2 f_1^{\bar{a}}(x_2)}\,, \tag{7}$$

where $x_{1,2} = (Q^2/s)^{1/2}\, e^{\pm y}$ with $s = (p_1 + p_2)^2$, $Q^2 = (k_1 + k_2)^2$, and $y = \frac{1}{2} \ln \frac{p_1 \cdot (k_1+k_2)}{p_2 \cdot (k_1+k_2)}$. The momenta of the incoming proton (outgoing lepton) pair are denoted by $p_{1/2}$ ($k_{1/2}$). The Gaussian factor reads

$$a_{\text{Gauss}}^{\text{DY}} = \frac{\sqrt{\pi}}{2} \frac{M_N}{\sqrt{p_{\text{Siv}}^2 + p_{\text{unp}}^2}}\,. \tag{8}$$

Considering the sign reversal[11] we obtain the results shown in Fig. 2. In our estimate we assume the ratio $\sum_a e_a^2 f_{1T}^{\perp(1)a} f_1^{\bar{a}} / \sum_b e_b^2 f_1^b f_1^{\bar{b}}$ to be weakly scale-dependent, and roughly simulate Sudakov dilution by assuming that $p_{\text{unp/Siv}}^2$ increase by a factor of two from HERMES to RHIC energies. Notice that SSA weighted appropriately with the transverse dilepton momentum[30] were argued to be less sensitive to Sudakov suppression.[46]

In the region of positive rapidities $1 \leq y \leq 2$ the Sivers SSA at RHIC is well constrained by the SIDIS data[2] and shows little sensitivity to the unknown $\bar{q}$-Sivers distributions. Thus, in this region STAR and PHENIX can also test the sign reversal[11] of the *quark* Sivers function.

For negative y the Sivers SSA is strongly sensitive to the antiquark Sivers function — with the effect being more pronounced at larger dilepton masses Q.[33] This reveals the unique feature of RHIC, which — in contrast to COMPASS and PAX — can also provide information on the antiquark Sivers distribution.

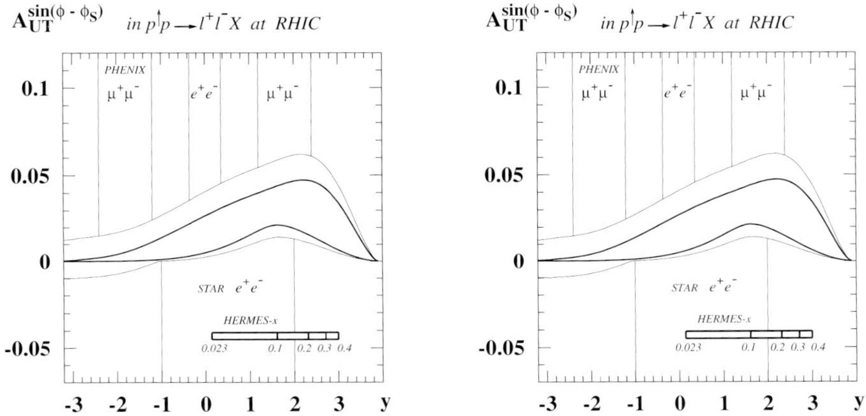

Fig. 2. The Sivers SSA $A_{UT}^{\sin(\phi-\phi_S)}$ in $p^\uparrow p \to l^+l^-X$ as function of y for the kinematics of the RHIC experiment with $\sqrt{s} = 200\,\text{GeV}$, and $Q^2 = (4\,\text{GeV})^2$. The inner error band (thick lines) shows the 1-σ uncertainty of the fit.[33] The x-region explored at HERMES is included to indicate where the SIDIS data constrain the prediction. The outer error band (thin lines) arises from assuming Sivers $\bar{q}$-distribution functions according to model I (left) and model II (right) in Eq. (6). Also shown are the regions in which PHENIX and STAR can detect $\mu^+\mu^-$ or e^+e^- pairs.

8. Conclusions

We reviewed our studies[30,33] of the HERMES and COMPASS data on the Sivers effect in SIDIS.[1–4] The data from various targets are compatible with each other and can be well described assuming a Gaussian distribution of parton transverse momenta in the distribution and fragmentation functions. The parameters in the Gaussian ansatz were constrained by HERMES data and are compatible with results obtained from studies of the Cahn effect.[31]

The data[1–4] confirm the predictions from the large-N_c limit on the flavour dependence of the Sivers function.[36] The sign of the extracted quark Sivers functions is in agreement with the intuitive picture discussed in Ref. 47. Results by other groups[31,32] confirm these findings — see the detailed comparison in Ref. 48.

The information on the quark Sivers distributions extracted from SIDIS is required for reliable estimates of the Sivers effect in DY for current or planned experiments. We estimated that the Sivers SSA in DY are sizeable enough to be observed at RHIC, COMPASS and PAX [30,33,34] allowing one to test the QCD prediction [11] that the Sivers function should appear with opposite signs in SIDIS and in DY. In addition, RHIC can provide information on the antiquark Sivers distributions.[34]

Acknowledgments.

The work is part of the European Integrated Infrastructure Initiative Hadron Physics project under contract number RII3-CT-2004-506078. A.E. is supported by grants RFBR 03-02-16816 and DFG-RFBR 03-02-04022. J.C.C. is supported in part by the U.S. D.O.E., and by a Mercator Professorship of DFG.

References

1. N.C. Makins [HERMES], *"Transversity Workshop", 6–7 Oct. 2003, Athens*; R. Seidl [HERMES], proc. of *DIS 2004, 13–18 April 2004, Štrbské Pleso*; I.M. Gregor [HERMES], *Acta Phys. Polon.* **B36** (2005) 209.
2. A. Airapetian *et al.* [HERMES], *Phys. Rev. Lett.* **94** (2005) 012002.
3. V.Y. Alexakhin *et al.* [COMPASS], *Phys. Rev. Lett.* **94** (2005) 202002.
4. M. Diefenthaler [HERMES], arXiv:hep-ex/0507013.
5. D. Boer and P.J. Mulders, *Phys. Rev.* **D57** (1998) 5780.
6. J.C. Collins and D.E. Soper, *Nucl. Phys.* **B193** (1981) 381 [*Erratum-ibid.* **B213** (1983) 545].
7. X.D. Ji, J.P. Ma and F. Yuan, *Phys. Rev.* **D71** (2005) 034005; *Phys. Lett.* **B597** (2004) 299.
8. J.C. Collins and A. Metz, *Phys. Rev. Lett.* **93** (2004) 252001.
9. D.W. Sivers, *Phys. Rev.* **D41** (1990) 83; *Phys. Rev.* **D43** (1991) 261.
10. S.J. Brodsky, D.S. Hwang and I. Schmidt, *Phys. Lett.* **B530** (2002) 99; *Nucl. Phys.* **B642** (2002) 344.
11. J.C. Collins, *Phys. Lett.* **B536** (2002) 43.
12. A.V. Belitsky, X. Ji and F. Yuan, *Nucl. Phys.* **B656** (2003) 165; X.D. Ji and F. Yuan, *Phys. Lett.* **B543** (2002) 66; D. Boer, P.J. Mulders and F. Pijlman, *Nucl. Phys.* **B667** (2003) 201.
13. J.C. Collins, *Nucl. Phys.* **B396** (1993) 161.
14. D.L. Adams *et al.*, *Phys. Lett.* **B261** (1991) 201 and **B264** (1991) 462; *Z. Phys.* **C56** (1992) 181.
15. H. Avakian [HERMES], *Nucl. Phys. Proc. Suppl.* **79** (1999) 523.
16. A. Airapetian *et al.* [HERMES], *Phys. Rev. Lett.* **84** (2000) 4047; *Phys. Rev.* **D64** (2001) 097101.
17. A. Airapetian *et al.* [HERMES], *Phys. Lett.* **B562** (2003) 182.

18. H. Avakian *et al.* [CLAS], *Phys. Rev.* **D69** (2004) 112004;
E. Avetisyan, A. Rostomyan and A. Ivanilov [HERMES], hep-ex/0408002.
19. A. Airapetian *et al.* [HERMES], *Phys. Lett.* **B622** (2005) 14.
20. P.J. Mulders and R.D. Tangerman, *Nucl. Phys.* **B461** (1996) 197 [*Erratum-ibid.* **B484** (1997) 538].
21. A. Afanasev and C.E. Carlson, arXiv:hep-ph/0308163; A. Metz and M. Schlegel, *Eur. Phys. J.* **A22** (2004) 489; *Annalen Phys.* **13** (2004) 699; K. Goeke, A. Metz and M. Schlegel, *Phys. Lett.* **B618** (2005) 90.
22. M. Anselmino, M. Boglione and F. Murgia, *Phys. Lett.* **B362** (1995) 164.
23. M. Anselmino and F. Murgia, *Phys. Lett.* **B442** (1998) 470.
24. U. D'Alesio and F. Murgia, *Phys. Rev.* **D70** (2004) 074009.
25. M. Anselmino *et al.*, *Phys. Rev.* **D71** (2005) 014002;
B.Q. Ma, I. Schmidt and J.J. Yang, *Eur. Phys. J.* **C40** (2005) 63.
26. A.V. Efremov, K. Goeke, P. Schweitzer, *Phys. Lett.* **B522** (2001) 37 and **B544** (2002) 389E; *Eur. Phys. J.* **C24** (2002) 407; *Acta Phys. Polon.* **B33** (2002) 3755; P. Schweitzer and A. Bacchetta, *Nucl. Phys.* **A732** (2004) 106.
27. E. De Sanctis, W.D. Nowak and K.A. Oganessian, *Phys. Lett.* **B483** (2000) 69; K.A. Oganessian *et al.*, *Nucl. Phys.* **A689** (2001) 784. B.Q. Ma, I. Schmidt and J.J. Yang, *Phys. Rev.* **D66** (2002) 094001.
28. A.V. Efremov, K. Goeke and P. Schweitzer, *Phys. Lett.* **B568** (2003) 63.
29. A.V. Efremov, K. Goeke and P. Schweitzer, hep-ph/0412420.
30. A.V. Efremov, K. Goeke, S. Menzel, A. Metz and P. Schweitzer, *Phys. Lett.* **B612** (2005) 233.
31. M. Anselmino *et al.*, *Phys. Rev.* **D71** (2005) 074006 and hep-ph/0507181.
32. W. Vogelsang and F. Yuan, hep-ph/0507266.
33. J.C. Collins, A.V. Efremov, K. Goeke, S. Menzel, A. Metz and P. Schweitzer, hep-ph/0509076.
34. J.C. Collins, A.V. Efremov, K. Goeke, M. Grosse Perdekamp, S. Menzel, B. Meredith, A. Metz and P. Schweitzer, "Sivers effect in the RHIC experiment", in preparation.
35. A.V. Efremov, K. Goeke and P. Schweitzer, talk at SIR'05.
36. P.V. Pobylitsa, hep-ph/0301236.
37. M. Glück, E. Reya and A. Vogt, *Eur. Phys. J.* **C5** (1998) 461.
38. S. Kretzer, E. Leader and E. Christova, *Eur. Phys. J.* **C22** (2001) 269.
39. J.C. Collins, *Acta Phys. Polon.* **B34** (2003) 3103.
40. M. Arneodo *et al.* [European Muon Collaboration], *Z. Phys.* **C34** (1987) 277.
41. R.N. Cahn, *Phys. Lett.* **B78** (1978) 269.
42. A. Bacchetta *et al.*, *Phys. Rev. Lett.* **85** (2000) 712.
43. The COMPASS Collaboration, CERN/SPLC 96-14, SPSC/P 297.
44. F. Rathmann *et al.*, *Phys. Rev. Lett.* **94** (2005) 014801; hep-ex/0505054.
45. A.V. Efremov *et al.*, *Eur. Phys. J.* **C35** (2004) 207; and hep-ph/0412427.
M. Anselmino *et al.*, *Phys. Lett.* **B594** (2004) 97.
46. D. Boer, *Nucl. Phys.* **B603** (2001) 195 [arXiv:hep-ph/0102071].
47. M. Burkardt, *Phys. Rev.* **D66** (2002) 114005; *Nucl. Phys.* **A735** (2004) 185.
48. M. Anselmino *et al.*, "Comparing extractions Sivers functions", these proceedings.

HELICITY FORMALISM AND SPIN ASYMMETRIES IN HADRONIC PROCESSES

M. Anselmino, [a] M. Boglione, [a] U. D'Alesio, [b]
E. Leader, [c] S. Melis, [b] F. Murgia [b]

[a] *Dipartimento di Fisica Teorica, Università di Torino and INFN, Sezione di Torino, Via P. Giuria 1, 10125 Torino, Italy*
[b] *Dipartimento di Fisica, Università di Cagliari and INFN, Sezione di Cagliari, C.P. 170, 09042 Monserrato (CA), Italy*
[c] *Imperial College London, Prince Consort Road, London SW7 2BW, U.K.*

We present a generalized QCD factorization scheme for the high energy inclusive polarized process, $(A, S_A) + (B, S_B) \to C + X$, including all intrinsic partonic motions. This introduces many non-trivial azimuthal phases and several new spin and $\boldsymbol{k}_\perp$ dependent soft functions. The formal expressions for single and double spin asymmetries are discussed. Numerical results for $A_N(p^\uparrow\, p \to \pi\, X)$ are presented.

1. Introduction and formalism

Recently [1–3] we have developed an approach to study (un)polarized cross sections for inclusive particle production in hadronic collisions at high energy and moderately large p_T and semi-inclusive deeply inelastic scattering (SIDIS).[4] Assuming that factorization is preserved, this approach generalizes the usual Leading Order (LO), collinear perturbative QCD formalism by including spin and intrinsic transverse momentum, $\boldsymbol{k}_\perp$, effects both in the soft contributions (parton distribution (PDF) and fragmentation (FF) functions) and in the elementary processes. Helicity formalism is adopted and exact non collinear kinematics is fully taken into account. Unpolarized cross sections and transverse single spin asymmetries (SSA), with emphasis on the Sivers [5] and Collins [6] effects, were already discussed in Refs. 1 and 2. Here we report on the most complete case of unpolarized cross sections and single and double spin asymmetries for the process $(A, S_A) + (B, S_B) \to C + X$.[3] The cross section for this process can be given as a LO (factorized) convolution of all possible hard elementary QCD processes, $ab \to cd$, with soft, leading twist, spin and $\boldsymbol{k}_\perp$ dependent PDF and

FF (see Eq. (8) of Ref. 2):

$$\frac{E_C\, d\sigma^{(A,S_A)+(B,S_B)\to C+X}}{d^3\boldsymbol{p}_C} =$$

$$\sum_{a,b,c,d,\{\lambda\}} \int \frac{dx_a\, dx_b\, dz}{16\pi^2 x_a x_b z^2 s}\, d^2\boldsymbol{k}_{\perp a}\, d^2\boldsymbol{k}_{\perp b}\, d^3\boldsymbol{k}_{\perp C}\, \delta(\boldsymbol{k}_{\perp C}\cdot\hat{\boldsymbol{p}}_c)\, J(\boldsymbol{k}_{\perp C})$$
$$\times\, \rho^{a/A,S_A}_{\lambda_a,\lambda'_a}\, \hat{f}_{a/A,S_A}(x_a,\boldsymbol{k}_{\perp a})\, \rho^{b/B,S_B}_{\lambda_b,\lambda'_b}\, \hat{f}_{b/B,S_B}(x_b,\boldsymbol{k}_{\perp b}) \qquad (1)$$
$$\times\, \hat{M}_{\lambda_c,\lambda_d;\lambda_a,\lambda_b}\, \hat{M}^*_{\lambda'_c,\lambda_d;\lambda'_a,\lambda'_b}\, \delta(\hat{s}+\hat{t}+\hat{u})\, \hat{D}^{\lambda_C,\lambda_C}_{\lambda_c,\lambda'_c}(z,\boldsymbol{k}_{\perp C})\,,$$

where A, B are initial, spin 1/2 hadrons in pure spin states S_A and S_B; C is the unpolarized observed hadron; $J(\boldsymbol{k}_{\perp C})$ is a phase-space kinematical factor;[1] $\rho^{a/A,S_A}_{\lambda_a,\lambda'_a}\, \hat{f}_{a/A,S_A}(x_a,\boldsymbol{k}_{\perp a})$ contains all information on parton a and its polarization state, through its helicity density matrix and the spin and $\boldsymbol{k}_\perp$ dependent PDF (analogously for parton b); $\hat{M}_{\lambda_c,\lambda_d;\lambda_a,\lambda_b}$ are the LO helicity amplitudes for the elementary process $ab\to cd$; $D^{\lambda_C,\lambda_C}_{\lambda_c,\lambda'_c}(z,\boldsymbol{k}_{\perp C})$ is a product of soft helicity fragmentation amplitudes for the $c\to C+X$ process. The remaining notation, in particular for kinematical variables, should be obvious.[1,2] Let us stress here that a formal proof of factorization for the $AB\to C+X$ process in the non collinear case is still missing; universality and evolution properties of the new spin and $\boldsymbol{k}_\perp$ dependent PDF and FF are not established or well known yet; a consistent account of all higher-twist effects is still missing. In the sequel, we will discuss in more detail the basic ingredients of Eq. (1).

2. Spin and $\boldsymbol{k}_\perp$ dependent PDF and FF (leading twist)

The most general expression for the helicity density matrix of quark a inside hadron A with polarization state S_A is

$$\rho^{a/A,S_A}_{\lambda_a,\lambda'_a} = \frac{1}{2}\begin{pmatrix} 1+P^a_z & P^a_x - iP^a_y \\ P^a_x + iP^a_y & 1-P^a_z \end{pmatrix}_{A,S_A} = \frac{1}{2}\begin{pmatrix} 1+P^a_L & P^a_T\, e^{-i\phi_{s_a}} \\ P^a_T\, e^{i\phi_{s_a}} & 1-P^a_L \end{pmatrix}_{A,S_A}, \qquad (2)$$

where P^a_j is the j-th component of the quark polarization vector in its helicity frame. Introducing soft, nonperturbative helicity amplitudes for the inclusive process $A\to a+X$, $\hat{\mathcal{F}}_{\lambda_a,\lambda_{X_A};\lambda_A}$, we can write

$$\rho^{a/A,S_A}_{\lambda_a,\lambda'_a}\, \hat{f}_{a/A,S_A}(x_a,\boldsymbol{k}_{\perp a}) = \sum_{\lambda_A,\lambda'_A} \rho^{A,S_A}_{\lambda_A,\lambda'_A} \not\!\!\int_{X_A,\lambda_{X_A}} \hat{\mathcal{F}}_{\lambda_a,\lambda_{X_A};\lambda_A}\, \hat{\mathcal{F}}^*_{\lambda'_a,\lambda_{X_A};\lambda'_A}$$
$$\equiv \sum_{\lambda_A,\lambda'_A} \rho^{A,S_A}_{\lambda_A,\lambda'_A}\, \hat{F}^{\lambda_a,\lambda'_a}_{\lambda_A,\lambda'_A}\,, \qquad (3)$$

where $\rho^{A,S_A}_{\lambda_A,\lambda'_A}$ is in turn the helicity density matrix of hadron A

$$\rho^{A,S_A}_{\lambda_A,\lambda'_A} = \frac{1}{2}\begin{pmatrix} 1+P^A_Z & P^A_X - iP^A_Y \\ P^A_X + iP^A_Y & 1-P^A_Z \end{pmatrix} = \frac{1}{2}\begin{pmatrix} 1+P^A_L & P^A_T\, e^{-i\phi_{S_A}} \\ P^A_T\, e^{i\phi_{S_A}} & 1-P^A_L \end{pmatrix}, \quad (4)$$

and P^A_J is the J-th component of the hadron polarization vector in its helicity rest frame. The definition of $\hat{F}^{\lambda_a,\lambda'_a}_{\lambda_A,\lambda'_A}$ in terms of the helicity amplitudes $\hat{\mathcal{F}}_{\lambda_a,\lambda_{X_A};\lambda_A}$ can be deduced from Eq. (3). One can see that, due to rotational invariance and parity properties, the following relations hold:[7]

$$\begin{aligned} \hat{F}^{\lambda'_a,\lambda_a}_{\lambda'_A,\lambda_A} &= \left(\hat{F}^{\lambda_a,\lambda'_a}_{\lambda_A,\lambda'_A}\right)^*, \\ \hat{F}^{\lambda_a,\lambda'_a}_{\lambda_A,\lambda'_A}(x_a,\boldsymbol{k}_{\perp a}) &= F^{\lambda_a,\lambda'_a}_{\lambda_A,\lambda'_A}(x_a,k_{\perp a})\,\exp\left[\,i(\lambda_A-\lambda'_A)\phi_a\,\right], \\ F^{-\lambda_a,-\lambda'_a}_{-\lambda_A,-\lambda'_A} &= (-1)^{2(S_A-s_a)}\,(-1)^{(\lambda_A-\lambda_a)+(\lambda'_A-\lambda'_a)}\,F^{\lambda_a,\lambda'_a}_{\lambda_A,\lambda'_A}. \end{aligned} \quad (5)$$

Using Eq. (5) one can easily associate the eight functions

$$F^{++}_{++},\quad F^{++}_{--},\quad F^{+-}_{+-},\quad F^{+-}_{-+},\quad F^{++}_{+-},\quad F^{--}_{+-},\quad F^{+-}_{++},\quad F^{+-}_{--}, \quad (6)$$

to the eight leading twist, spin and $\boldsymbol{k}_\perp$ dependent PDF: $F^{++}_{++} \pm F^{++}_{--}$ are respectively related to the unpolarized and longitudinally polarized PDF, $f_{a/A}$ and $\Delta_L f_{a/A}$; $F^{+-}_{+-} \pm F^{+-}_{-+}$ are related with the two possible contributions to the transversity PDF; $F^{++}_{+-} \pm F^{--}_{+-}$ are respectively related to the probability of finding an unpolarized (longitudinally polarized) parton inside a transversely polarized hadron, the Sivers function,[5] $\Delta^N f_{a/A^\uparrow}$ (the $g^\perp_{1T}$ PDF); $F^{+-}_{++} \pm F^{+-}_{--}$ are related to the probability of finding a transversely polarized parton respectively inside an unpolarized hadron (the so-called Boer–Mulders function,[8] $\Delta^N f_{a^\uparrow/A}$) and inside a longitudinally polarized hadron, the $h^\perp_{1L}$ PDF. More precisely, the relations are the following:[3]

$$\begin{aligned} \hat{f}_{a/A} = \hat{f}_{a/A,S_L} &= \left(F^{++}_{++} + F^{++}_{--}\right) \\ \hat{f}_{a/A,S_T} &= \hat{f}_{a/A} + \tfrac{1}{2}\,\Delta\hat{f}_{a/S_T} = \left(F^{++}_{++} + F^{++}_{--}\right) + 2\,\mathrm{Im}F^{++}_{+-}\sin(\phi_{S_A}-\phi_a) \\ P^a_x\,\hat{f}_{a/A,S_L} &= \Delta\hat{f}_{s_x/S_L} = 2\,\mathrm{Re}F^{+-}_{++} \\ P^a_x\,\hat{f}_{a/A,S_T} &= \Delta\hat{f}_{s_x/S_T} = \left(F^{+-}_{+-} + F^{-+}_{+-}\right)\cos(\phi_{S_A}-\phi_a) \\ P^a_y\,\hat{f}_{a/A,S_L} &= P^a_y\,\hat{f}_{a/A} = \Delta\hat{f}_{s_y/S_L} = -2\,\mathrm{Im}F^{+-}_{++} \\ P^a_y\,\hat{f}_{a/A,S_T} &= \Delta\hat{f}_{s_y/S_T} = -2\,\mathrm{Im}F^{+-}_{++} + \left(F^{+-}_{+-} - F^{-+}_{+-}\right)\sin(\phi_{S_A}-\phi_a) \\ P^a_z\,\hat{f}_{a/A,S_L} &= \Delta\hat{f}_{s_z/S_L} = \left(F^{++}_{++} - F^{++}_{--}\right) \\ P^a_z\,\hat{f}_{a/A,S_T} &= \Delta\hat{f}_{s_z/S_T} = 2\,\mathrm{Re}F^{++}_{+-}\cos(\phi_{S_A}-\phi_a), \end{aligned} \quad (7)$$

where ϕ_{S_A} and ϕ_a are respectively the azimuthal angle of the hadron spin polarization vector and of the parton a transverse momentum, $\boldsymbol{k}_{\perp a}$, in the hadronic c.m. frame. We have also used the notation $\Delta\hat{f}_{s_i/S_J} = \hat{f}_{s_i/S_J} - \hat{f}_{-s_i/S_J}$. More details and relations with the notation of the Amsterdam group [8] can be found in Ref. 3. Since the helicity density matrix for a massless gluon can be formally written similarly to that of a quark,

$$\rho^{g/A,S_A}_{\lambda_g,\lambda'_g} = \frac{1}{2}\begin{pmatrix} 1+P^g_z & \mathcal{T}^g_1 - i\mathcal{T}^g_2 \\ \mathcal{T}^g_1 + i\mathcal{T}^g_2 & 1-P^g_z \end{pmatrix}_{A,S_A} = \frac{1}{2}\begin{pmatrix} 1+P^g_{\rm circ} & -P^g_{\rm lin}\, e^{-2i\phi} \\ -P^g_{\rm lin}\, e^{2i\phi} & 1-P^g_{\rm circ} \end{pmatrix}_{A,S_A}, \quad (8)$$

where $\mathcal{T}^g_1$ and $\mathcal{T}^g_2$ are related to the degree of linear polarization of the gluon, relations analogous to those shown for quarks hold also for gluons.[3]

Analogously, introducing soft nonperturbative helicity fragmentation amplitudes for the process $c \to C + X$, and limiting to the case of unpolarized hadron C, properties similar to those shown for PDF in Eq. (5) hold.[2,3] From these relations one can easily see that, for each parton, only two independent FF survive: the usual unpolarized FF; the well known Collins function [6] for quarks, $\Delta^N\hat{D}_{C/q^\uparrow}(z, k_{\perp C})$, and a Collins-like function for (linearly) polarized gluons, $\Delta^N\hat{D}_{C/\mathcal{T}^g_1}(z, k_{\perp C})$.[3]

3. Helicity amplitudes for the elementary process *ab* → *cd*

Since intrinsic parton motions are fully taken into account in our approach, all soft processes, $A(B) \to a(b) + X$ and $c \to C + X$, and the elementary process $ab \to cd$ take place out of the hadronic production plane (the $XZ_{\rm cm}$ plane). The relation between the elementary helicity amplitudes given in the hadronic c.m. frame, $\hat{M}_{\lambda_c,\lambda_d;\lambda_a,\lambda_b}$, and those given in the canonical partonic c.m. frame (no azimuthal phases), $\hat{M}^0_{\lambda_c,\lambda_d;\lambda_a,\lambda_b}$, has been given in Ref. 2. In summary,

$$\hat{M}_{\lambda_c,\lambda_d;\lambda_a,\lambda_b} = \hat{M}^0_{\lambda_c,\lambda_d;\lambda_a,\lambda_b}\, e^{-i(\lambda_a\xi_a+\lambda_b\xi_b-\lambda_c\xi_c-\lambda_d\xi_d)} \times e^{-i[(\lambda_a-\lambda_b)\tilde{\xi}_a-(\lambda_c-\lambda_d)\tilde{\xi}_c]}\, e^{i(\lambda_a-\lambda_b)\phi''_c}, \quad (9)$$

where ξ_j, $\tilde{\xi}_j$ $(j = a, b, c, d)$, ϕ''_c are phases which depend on the initial kinematical configuration in the hadronic c.m. frame.[2] Parity properties for the canonical helicity amplitudes $\hat{M}^0$ are well known,[7] and so are the relations between a given canonical helicity amplitude and those obtained by exchanging the two initial (final) partons.[7] For massless partons there are only three independent helicity amplitudes, $\hat{M}_{++;++} = \hat{M}^0_1 \exp(i\varphi_1)$, $\hat{M}_{-+;-+} = \hat{M}^0_2 \exp(i\varphi_2)$, $\hat{M}_{-+;+-} = \hat{M}^0_3 \exp(i\varphi_3)$, where φ_j $(j = 1, 2, 3)$

are the corresponding phases given in Eq. (9). At LO there are eight elementary contributions $ab \to cd$ which must be considered separately, since they involve different combinations of PDF and FF in Eq. (1): $q_a q_b \to q_c q_d$, $g_a g_b \to g_c g_d$, $qg \to qg$, $gq \to gq$, $qg \to gq$, $gq \to qg$, $g_a g_b \to q\bar{q}$, $q\bar{q} \to g_c g_d$ (the first contribution includes all quark and antiquark cases).

4. Kernels for the process $A(S_A) + B(S_B) \to C + X$

In the previous sections we have presented all the ingredients required for the evaluation of the convolution integral for the double polarized cross section, Eq. (1). We can then derive the expression of all the physically relevant single and double spin asymmetries for the process $A(S_A)+B(S_B) \to C+X$, and, with appropriate modifications, for other inclusive production processes. Defining kernels as:

$$\Sigma(S_A, S_B)^{ab\to cd} = \sum_{\{\lambda\}} \rho^{a/A,S_A}_{\lambda_a,\lambda'_a} \hat{f}_{a/A,S_A}(x_a, \boldsymbol{k}_{\perp a})\, \rho^{b/B,S_B}_{\lambda_b,\lambda'_b} \hat{f}_{b/B,S_B}(x_b, \boldsymbol{k}_{\perp b}) \times \hat{M}_{\lambda_c,\lambda_d;\lambda_a,\lambda_b} \hat{M}^*_{\lambda'_c,\lambda_d;\lambda'_a,\lambda'_b} \hat{D}^{\lambda_C,\lambda_C}_{\lambda_c,\lambda'_c}(z, \boldsymbol{k}_{\perp C}), \quad (10)$$

we present here, as an example, the kernel for the $q_a q_b \to q_c q_d$ process:

$$\begin{aligned}\Sigma(S_A, S_B)^{q_a q_b \to q_c q_d} &= \tfrac{1}{2}\, \hat{D}_{C/c}(z, k_{\perp C})\, \hat{f}_{a/S_A}(x_a, \boldsymbol{k}_{\perp a})\, \hat{f}_{b/S_B}(x_b, \boldsymbol{k}_{\perp b}) \\ &\times \Big\{ \left(|\hat{M}_1^0|^2 + |\hat{M}_2^0|^2 + |\hat{M}_3^0|^2\right) + P_z^a\, P_z^b \left(|\hat{M}_1^0|^2 - |\hat{M}_2^0|^2 - |\hat{M}_3^0|^2\right) \\ &+ \hat{M}_2^0\, \hat{M}_3^0 \Big[\left(P_x^a P_x^b + P_y^a P_y^b\right) \cos(\varphi_3 - \varphi_2) - \left(P_x^a P_y^b - P_y^a P_x^b\right) \sin(\varphi_3 - \varphi_2) \Big] \Big\} \\ &- \tfrac{1}{2}\, \Delta^N \hat{D}_{C/c^\uparrow}(z, k_{\perp C})\, \hat{f}_{a/S_A}(x_a, \boldsymbol{k}_{\perp a})\, \hat{f}_{b/S_B}(x_b, \boldsymbol{k}_{\perp b}) \qquad (11) \\ &\times \Big\{ \hat{M}_1^0\, \hat{M}_2^0 \left[P_x^a \sin(\varphi_1 - \varphi_2 + \phi_C^H) - P_y^a \cos(\varphi_1 - \varphi_2 + \phi_C^H) \right] \\ &+ \hat{M}_1^0\, \hat{M}_3^0 \left[P_x^b \sin(\varphi_1 - \varphi_3 + \phi_C^H) - P_y^b \cos(\varphi_1 - \varphi_3 + \phi_C^H) \right] \Big\},\end{aligned}$$

where ϕ_C^H is the azimuthal angle of hadron C three-momentum in the helicity frame of parton c. A more complete list of kernels for all partonic contributions and more details can be found in Ref. 3.

5. Cross section and SSA for the process $pp \to \pi + X$

The formalism described in the previous sections is very general and can be applied to the calculation of unpolarized cross sections, single and double spin asymmetries for inclusive particle production in hadronic collisions. As an explicit example, we now discuss the transverse single-spin asymmetry for inclusive pion production in proton–proton collisions, $A_N(pp \to \pi + X) = (d\sigma^\uparrow - d\sigma^\downarrow)/(d\sigma^\uparrow + d\sigma^\downarrow)$, where $d\sigma^{\uparrow,\downarrow}$ stands for the cross section of Eq. (1) with $S_A = \uparrow, \downarrow$, $S_B = 0$. We present, limiting ourselves to the case $q_a q_b \to q_c q_d$, the combination of kernels appearing in the numerator and denominator of the SSA (dependence on $x_{a,b}$, $\boldsymbol{k}_{\perp a,b}$ in PDF and on z, $\boldsymbol{k}_{\perp C}$ in FF is understood):

$$
\begin{aligned}
&[\Sigma(\uparrow,0) - \Sigma(\downarrow,0)]^{q_a q_b \to q_c q_d} = \\
&\tfrac{1}{2}\,\Delta \hat{f}_{a/A^\uparrow}\,\hat{f}_{b/B}\left[|\hat{M}_1^0|^2 + |\hat{M}_2^0|^2 + |\hat{M}_3^0|^2\right]\hat{D}_{C/c} \\
&+ 2\left[\Delta^- \hat{f}^a_{s_y/\uparrow}\cos(\varphi_3 - \varphi_2) - \Delta \hat{f}^a_{s_x/\uparrow}\sin(\varphi_3 - \varphi_2)\right]\Delta \hat{f}^b_{s_y/B}\,\hat{M}_2^0\,\hat{M}_3^0\,\hat{D}_{C/c} \\
&+ \left[\Delta^- \hat{f}^a_{s_y/\uparrow}\cos(\varphi_1 - \varphi_2 + \phi_C^H) - \Delta \hat{f}^a_{s_x/\uparrow}\sin(\varphi_1 - \varphi_2 + \phi_C^H)\right] \\
&\quad \times \hat{f}_{b/B}\,\hat{M}_1^0\,\hat{M}_2^0\,\Delta^N \hat{D}_{C/c^\uparrow} \\
&+ \tfrac{1}{2}\,\Delta \hat{f}_{a/A^\uparrow}\,\Delta \hat{f}^b_{s_y/B}\cos(\varphi_1 - \varphi_3 + \phi_C^H)\,\hat{M}_1^0\,\hat{M}_3^0\,\Delta^N \hat{D}_{C/c^\uparrow}\,,
\end{aligned}
\tag{12}
$$

$$
\begin{aligned}
&[\Sigma(\uparrow,0) + \Sigma(\downarrow,0)]^{q_a q_b \to q_c q_d} = \\
&\qquad \hat{f}_{a/A}\,\hat{f}_{b/B}\left[|\hat{M}_1^0|^2 + |\hat{M}_2^0|^2 + |\hat{M}_3^0|^2\right]\hat{D}_{C/c} \\
&\qquad + 2\,\Delta \hat{f}^a_{s_y/A}\,\Delta \hat{f}^b_{s_y/B}\cos(\varphi_3 - \varphi_2)\,\hat{M}_2^0\,\hat{M}_3^0\,\hat{D}_{C/c} \\
&\qquad + \left[\hat{f}_{a/A}\,\Delta \hat{f}^b_{s_y/B}\cos(\varphi_1 - \varphi_3 + \phi_C^H)\,\hat{M}_1^0\,\hat{M}_3^0\right. \\
&\qquad \left. + \,\Delta \hat{f}^a_{s_y/A}\,\hat{f}_{b/B}\cos(\varphi_1 - \varphi_2 + \phi_C^H)\,\hat{M}_1^0\hat{M}_2^0\right]\Delta^N \hat{D}_{C/c^\uparrow}\,.
\end{aligned}
\tag{13}
$$

There are four terms contributing to the numerator of the SSA, Eq. (12): the Sivers contribution (2nd line); the transversity⊗Boer–Mulders contribution (3rd line); the transversity⊗Collins contribution (4th and 5th lines); the Sivers⊗Boer–Mulders⊗Collins contribution (last line). Similarly, there are three terms contributing to the denominator of the SSA, Eq. (13): the usual term involving only unpolarized quantities (2nd line); the Boer–Mulders⊗Boer–Mulders contribution (3rd line); the Boer–Mulders⊗Collins contribution (last two lines). Similar considerations apply to all other partonic contributions. Whenever gluons are involved, functions analogous to the transversity, Boer–Mulders, Collins functions for quarks, but describing linearly polarized gluons appear, see Ref. 3. Let us now present some numerical results on unpolarized cross sections and SSA for the $pp \to \pi + X$ process. Our aim here is basically that of showing the (maximized) possible contributions of all terms appearing in Eqs. (12) and (13), which involve

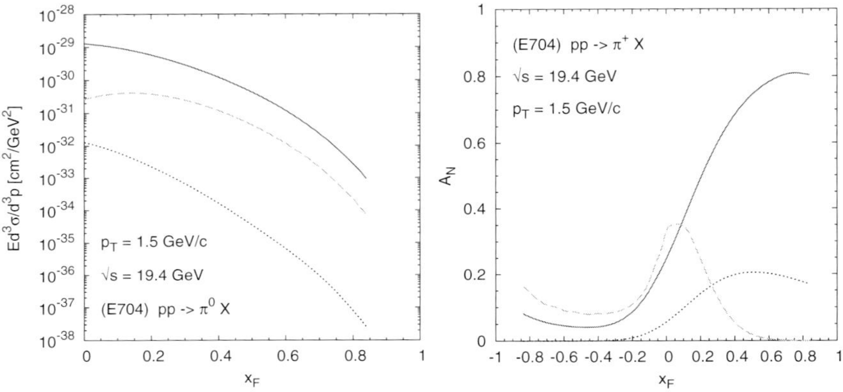

Fig. 1. LEFT: (maximized) contributions to the unp. cross section for $pp \to \pi^0 + X$ process and E704 kinematics; solid line: usual contribution; dashed line: Boer–Mulders⊗Collins; dotted line: Boer–Mulders⊗Boer–Mulders; RIGHT: (maximized) contributions to $A_N(p^\uparrow p \to \pi^+ + X)$ for E704 kinematics; solid line: quark Sivers contribution; dashed line: gluon Sivers; dotted line: transversity⊗Collins; all other contributions are much smaller.

several unknown or poorly known functions. To this end we saturate known positivity bounds for the Collins and Sivers functions, replacing all other $k_\perp$ dependent polarized PDF with the corresponding unpolarized ones (keeping trace of azimuthal phases); we sum all possible contributions with the same sign; for all PDF we assume an x and flavour independent gaussian shape vs. $k_\perp$, taking $\langle k_\perp \rangle = 0.8\,\mathrm{GeV}/c$, while for FF $\langle k_{\perp C}(z) \rangle$ is taken as in Ref. 1; for the unpolarized, $\boldsymbol{k}_\perp$-integrated PDF and FF, we take respectively the MRST01 [9] and the KKP [10] sets. See Ref. 1 for all other details on numerical calculations. In Fig. 1 (left) we show the maximized contributions to the unpolarized differential cross section for the $pp \to \pi^0 + X$ process for the kinematical regime of the E704 data on SSA. The usual contribution clearly dominates, the other two being suppressed by the azimuthal phases. In Fig. 1 (right) we show the contributions to the SSA, $A_N(p^\uparrow p \to \pi^+ + X)$, in the same kinematical regime (including the negative x_F region). Full $\boldsymbol{k}_\perp$ treatment and azimuthal phases considerably suppress the Collins effect;[2] the same is not true for Sivers contribution. In Fig. 2 (left) we plot $A_N(p^\uparrow p \to \pi^0 + X)$ for the kinematical regime of the STAR experiment at RHIC. In Fig. 2 (right) we show $A_N(p^\uparrow \bar{p} \to \pi^+ + X)$, in the kinematical regime of the proposed PAX experiment at GSI. These last results are particularly interesting for the gluon Sivers function. Many other applications of our formalism are possible and some of them are un-

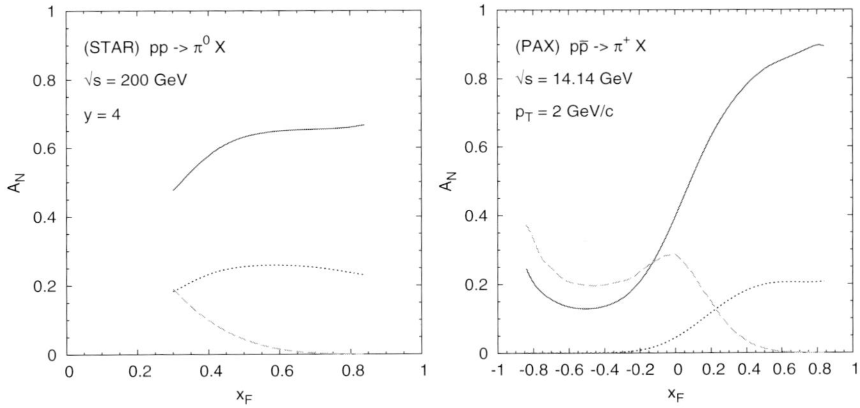

Fig. 2. (Maximized) contributions to: (left) $A_N(p^\uparrow p \to \pi^0 + X)$ for STAR kinematics (at $x_F < 0$ all contributions are vanishingly small); (right) $A_N(p^\uparrow \bar{p} \to \pi^+ + X)$ for PAX kinematics; lines are as in Fig. 1.

der investigation, like the study of the double longitudinal asymmetry, A_{LL}, for pion production; the transverse Λ polarization in unpolarized hadronic reactions; the study of the Collins effect in polarized SIDIS; the Drell–Yan process; inclusive particle production in pion–proton collisions and so on. Hopefully, this thorough phenomenological analysis of present and forthcoming experimental results on spin asymmetries will help in clarifying the role of spin effects in high energy hadronic reactions.

References

1. U. D'Alesio and F. Murgia, *Phys. Rev.* **D70** (2004) 074009.
2. M. Anselmino, M. Boglione, U. D'Alesio, E. Leader and F. Murgia, *Phys. Rev.* **D71** (2005) 014002.
3. M. Anselmino, M. Boglione, U. D'Alesio, E. Leader, S. Melis and F. Murgia, hep-ph/0509035.
4. M. Anselmino, M. Boglione, U. D'Alesio, A. Kotzinian, F. Murgia and A. Prokudin, *Phys. Rev.* **D71** (2005) 074006; hep-ph/0507181, *Phys. Rev.* **D**, in press.
5. D. Sivers, *Phys. Rev.* **D41** (1990) 83; **D43** (1991) 261.
6. J.C. Collins, *Nucl. Phys.* **B396** (1993) 161.
7. E. Leader, *Spin in Particle Physics*, Cambridge University Press, 2001.
8. D. Boer and P.J. Mulders, *Phys. Rev.* **D57** (1998) 5780.
9. A.D. Martin, R.G. Roberts, W.J. Stirling and R.S. Thorne, *Phys. Lett.* **B531** (2002) 216.
10. B.A. Kniehl, G. Kramer, and B. Pötter, *Nucl. Phys.* **B582** (2000) 514.

INCLUDING CAHN AND SIVERS EFFECTS INTO EVENT GENERATORS

A. Kotzinian

Dipartimento di Fisica Generale, Università di Torino;
INFN, Sezione di Torino, Via P. Giuria 1, I-10125 Torino, Italy;
Yerevan Physics Institute, 375036 Yerevan, Armenia;
and JINR, 141980 Dubna, Russia
E-mail: aram.kotzinian@cern.ch

It is demonstrated that event generators LEPTO and PYTHIA can be modified to describe some azimuthal modulations. The comparisons of results obtained with modified LEPTO with existing data in the current fragmentation region of SIDIS are presented for Cahn and Sivers effects as well as the predictions for the target fragmentation region. The predictions for Sivers effect in the Drell–Yan process obtained with modified PYTHIA are also presented. The concept of hadronization function is discussed.

1. Introduction

The spin (in)dependent azimuthal asymmetries arising in high energy reactions allow us to study the dynamical effects related to spin and transverse momentum of partons in target nucleon and in hadronization.

The Drell–Yan lepton pair production is the simplest process which does not include the hadronization dynamics and within QCD can be described using as a nonperturbative input only the parton distribution functions[a]:

$$d\sigma^{h_1+h_2\to\ell^++\ell^-+X} \sim \sum_q (f_{\bar{q}/h_1} f_{q/h_2} + 1 \leftrightarrow 2) \otimes d\hat{\sigma}^{q+\bar{q}\to\ell^++\ell^-}. \quad (1)$$

The predictions for the Sivers[2] effect for this process recently have been presented in Refs. 3 and 4. Here it will be demonstrated that similar results are obtained using modified PYTHIA event generator.

More nonperturbative inputs are needed to describe the semi-inclusive DIS (SIDIS) processes within the QCD formalism. Namely, for the particles

[a]In the following the notations of Ref. 1 are used.

produced in the current fragmentation region (CFR) of SIDIS one needs to introduce fragmentation function, $D_q^h(z)$:

$$d\sigma^{\ell p\to\ell hX} \sim \sum_q f_q \otimes d\hat{\sigma}^{\ell q\to\ell q} \otimes D_q^h. \tag{2}$$

For the particles produced in the target fragmentation region (TFR) other nonperturbative input — the fracture functions, $M_{q/N}^h(x,z)$, are needed:

$$d\sigma^{\ell p\to\ell hX} \sim \sum_q d\hat{\sigma}^{\ell q\to\ell q} \otimes M_{q/N}^h. \tag{3}$$

In practice it is not easy to separate the two regions at moderate beam energies and final hadronic state invariant masses.[5] An alternative approach which is able to describe the particles production in the whole phase space is based on the LUND string fragmentation model and adopted in the Monte Carlo event generators.[6,7] Here the SIDIS cross section can be represented as [8]

$$d\sigma^{\ell p\to\ell hX} \sim \sum_q f_q \otimes d\hat{\sigma}^{\ell q\to\ell q} \otimes H_{q/N}^h, \tag{4}$$

where the hadronization function, $H_{q/N}^h$, describes the particle production from the system formed by the struck quark and target remnant.

In Refs. 9 and 1 the role of parton intrinsic motion in SIDIS processes in CFR within QCD parton model has been considered at leading order; intrinsic $\mathbf{k}_\perp$ is fully taken into account in quark distribution functions and in the elementary processes as well as the hadron transverse momentum, $\mathbf{p}_\perp$, with respect to fragmenting quark momentum.

The average values of $k_\perp$ for quarks inside protons and $p_\perp$ for final hadrons inside the fragmenting quark jet were fixed by a comparison with data on Cahn effect [10] – the dependence of the unpolarized cross section on the azimuthal angle between the leptonic and the hadronic planes. The single-spin asymmetry (SSA) $A_{UT}^{\sin(\phi_\pi-\phi_S)}$ recently observed by HERMES [11] and COMPASS [12] Collaborations was successfully described by Sivers mechanism.

Here it will be demonstrated that both Cahn and Sivers effects can be implemented [13] into Monte Carlo event generators.

2. Including Cahn effect in LEPTO

In the simplest case, corresponding to LO approximation of parton model, event generation in LEPTO proceeds in several steps:

(1) the hard scattering kinematics is generated,
(2) the active quark inside the nucleon is chosen according to the quark density function $f_q(x, Q^2)$,
(3) the transverse momentum of the final quark is simulated with Gaussian $k_\perp$ and flat φ distributions. Note that the transverse momentum of the final quark is equal to that of initial quark for leading order hard subprocess.
(4) the string fragmentation machinery of `JETSET` program [7] is applied to form the final hadrons.

The Cahn effect [10] is a kinematical effect arising due to the presence of nonzero intrinsic transverse momentum of quarks in the nucleon. In the general case of non collinear kinematics Mandelstam variables depend on the quark transverse momentum and its azimuthal angle and at order $\mathcal{O}(k_\perp/Q)$ one has

$$d\hat{\sigma}^{\ell q\to\ell q} \propto 1 - \frac{(2+y)\sqrt{1-y}}{1+(1-y)^2}\frac{k_\perp}{Q}\cos\varphi. \tag{5}$$

Eq. (5) shows that the azimuthal angle of the final quark (and of the string's end associated with the struck quark) is now modulated with amplitude depending on y, Q and $k_\perp$.

This effect can be introduced in the `LEPTO` event generator at the step (3) of the event generation, when the transverse momentum and azimuthal angle of the scattered quark are generated. To do this the generation of the quark transverse momentum, $k_\perp$, is left unchanged and then the azimuthal angle is generated according to Eq. (5). This leads to azimuthal modulation of the string axis. The momentum conservation means that the transverse momentum of the quark is balanced by that of the target remnant, which in turn means that the azimuthal angle of the target remnant $\varphi_{qq} = \varphi + \pi$. Hence, one expects that the azimuthal angle of the hadrons in the target fragmentation region (TFR), $x_F < 0$, will be modulated with a phase shifted by π with respect to that in CFR.

Data on azimuthal dependencies of SIDIS covering a large x_F range have been obtained by the EMC Collaboration [14] for a beam energy of 280 GeV. The x_F dependence of $\langle\cos\phi_h\rangle/w_1(y)$, where $w_1(y) = (2-y)\sqrt{1-y}/\left(1+(1-y)^2\right)$, obtained by using modified `LEPTO` for EMC kinematics are presented in Fig. 1 together with data points from Ref. 14 (left panel). The simulation has been done with LO setting of `LEPTO` (LST(8)=0) and with values of the parameters describing intrinsic k_T (PARL(3)=0.5) and fragmentation p_T (PARL(21)=0.45) as adopted in Ref. 1.

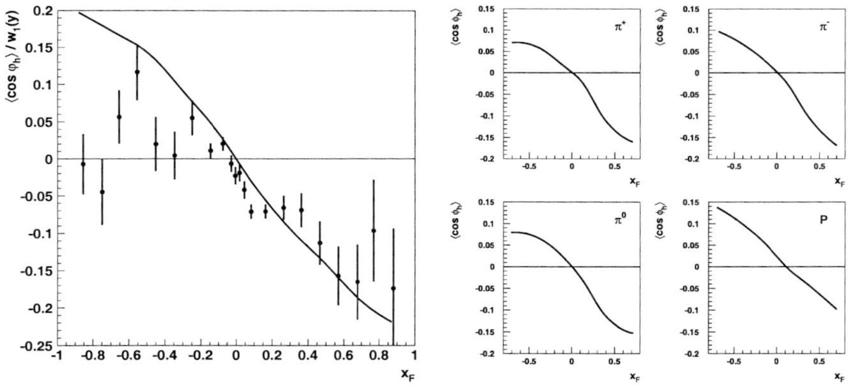

Fig. 1. Left: The x_F dependence of $\langle\cos\phi_h\rangle/w_1(y)$ for charged hadrons compared with EMC data. Right: Predictions of modified LEPTO for x_F dependence of $\langle\cos\phi_h\rangle$ for different hadrons produced in 12 GeV unpolarized SIDIS process.

The predictions of modified LEPTO for $\langle\cos\phi_h\rangle$ of different hadron (π^+, π^-, π^0 and p) produced in SIDIS on a proton target at future CEBAF 12 GeV facility at JLab[15] are presented in Fig. 1 (right panel). One can see from Fig. 1 and that the predicted mean value of $\cos\phi_h$ in the CFR is negative $\langle\cos\phi_h\rangle_{rmCFR} < 0$, while in the TFR is positive $\langle\cos\phi_h\rangle_{\rm TFR} > 0$, as suggested by arguments based on transverse momentum conservation.

3. Including Sivers effect in LEPTO

The azimuthal modulation of the string transverse momentum in the previous section was due to Cahn effect – the dependence of the non planar lepton–quark scattering cross section on the quark azimuth. The quark distribution, $f_q(x, k_\perp)$ itself is independent of quark azimuthal angle.

The situation is different when one considers SIDIS on a transversely polarized nucleon. Now a correlation between transverse momentum of quark, $\boldsymbol{k}_\perp$ and target transverse polarization $\boldsymbol{S}_T$ of type $\boldsymbol{S}_T \cdot [\hat{\boldsymbol{P}} \times \hat{\boldsymbol{k}}_\perp]$ is possible – the so-called Sivers effect.[2]

The unpolarized quark (and gluon) distributions inside a transversely polarized proton can be written as:

$$f_{q/p^\uparrow}(x, \boldsymbol{k}_\perp) = f_{q/p}(x, k_\perp) + \frac{1}{2}\,\Delta^N f_{q/p^\uparrow}(x, k_\perp)\, \boldsymbol{S}_T \cdot (\hat{\boldsymbol{P}} \times \hat{\boldsymbol{k}}_\perp)\,. \qquad (6)$$

Eq. (6) implies

$$\begin{aligned} f_{q/p^\uparrow}(x, \boldsymbol{k}_\perp) + f_{q/p^\downarrow}(x, \boldsymbol{k}_\perp) &= 2 f_{q/p}(x, k_\perp)\,, \\ f_{q/p^\uparrow}(x, \boldsymbol{k}_\perp) - f_{q/p^\downarrow}(x, \boldsymbol{k}_\perp) &= \Delta^N f_{q/p^\uparrow}(x, k_\perp)\, \boldsymbol{S}_T \cdot (\hat{\boldsymbol{P}} \times \hat{\boldsymbol{k}}_\perp)\,, \end{aligned} \qquad (7)$$

where $f_{q/p}(x, k_\perp)$ is the unpolarized parton density and $\Delta^N f_{q/p^\uparrow}(x, k_\perp)$ is referred to as the Sivers function. Notice that, as requested by parity invariance, the scalar quantity $\boldsymbol{S}_{_T} \cdot (\hat{\boldsymbol{P}} \times \hat{\boldsymbol{k}}_\perp)$ singles out the polarization component perpendicular to the $\boldsymbol{P} - \boldsymbol{k}_\perp$ plane. For a proton moving along $-z$ and a generic transverse polarization vector $\boldsymbol{S}_{_T} = |\boldsymbol{S}_{_T}|\,(\cos\phi_S, \sin\phi_S, 0)$ one has:

$$\boldsymbol{S}_{_T} \cdot (\hat{\boldsymbol{P}} \times \hat{\boldsymbol{k}}_\perp) = |\boldsymbol{S}_{_T}|\,\sin(\varphi - \phi_S) \equiv |\boldsymbol{S}_{_T}|\,\sin\phi_{\rm Siv}\,, \tag{8}$$

where $(\varphi - \phi_S) = \phi_{\rm Siv}$ is the Sivers angle.

The Sivers function for each light quark flavor $q = u, d$ are parametrized in the following factorized form:[1]

$$\Delta^N f_{q/p^\uparrow}(x, k_\perp) = 2\,\mathcal{N}_q(x)\,h(k_\perp)\,f_{q/p}(x, k_\perp)\,, \tag{9}$$

where

$$\mathcal{N}_q(x) = N_q\,x^{a_q}(1-x)^{b_q}\,\frac{(a_q+b_q)^{(a_q+b_q)}}{a_q^{a_q}b_q^{b_q}}\,, \tag{10}$$

$$h(k_\perp) = \sqrt{2e}\,\frac{k_\perp}{M}\,e^{-k_\perp^2/M^2}\,, \tag{11}$$

where N_q, a_q, b_q and M (GeV/c) are parameters. Then Eq. (6)can be rewritten as

$$f_{q/p^\uparrow}(x, \boldsymbol{k}_\perp) = f_{q/p}(x, k_\perp)\,[1 + |\boldsymbol{S}_{_T}|\,\mathcal{N}_q(x)\,h(k_\perp)\,\sin\phi_{\rm Siv}]. \tag{12}$$

Again, the Sivers effect is incorporated into `LEPTO` at the stage 3) of the event generation in the same way as for the Cahn effect but now the azimuthal angle is generated according to Eq. (12). For simulations the following set of parameters compatible with those obtained in Refs. 1 and 3 have been used: $N_u = N_{\bar{u}} = 0.5$, $N_d = N_{\bar{d}} = -0.2$, $a_q = 0.3$, $b_q = 2$ and $M^2 = 0.36\,({\rm GeV}/c)^2$.

In Fig. 2 the results of simulation for HERMES experimental conditions are compared with observed Sivers asymmetries [11] (left panel).

Future facilities as Electron Ion Colliders or upgraded JLab will have larger kinematic coverage and will offer the possibility of studying the Sivers effect also with hadrons produced in the TFR. As an example, the simulations have been done for 12 GeV electron SIDIS of a proton target. The DIS cut $Q^2 > 1\,({\rm GeV}/c)^2$ and $W^2 > 4\,{\rm GeV}^2$ and a cut on the produced hadron transverse momentum $P_T > 0.05\,{\rm GeV}/c$ was imposed. The predictions for x_F dependence for JLab kinematics is presented in Fig. 2 (right panel). The x and P_T dependencies in the TFR are presented Fig. 3.

In Fig. 4 the results for Sivers asymmetry in the Drell–Yan process obtained with modified `PYTHIA` generator are presented for two different

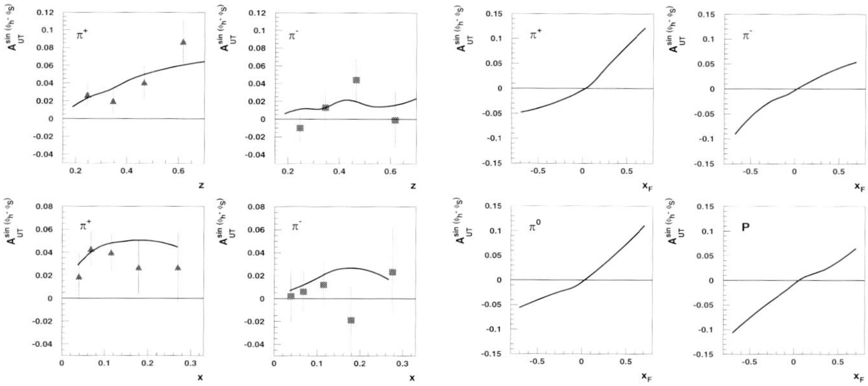

Fig. 2. Left: HERMES data on $A_{UT}^{\sin(\phi_\pi - \phi_S)}$ for scattering off a transversely polarized proton target. The curves are the results of simulations obtained with modified `LEPTO`; Right: Predicted dependence of $A_{UT}^{\sin(\varphi_h - \varphi_S)}$ on x_F for different hadrons produced in SIDIS of 12 GeV electrons off a transversely polarized proton target.

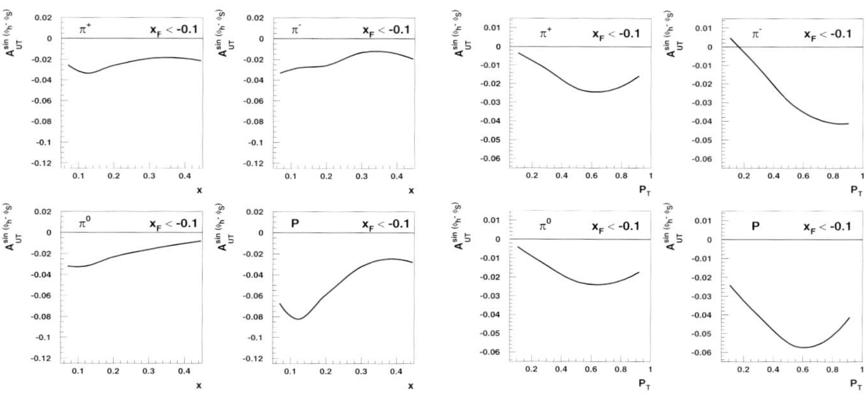

Fig. 3. Predicted dependence of $A_{UT}^{\sin(\varphi_h - \varphi_S)}$ on x, left panel, (p_T, right panel) for different hadrons produced in the TFR ($x_F < -0.1$) of SIDIS of 12 GeV electrons off a transversely polarized proton target.

energies: planned GSI $\bar{p}p$ collider with $\sqrt{s} = 14.4\,\text{GeV}$ (left panel) and RHIC $\sqrt{s} = 200\,\text{GeV}$ (right panel). The results[b] are similar to that obtained in Refs. 3 and 4.

[b]Note, that here the sign convention of Ref. 4 is adopted.

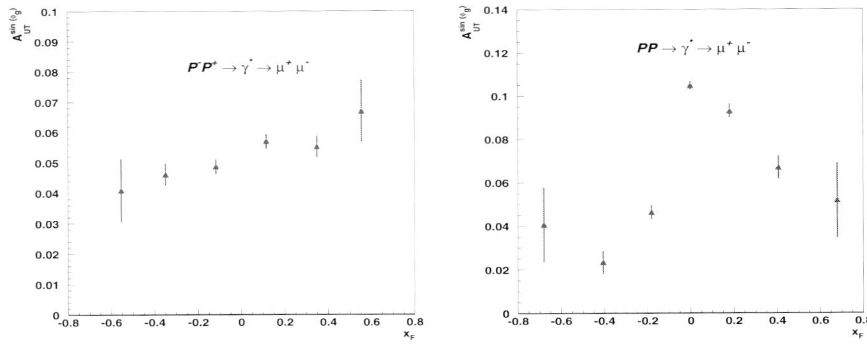

Fig. 4. Predicted dependence of A_N for Drell–Yan process on x_F. Left: $\sqrt{s} = 14.4\,\mathrm{GeV}$, right: $\sqrt{s} = 200\,\mathrm{GeV}$

4. Discussion and Conclusions

The advantage of this MC based approach compared to standard QCD factorized approach is the full coverage of produced hadron phase space. Figs. 1 and 2 demonstrate that the modified LEPTO event generator well describes the data in the CFR both for Cahn and Sivers asymmetries. One can notice in Fig. 1 that the integrated experimental value of $\langle \cos\phi_h \rangle$ for charged hadrons in the CFR is not compensated by that in TFR. It seems improbable that this imbalance can be compensated by larger values of $\langle \cos\phi_h \rangle$ of neutral hadrons at $x_F \simeq -1$.

Note, that in present approach a possible modifications of hadronization in the case of polarized target have been ignored. In Ref. 8 it was shown that the hadronization functions in principle depends on polarization states of struck quark and target nucleon even for production of (pseudo)scalar or unpolarized particles. This dependence cannot be neglected at moderate energies and new nonperturbative input — the polarized hadronization functions, $\Delta H^h_{q/N}$, are needed to describe the polarized SIDIS. The expression for the SIDIS helicity asymmetry is then looks as:

$$A_1^h(x,z,Q^2) = \frac{\sum_q e_q^2\, q(x,Q^2) H^h_{q/N}(x,z,Q^2)\left[\frac{\Delta q(x,Q^2)}{q(x,Q^2)} + \frac{\Delta H^h_{q/N}(x,z,Q^2)}{H^h_{q/N}(x,z,Q^2)}\right]}{\sum_q e_q^2\, q(x,Q^2)\, H^h_{q/N}(x,z,Q^2)\left[1 + \frac{\Delta q(x,Q^2)\,\Delta H^h_{q/N}(x,z,Q^2)}{q(x,Q^2)\,H^h_{q/N}(x,z,Q^2)}\right]}. \tag{13}$$

Since the hadronization functions depend on target nucleon, active quark and produced hadron variables the new correlations as (**a**): $\boldsymbol{S}_L \cdot [\mathbf{p}_\perp \times \hat{\boldsymbol{k}}_\perp]$, (**b**): $\boldsymbol{s}_L \cdot [\mathbf{p}_\perp \times \hat{\boldsymbol{k}}_\perp]$, (**c**): $[\boldsymbol{S}_T \times \mathbf{p}_\perp] \cdot [\mathbf{p}^h_T \times \hat{\mathbf{s}}_T]$ *etc.* are possible. This correlations cannot be present separately in the distribution and fragmentation functions. They will induce the azimuthal asymmetries in the SIDIS of (**a**): unpolarized lepton off longitudinally polarized target, (**b**): longitudinally polarized lepton off unpolarized target, (**c**): unpolarized lepton off transversely polarized target *etc.*

The new high statistic measurements in both CFR and TFR of SIDIS will allow to check the predictions of the approach presented here and better understand the effects of the quark intrinsic transverse momentum and hadronization mechanism in SIDIS. The study of single spin asymmetries in Drell–Yan process will provide an additional test of our understanding od spin dependent phenomena.

Acknowledgements

The author expresses his gratitude to M. Anselmino, A. Prokudin for discussions and to P. Ratcliffe for the kind invitation to the Transversity 2005 workshop.

References

1. M. Anselmino *et al.*, *Phys. Rev.* **D72** (2005) 054028, arXiv:hep-ph/0501196.
2. D. Sivers, *Phys. Rev.* **D41** (1990) 83; **D43** (1991) 261.
3. M. Anselmino *et al.*, arXiv:hep-ph/0507181.
4. W. Vogelsang and F. Yuan, *Phys. Rev.* **D71** (2005) 074006, arXiv:hep-ph/0507266.
5. E. Berger, proc. of the NPAS Workshop on Electronuclear Physics with Internal Targets (SIAC, 1987), SLAC report 316, eds. R.G. Arnold and R.C. Mmehart, p. 82; Preprint ANL-HEP-CP-87-45, April 30, 1987.
6. G. Ingelman, A. Edin and J. Rathsman, *Comput. Phys. Commun.* **101** (1997) 108.
7. T. Sjöstrand, *Comput. Phys. Commun.* **39** (1986) 347; **43** (1987) 367;
   T. Sjöstrand, *PYTHIA 5.7 and JETSET 7.4: Physics and Manual*, arXiv:hep-ph/9508391;
   T. Sjöstrand *et al.*, *Comput. Phys. Commun.* **135** (2001) 238.
8. A. Kotzinian, *Eur. Phys. J.* **C44** (2005) 211, arXiv:hep-ph/0410093.
9. A. Kotzinian, *Nucl. Phys.* **B441** (1995) 234, arXiv:hep-ph/9412283.
10. R.N. Cahn, *Phys. Lett.* **B78** (1978) 269; *Phys. Rev.* **D40** (1989) 3107.
11. HERMES Collaboration, A. Airapetian *et al.*, *Phys. Rev. Lett.* **94** (2005) 012002, arXive:hep-ex/0408013; M. Diefenthaler, arXive:hep-ex/0507013.
12. COMPASS Collaboration, V.Y. Alexakhin *et al.*, *Phys. Rev. Lett.* **94** (2005) 202002, arXiv:hep-ex/0503002.
13. A. Kotzinian, arXiv:hep-ph/0504081.
14. EMC Collaboration, M. Arneodo *et al.*, *Z. Phys.* **C34** (1987) 277.
15. Pre-conceptual Design Report,
    `http://www.jlab.org/12GeV/collaboration.html`.

COMPARING EXTRACTIONS OF SIVERS FUNCTIONS

M. Anselmino[1], M. Boglione[1], J.C. Collins[2], U. D'Alesio[3], A.V. Efremov[4], K. Goeke[5], A. Kotzinian[1], S. Menzel[5], A. Metz[5], F. Murgia[3], A. Prokudin[1], P. Schweitzer[5], W. Vogelsang[6,7] and F. Yuan[7]

[1] *Università di Torino and INFN, Sezione di Torino, Italy*
[2] *Penn State University, 104 Davey Lab, University Park PA 16802, U.S.A.*
[3] *INFN, Sezione di Cagliari and Università di Cagliari, Italy*
[4] *Joint Institute for Nuclear Research, Dubna, 141980 Russia*
[5] *Institut für Theoretische Physik II, Ruhr-Universität Bochum, Germany*
[6] *Physics Department, Brookhaven National Laboratory, Upton, NY 11973, U.S.A.*
[7] *RIKEN BNL Research Center, Building 510A, BNL, Upton, NY 11973, U.S.A.*

A comparison is given of the various recently published extractions of the Sivers functions from the HERMES and COMPASS data on single-transverse spin asymmetries in semi-inclusive deeply inelastic scattering.

1. Introduction

Single-spin asymmetries (SSA) in semi-inclusive deeply inelastic scattering (SIDIS) off transversely polarized nucleon targets have been under intense experimental investigation over the past few years.[1–6]. Substantial asymmetries have been reported in some cases, in particular, with best statistics, by the HERMES collaboration for scattering off a proton target.

The importance of SSA lies in the fact that they provide new insights into QCD and nucleon structure.[7–14] For instance, the asymmetry in SIDIS may contain an angular dependence of the form $\sin(\phi - \phi_S)$, where ϕ and ϕ_S denote respectively the azimuthal angles of the produced hadron and the target polarization vector with respect to the axis defined by the hard virtual photon.[11] This angular dependence arises from the so-called Sivers effect[7] tightly related to notions of an intrinsic asymmetry in the parton transverse momentum distribution and angular momenta. Factorization theorems[15–17] proven to leading power in the photon virtuality Q provide the basis for a QCD description of the process, and allow to extract the Sivers function from SIDIS data[1–4] and to use it for predictions for the SSA in the Drell–Yan (DY) process, hopefully to be explored experimentally at RHIC, COMPASS and the GSI. Comparisons of SIDIS and the DY process will be particularly important for testing our understanding of the underlying physics, since it has been predicted [13,14] that the Sivers functions appear with opposite signs in these two processes. The approach just outlined has been followed recently in Refs. 18–23 In this note we compare the results of these papers for the extracted Sivers functions

$$\Delta^N f_{q/p^\uparrow}(x,\mathbf{p}_T^2) \equiv -\frac{2|\mathbf{p}_T|}{M_N}\, f_{1T}^{\perp a}(x,\mathbf{p}_T^2) \equiv -\frac{2|\mathbf{p}_T|}{M_N}\, q_T(x,\mathbf{p}_T^2)\,. \tag{1}$$

In the extractions of the Sivers functions from SIDIS several simplifying approximations were common between the groups, namely the neglect of the so-called "soft factor" [16,17] and the Sivers antiquark functions. Different approaches were, however, followed in Refs. 19–23 concerning the treatment of the dependence of the distributions on transverse parton momenta.

The Sivers SSA is obtained [2,24] by weighting the events entering the spin asymmetry with $\sin(\phi-\phi_S)$. When analyzed in this way, however, specific models for the dependence on parton transverse momenta need to be made in the theoretical expression. By assuming that the transverse momentum dependence of the Sivers function is of the form $f_{1T}^{\perp a}(x,\mathbf{p}_T^2) = f_{1T}^{\perp a}(x)\, G(\mathbf{p}_T^2)$ and/or similarly for other distribution or fragmentation functions, the Sivers SSA as defined at HERMES [2] can be written generically as

$$A_{UT}^{\sin(\phi-\phi_S)} = (-2)\,\frac{\sum_a e_a^2\, x F_{\rm Siv}^a(x)\, D_1^{a/\pi}(z)}{\sum_a e_a^2\, x f_1^a(x)\, D_1^{a/\pi}(z)}\,. \tag{2}$$

The factor (-2) is due to conventions [24] and $F_{\rm Siv}^q(x)$ is some functional depending on $f_{1T}^{\perp}$ and the model used for parton transverse momenta.

Notice that by including in addition a factor of $P_{h\perp}/M_N$ into the weight in (2) the resulting SSA can be interpreted model-independently in terms of the transverse moment of the Sivers function [11]

$$f_{1T}^{\perp(1)a}(x) \equiv \int \mathrm{d}^2\mathbf{p}_T\, \frac{\mathbf{p}_T^2}{2M_N^2}\, f_{1T}^{\perp a}(x,\mathbf{p}_T^2) = -\int \mathrm{d}^2\mathbf{p}_T\, \frac{|\mathbf{p}_T|}{4M_N}\, \Delta^N f_{q/p^\uparrow}(x,\mathbf{p}_T^2)\,. \tag{3}$$

Such weighted SSA were argued to be less sensitive to Sudakov suppression which can be important for predictions involving the Sivers function. *Preliminary* HERMES data for such SSA are available [1] and were studied in Ref. 18, where a first fit for the transverse moment of the Sivers function (3) was obtained. The result of Ref. 18 is in good agreement with the studies of SSA analyzed *without* a power of $P_{h\perp}$ in the weight [2–4] reported in Refs. 19–23.

The next Sections review and compare the fit results for the Sivers functions extracted in the different approaches in Refs. 19–23.

2. The approach of Refs. 19 and 20

In Ref. 19 the azimuthal angular dependence (Cahn effect) of the SIDIS unpolarized cross section was used to extract the widths of the Gaussian p_T-dependent parton distribution (pdf) and fragmentation (ff) functions respectively as $\langle p_T^2\rangle = 0.25\,(\mathrm{GeV}/c)^2$ and $\langle K_T^2\rangle = 0.2\,(\mathrm{GeV}/c)^2$. A first estimate of the Sivers functions was then obtained by fitting the data on $A_{UT}^{\sin(\phi-\phi_S)}$ observed by HERMES collaboration.[1,2] In Ref. 20 a novel fit on the new HERMES data[4] together with data from the COMPASS collaboration[3] was performed. In both fits the full exact kinematics was always adopted. The Sivers function (u, d quarks) was parametrized as:

$$\Delta^N f_{q/p^\uparrow}(x,\mathbf{p}_T^2) = 2\,\mathcal{N}_q(x)\,f_{q/p}(x)\,g(\mathbf{p}_T^2)\,h(\mathbf{p}_T^2)\,, \tag{4}$$

$$\mathcal{N}_q(x) = N_q\,x^{a_q}(1-x)^{b_q}\,\frac{(a_q+b_q)^{(a_q+b_q)}}{a_q^{a_q}\,b_q^{b_q}}\,, \quad g(\mathbf{p}_T^2) = \frac{e^{-p_T^2/\langle p_T^2\rangle}}{\pi\langle p_T^2\rangle}\,. \tag{5}$$

Two options for the $h(\mathbf{p}_T^2)$ function were considered, namely:

$$\text{(a)}\;\; h(\mathbf{p}_T^2) = \frac{2p_T M_0}{p_T^2+M_0^2}\,, \qquad \text{(b)}\;\; h(\mathbf{p}_T^2) = \sqrt{2e}\,\frac{p_T}{M'}\,e^{-p_T^2/M'^2}\,, \tag{6}$$

the latter allowing, at leading order in p_T/Q, to give for F^a_{Siv} in Eq. (2):

$$F^a_{\mathrm{Siv}}(x) = \frac{\sqrt{\pi}}{2}\,\frac{M_N}{\sqrt{\widehat{\langle p_T^2\rangle} + \langle K_T^2\rangle/z^2}}\,f_{1T}^{\perp(1)a}(x)\;\;\text{with}\;\;\widehat{\langle p_T^2\rangle} = \frac{\langle p_T^2\rangle}{1+\langle p_T^2\rangle/M'^2}\,. \tag{7}$$

In the fits, $f_{q/p}(x)$ was taken from the LO MRST01 set,[25] whereas Kretzer's set[26] for the LO ff was used. The 7 parameters were then extracted as:[20]

$$\begin{aligned} &N_u = 0.32 \pm 0.11\,, \;\; a_u = 0.29 \pm 0.35\,, \;\; b_u = 0.53 \pm 3.58\,, \\ &N_d = -1.0 \pm 0.12\,, \;\; a_d = 1.16 \pm 0.47\,, \;\; b_d = 3.77 \pm 2.59\,, \\ &M'^2 = 0.55 \pm 0.38\;(M_0^2 = 0.32 \pm 0.25)\;(\mathrm{GeV}/c)^2\,, \end{aligned} \tag{8}$$

with a χ^2 per degree of freedom (χ^2_{dof}) of 1.06. The one-sigma band shown in Fig. 1 (Eq. (6b)) takes into account the errors with their correlations.

These results were then used to give predictions for SSA measurable in SIDIS and DY processes for various kinematical configurations.

These effects were also invoked[9,10,27,28] to generate SSA for other processes in hadron–hadron collisions[29,30] although the status of factorization is less clear in this case. Here we only point out that the SIDIS data are sensitive to much smaller x values than the E704 (STAR) ones.

3. The approach of Ref. 21

In Ref. 21 it was assumed that the final hadron's transverse momentum is entirely due to the transverse-momentum dependence in the Sivers function. There is then no further assumption on the particular form of this dependence; rather it is integrated out in order to compare to the experimental data. The transverse momenta contributed by the other factors in the factorization formula will give some smearing effects which may be viewed as "sub-dominant". (However, we emphasize that this will not be true toward small z where the transverse momentum in the fragmentation functions will become important, likely resulting in a suppression of the asymmetry at small z.) The "1/2-moments" of the Sivers functions were then introduced in Ref. 21 in the fit to the experimental data:

$$q_T^{(1/2)}(x) \equiv \int d^2\mathbf{p_T} \frac{|\mathbf{p_T}|}{M_N} f_{1T}^{\perp q}(x, \mathbf{p}_T^2) \; . \tag{9}$$

These appear in an expression of the form (2) for the Sivers asymmetry, where

$$F_{\mathrm{Siv}}^q(x) = \frac{1}{2}\, q_T^{(1/2)}(x) \; . \tag{10}$$

In the actual fit to the HERMES data in Ref. 21 the functions $q_T^{(1/2)}(x)$ were modeled in terms of the unpolarized u-quark distribution as

$$\frac{u_T^{(1/2)}(x)}{u(x)} = S_u\, x(1-x) \; , \qquad \frac{d_T^{(1/2)}(x)}{u(x)} = S_d\, x(1-x) \; , \tag{11}$$

where $u(x)$ was taken from the GRV LO parametrizations for the unpolarized parton distributions.[31] Furthermore, Kretzer's set for the LO fragmentation functions[26] was used. The fit to the new preliminary HERMES data gave

$$S_u = -0.81 \pm 0.07 \, , \qquad S_d = 1.86 \pm 0.28 \, , \tag{12}$$

with $\chi^2_{\mathrm{dof}} \approx 1.2$. A fit to the old published HERMES data gave instead $S_u = -0.55 \pm 0.37$ and $S_d = 1.1 \pm 1.6$, with a similar size of χ^2_{dof}. The COMPASS data were not included in the fit performed in Ref. 21, but a comparison of the fit with the data was given, showing good agreement. The results of the fit to the HERMES data were furthermore used for making predictions for the SSAs in the Drell–Yan process and in di-jet and jet–photon correlations at RHIC.

4. The approach of Refs. 22 and 23

In Refs. 22 and 23 the distributions of transverse parton momenta in f_1^a, $f_{1T}^{\perp a}$ and D_1^a were assumed to be Gaussian with the respective widths $\langle p_T^2 \rangle$, $\langle p_T^2 \rangle_{\rm Siv}$ and $\langle K_T^2 \rangle$ taken to be flavour- and x- or z-independent. In this model the $F_{\rm Siv}^a$ defined in (2) is given by the expression in Eq. (7) with $\widehat{\langle p_T^2 \rangle}$ replaced by $\langle p_T^2 \rangle_{\rm Siv}$.

The values $\langle K_T^2 \rangle = 0.16\,({\rm GeV}/c)^2$, $\langle p_T^2 \rangle = 0.33\,({\rm GeV}/c)^2$ were extracted [22] from the HERMES data [32] on $\langle P_{h\perp} \rangle$ and are similar to those discussed in Sec. 2, while $\langle p_T^2 \rangle_{\rm Siv} \in [0.01; 0.32]\,({\rm GeV}/c)^2$ remained poorly constrained by positivity [33] – still allowing an extraction of the *transverse moment* of the Sivers function (3).

In order to reduce the number of fit parameters the prediction [34] from the limit of a large number of colours N_c was imposed:

$$f_{1T}^{\perp u}(x, \mathbf{p}_T^2) = -f_{1T}^{\perp d}(x, \mathbf{p}_T^2) \quad \text{modulo } 1/N_c \text{ corrections.} \tag{13}$$

The best fit [22] (using parametrizations [35,36]) to the *published data* [2] is

$$x f_{1T}^{\perp(1)u}(x) \overset{\text{ansatz}}{=} A x^b (1-x)^5 \overset{\text{fit}}{=} -0.17 x^{0.66} (1-x)^5 \,, \tag{14}$$

with a $\chi^2_{\rm dof} \sim 0.3$, and a 1-$\sigma$ uncertainty of roughly $\pm 30\%$. This result agrees well with the fit to the preliminary $P_{h\perp}$-weighted HERMES data,[1] which were analyzed in a (transverse parton momentum) model-independent way [18]. The good agreement of the results in Refs. 18 and 22 is an important cross check for the applicability of the Gauss model to the description of SSA in SIDIS.

For sake of a better comparison to the results by the other groups [19–21] the above fit procedure was applied [23] to the most recent and more precise *preliminary* HERMES data.[4] The new fit has a $\chi^2_{\rm dof} \sim 2$ and is consistent [23] with that quoted in Eq. (14). One has to keep in mind that the large-N_c relation (13) is a useful constraint at the *present stage*, and will have to be relaxed when future more precise data will become available.

Note that for $\langle K_T^2 \rangle \to 0$ in (7) one obtains $F_{\rm Siv}^a(x) \to \frac{1}{2} f_{1T}^{\perp(1/2)a}(x)$ within the Gaussian model. This limit means that the produced hadron acquires no additional transverse momentum from the fragmentation process, i.e. $D_1^a(z, \mathbf{K}_T^2) = D_1^a(z)\,\delta^{(2)}(\mathbf{K}_T)$. In this sense, the approach of Ref. 21 discussed in Sec. 3, cf. Eq. (10), is contained as a limiting case in the Gauss ansatz.

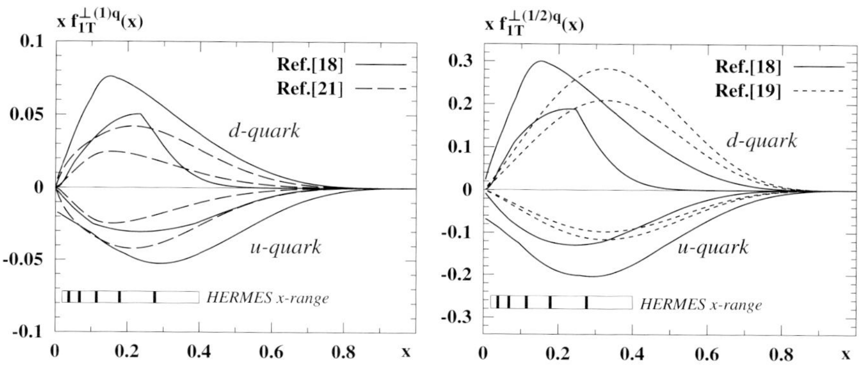

Fig. 1. The first and 1/2-transverse moments of the Sivers quark distribution functions, defined in Eqs. (3, 9), as extracted in Refs. 20, 21 and 23. The fits were constrained mainly (or solely) by the preliminary HERMES data [4] in the indicated x-range. The curves indicate the 1-σ regions of the various parametrizations.

5. Comparison of the results and Conclusions

It should be stressed that the various fit results, when used within the respective approaches, provide equally good descriptions of the HERMES and COMPASS data. Here we compare only those analyses [20,21,23] in which the most recent and more precise *preliminary* HERMES data [4] were used.

In Fig. 1a we compare the fits for $f_{1T}^{\perp(1)q}$ from Refs. 20 and 23, and in Fig. 1b the fits for $f_{1T}^{\perp(1/2)q}$ from Refs. 20 and 21. (A direct comparison of Refs. 21 and 23 is not possible.) In view of the different models assumed for the transverse parton momenta and the varying fit Ansätze, we observe a satisfactory *qualitative* agreement — in the x-region constrained by the HERMES data. However, a closer look reveals differences between the results in Fig. 1, which indicate the size of the systematic uncertainties of the three Sivers function fits mainly due to the use of different models for the parton transverse momenta. These uncertainties were not estimated in Refs. 20, 21 and 23.

We have presented a comparison of three extractions [20,21,23] of Sivers functions from HERMES and COMPASS data on single-transverse spin asymmetries in SIDIS. The three approaches somewhat differ, but they describe the data with similar quality. The fits are in good qualitative agreement, though there are differences with regard to the size and shape of the extracted Sivers functions. These differences reflect the model dependence of the fit results which gives rise to a certain theoretical systematic uncertainty of the fit results. The latter seems, however, less dominant than the statistical uncertainty of the fits at the present stage.

It is clear that further information from experiment will be vital. For now, one cannot really expect to obtain much more than a first qualitative picture of the Sivers functions. We also emphasize that it will be crucial for the future to experimentally confirm the leading-power nature of the observed spin asymmetries. For this, forthcoming COMPASS or JLab data for scattering off a proton target and studies of the Q^2-dependence of the asymmetries will be important.

The good *qualitative* agreement between the different approaches observed here means that the predictions [18–23] for the magnitude of the Sivers effect in DY are robust — in the kinematic region constrained by the HERMES data. This solidifies the conclusions [18–23] that the predicted sign reversal of the Sivers function between SIDIS and DY, can be tested in running or future experiments at RHIC, COMPASS and PAX.

Acknowledgments. U.D. and F.M. acknowledge partial support by MIUR under Cofinanziamento PRIN 2003. A.E. is supported by grants RFBR 03-02-16816 and DFG-RFBR 03-02-04022. J.C.C. is supported in part by the U.S. D.O.E., and by a Mercator Professorship of DFG. W.V. and F.Y. are grateful to RIKEN, Brookhaven National Laboratory and the U.S. Department of Energy (contract number DE-AC02-98CH10886) for providing the facilities essential for the completion of their work. The work is partially supported by the European Integrated Infrastructure Initiative Hadron Physics project under contract number RII3-CT-2004-506078.

References

1. N.C. Makins [HERMES], *"Transversity Workshop", 6–7 Oct. 2003, Athens*; R. Seidl [HERMES], proc. of *DIS 2004, 13–18 April 2004, Štrbské Pleso*; I.M. Gregor [HERMES], *Acta Phys. Polon.* **B36** (2005) 209.
2. A. Airapetian *et al.* [HERMES], *Phys. Rev. Lett.* **94** (2005) 012002.
3. V.Y. Alexakhin *et al.* [COMPASS], *Phys. Rev. Lett.* **94** (2005) 202002.
4. M. Diefenthaler [HERMES], arXiv:hep-ex/0507013.
5. A. Bravar [SMC], *Nucl. Phys.* **A666** (2000) 314.
6. H. Avakian [CLAS], talk presented at the RBRC workshop "Single-Spin Asymmetries", Brookhaven National Laboratory, Upton, New York, June 1–3, 2005, to appear in the proceedings.
7. D.W. Sivers, *Phys. Rev.* **D41** (1990) 83; *Phys. Rev.* **D43** (1991) 261.
8. J.C. Collins, *Nucl. Phys.* **B396** (1993) 161.
9. M. Anselmino, M. Boglione and F. Murgia, *Phys. Lett.* **B362** (1995) 164.

10. M. Anselmino and F. Murgia, *Phys. Lett.* **B442** (1998) 470.
11. D. Boer and P.J. Mulders, *Phys. Rev.* **D57** (1998) 5780.
12. S.J. Brodsky, D.S. Hwang and I. Schmidt, *Phys. Lett.* **B530** (2002) 99; *Nucl. Phys.* **B642** (2002) 344.
13. J.C. Collins, *Phys. Lett.* **B536** (2002) 43.
14. A.V. Belitsky, X.D. Ji and F. Yuan, *Nucl. Phys.* **B656** (2003) 165.
    X.D. Ji and F. Yuan, *Phys. Lett.* **B543** (2002) 66.
    D. Boer, P.J. Mulders and F. Pijlman, *Nucl. Phys.* **B667** (2003) 201.
15. J.C. Collins and D.E. Soper, *Nucl. Phys.* **B193** (1981) 381 [*Erratum-ibid.* **B213** (1983) 545].
16. X.D. Ji, J.P. Ma and F. Yuan, *Phys. Rev.* **D71** (2005) 034005; *Phys. Lett.* **B597** (2004) 299.
17. J.C. Collins and A. Metz, *Phys. Rev. Lett.* **93** (2004) 252001.
18. A.V. Efremov, K. Goeke, S. Menzel, A. Metz and P. Schweitzer, *Phys. Lett.* **B612** (2005) 233.
19. M. Anselmino, M. Boglione, U. D'Alesio, A. Kotzinian, F. Murgia and A. Prokudin, *Phys. Rev.* **D71** (2005) 074006.
20. M. Anselmino, M. Boglione, U. D'Alesio, A. Kotzinian, F. Murgia and A. Prokudin, to appear in *Phys. Rev.* **D**, arXiv:hep-ph/0507181; and these proceedings.
21. W. Vogelsang and F. Yuan, *Phys. Rev.* **D72** (2005) 054028.
22. J.C. Collins, A.V. Efremov, K. Goeke, S. Menzel, A. Metz and P. Schweitzer, arXiv:hep-ph/0509076.
23. J.C. Collins, A.V. Efremov, K. Goeke, M. Grosse Perdekamp, S. Menzel, B. Meredith, A. Metz and P. Schweitzer, these proceedings, arXiv:hep-ph/0510342, and work in progress.
24. A. Bacchetta, U. D'Alesio, M. Diehl and C.A. Miller, *Phys. Rev.* **D70** (2004) 117504.
25. A.D. Martin, R.G. Roberts, W.J. Stirling and R.S. Thorne, *Phys. Lett.* **B531** (2002) 216.
26. S. Kretzer, *Phys. Rev.* **D62** (2000) 054001.
27. U. D'Alesio and F. Murgia, *Phys. Rev.* **D70** (2004) 074009.
28. M. Anselmino, M. Boglione, U. D'Alesio, E. Leader and F. Murgia, *Phys. Rev.* **D71** (2005) 014002.
29. D.L. Adams *et al.*, *Phys. Lett.* **B261** (1991) 201 and **B264** (1991) 462; *Z. Phys.* **C56** (1992) 181.
30. J. Adams *et al.* [STAR], *Phys. Rev. Lett.* **92** (2004) 171801.
31. M. Glück, E. Reya and A. Vogt, *Z. Phys.* **C67** (1995) 433.
32. A. Airapetian *et al.* [HERMES], *Phys. Lett.* **B562** (2003) 182.
33. A. Bacchetta, M. Boglione, A. Henneman and P.J. Mulders, *Phys. Rev. Lett.* **85** (2000) 712.
34. P.V. Pobylitsa, arXiv:hep-ph/0301236.
35. M. Glück, E. Reya and A. Vogt, *Eur. Phys. J.* **C5** (1998) 461
36. S. Kretzer, E. Leader and E. Christova, *Eur. Phys. J.* **C22** (2001) 269.

ANOMALOUS DRELL–YAN ASYMMETRY FROM HADRONIC OR QCD VACUUM EFFECTS

Daniël Boer

Dept. of Physics and Astronomy, Vrije Universiteit Amsterdam,
De Boelelaan 1081, 1081 HV Amsterdam, The Netherlands
E-mail: D.Boer@few.vu.nl

The anomalously large $\cos(2\phi)$ asymmetry measured in the Drell–Yan process is discussed. Possible origins of this large deviation from the Lam–Tung relation are considered with emphasis on the comparison of two particular proposals: one that suggests it arises from a QCD vacuum effect and one that suggests it is a hadronic effect. Experimental signatures distinguishing these effects are discussed.

1. Introduction

Azimuthal asymmetries in the unpolarized Drell–Yan (DY) process differential cross section arise only in the following way

$$\frac{1}{\sigma}\frac{d\sigma}{d\Omega} \propto \left(1 + \lambda \cos^2\theta + \mu \sin 2\theta \, \cos\phi + \frac{\nu}{2}\sin^2\theta \cos 2\phi\right), \tag{1}$$

where ϕ is the angle between the lepton and hadron planes in the lepton center of mass frame (see Fig. 3 of Ref. 1). In the parton model (order α_s^0) quark-antiquark annihilation yields $\lambda = 1$, $\mu = \nu = 0$. The leading order (LO) perturbative QCD corrections (order α_s^1) lead to $\mu \neq 0$, $\nu \neq 0$ and $\lambda \neq 1$, such that the so-called Lam–Tung relation $1 - \lambda - 2\nu = 0$ holds. Beyond LO, small deviations from the Lam–Tung relation will arise. If one defines the quantity $\kappa \equiv -\frac{1}{4}(1 - \lambda - 2\nu)$ as a measure of the deviation from the Lam–Tung relation, it has been calculated [2,3] that at order α_s^2 κ is small and negative: $-\kappa \lesssim 0.01$, for values of the muon pair's transverse momentum Q_T of up to $3\,\mathrm{GeV}/c$.

Surprisingly, the data is incompatible with the Lam–Tung relation and with its small order-α_s^2 modification as well.[3] These data from CERN's NA10 Collaboration [4,5] and Fermilab's E615 Collaboration [6] are for $\pi^- N \to \mu^+\mu^- X$, with $N = D$ and W. The π^--beam energies range from 140 GeV up to 286 GeV and the invariant mass Q of the lepton pair is in the range $Q \sim 4 - 12$ GeV. The measured values for κ are an order of magnitude larger than the order-α_s^2 result and moreover, of opposite sign.

Several explanations have been put forward, but not all of them will be reviewed here. Some unlikely explanations would be: *i*) NNLO pQCD corrections could solve the discrepancy (but in that case the perturbative expansion itself would be questionable); *ii*) it could be a higher twist effect (but $Q^2 > 16\,\text{GeV}^2$ seems too high and according to the Fermilab data the deviation disappears at high x_π, contrary to higher twist expectation; also, one would expect $\mu > \nu$, whereas in the data $\nu \gg \mu \approx 0$); *iii*) it could be a nuclear effect, since $\sigma(Q_T)_W/\sigma(Q_T)_D$ is an increasing function of Q_T (but according to Ref. 5 $\nu(Q_T)$ shows no apparent nuclear dependence).

The two possible explanations that will be discussed and compared here are: *i*) a QCD vacuum effect;[3] *ii*) a hadronic effect, arising from noncollinear parton configurations.[1] The following will largely be based on a recent comparative study performed in collaboration with A. Brandenburg, O. Nachtmann and A. Utermann.[7]

2. Explanation in terms of a QCD vacuum effect

Usually the DY process at $Q \sim 4 - 12$ GeV is described by collinear factorization. Collinear quarks inside unpolarized hadrons are unpolarized themselves, implying a trivial quark-antiquark spin density matrix:

$$\rho^{(q,\bar{q})} = \frac{1}{4}\{\mathbf{1} \otimes \mathbf{1}\}. \tag{2}$$

The QCD vacuum may alter this. The gluon condensate leads to a chromomagnetic field strength (Savvidy; Shifman, Vainshtein, Zakharov; ...)

$$\langle g^2 \boldsymbol{B}^a(x) \cdot \boldsymbol{B}^a(x) \rangle \approx (700\,\text{MeV})^4, \tag{3}$$

with gluon fields having a typical correlation length $a \approx 0.35$ fm in Euclidean space. Taking this to be an invariant length in Minkowski space[8] leads to the picture of a fluctuating domain structure of the vacuum with typical domain size a, schematically depicted in Fig. 1. If a fast hadron, and with it a fast quark, traverses this domain structure, the time for traversing a vacuum domain is of the order of the correlation length: $t \approx a$. Due to the presence of a background chromomagnetic field the quark will acquire a transverse polarization (the Sokolov–Ternov effect). The time to build up transverse polarization is estimated[8,9] to be much shorter than the time it takes to traverse the domain, i.e. $t \ll a$. The radiated gluons/photons are

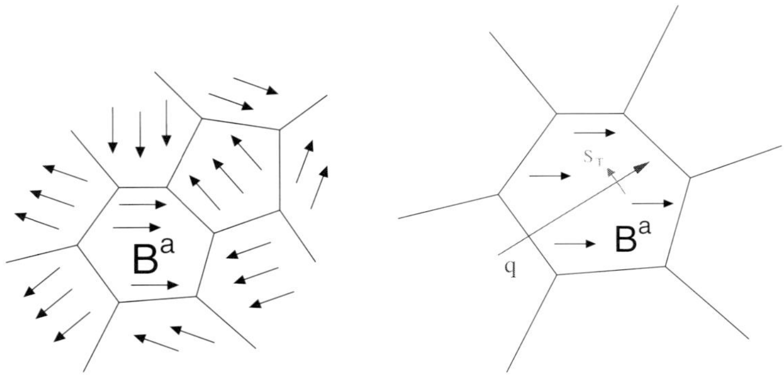

Fig. 1. Left figure: cartoon of the chromomagnetic field domain structure of the QCD vacuum. Right figure: a fast quark traversing a domain.

just part of the cloud of virtual particles; in other words, they are included in the wave function. There will be no average polarization. However, if the quark will annihilate with an antiquark in a high energy scattering experiment, such as DY, the polarization of the quark and the antiquark may be correlated if they annihilate within a certain domain. Therefore, the QCD

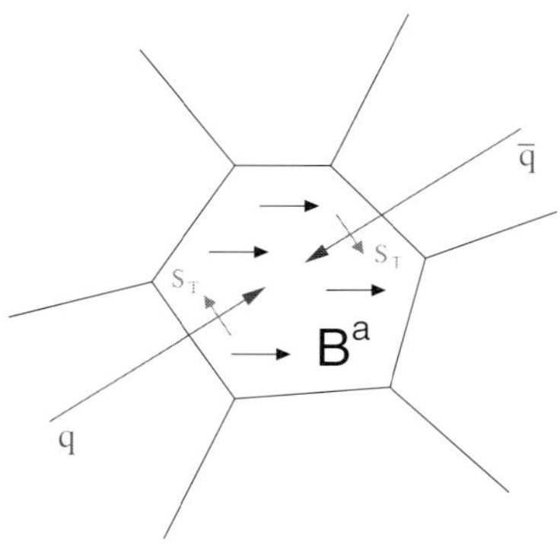

Fig. 2. Annihilation of a quark and an antiquark inside the same domain.

vacuum can induce a spin correlation between an annihilating $q\,\bar{q}$ pair. The quark–antiquark spin density matrix Eq. (2) will then be modified into

$$\rho^{(q,\bar{q})} = \frac{1}{4}\left\{\mathbf{1}\otimes\mathbf{1} + F_j\,\boldsymbol{\sigma}_j\otimes\mathbf{1} + G_j\,\mathbf{1}\otimes\boldsymbol{\sigma}_j + H_{ij}\,\boldsymbol{\sigma}_i\otimes\boldsymbol{\sigma}_j\right\}. \qquad (4)$$

Only if $H_{ij} = F_i G_j$, then the spin density matrix factorizes. But this is not necessarily so, in which case it could be called entangled. Brandenburg,

Nachtmann & Mirkes[3] demonstrated that the diagonal elements H_{11} and H_{22} can give rise to a deviation from the Lam–Tung relation:

$$\kappa \equiv -\frac{1}{4}(1-\lambda-2\nu) \approx \left\langle \frac{H_{22}-H_{11}}{1+H_{33}} \right\rangle. \tag{5}$$

A simple assumption for the transverse momentum dependence of $(H_{22} - H_{11})/(1+H_{33})$ produced a good fit to the data:

$$\kappa = \kappa_0 \, \frac{Q_T^4}{Q_T^4 + m_T^4}, \qquad \text{with } \kappa_0 = 0.17 \text{ and } m_T = 1.5 \text{ GeV}. \tag{6}$$

Note that for this Ansatz κ approaches a constant value (κ_0) for large Q_T. In other words, the vacuum effect could persist out to large values of Q_T. The Q^2 dependence of the vacuum effect is not known, but there is also no reason to assume that the spin correlation due to the QCD vacuum effect has to decrease with increasing Q^2.

3. Explanation as a hadronic effect

Usually if one assumes that factorization of soft and hard energy scales in a hard scattering process occurs, one implicitly also assumes factorization of the spin density matrix. In the present section this will indeed be assumed, but another common assumption will be dropped, namely that of *collinear* factorization. It will be investigated what happens if one allows for transverse momentum dependent parton distributions (TMDs). The spin density matrix of a noncollinear quark inside an unpolarized hadron can be nontrivial. In other words, the transverse polarization of a noncollinear quark inside an unpolarized hadron in principle can have a preferred direction and the TMD describing that situation is called $h_1^\perp$.[10] As pointed out in Ref. 1 nonzero $h_1^\perp$ leads to a deviation from Lam–Tung relation. It offers a parton model explanation of the DY data (i.e. with $\lambda = 1$ and $\mu = 0$): $\kappa = \frac{\nu}{2} \propto h_1^\perp(\pi)\, h_1^\perp(N)$. In this way a good fit to data was obtained by assuming Gaussian transverse momentum dependence. The reason for this choice of transverse momentum dependence is that in order to be consistent with the factorization of the cross section in terms of TMDs, the transverse momentum of partons should not introduce another large scale. Therefore, explaining the Lam–Tung relation within this framework necessarily implies that $\kappa = \frac{\nu}{2} \to 0$ for large Q_T. This offers a possible way to distinguish between the hadronic effect and the QCD vacuum effect.

It may be good to mention that not only a fit of $h_1^\perp$ to data has been made (under certain assumptions), but also several model calculations of

$h_1^\perp$ and some of its resulting asymmetries have been performed,[11–13] based on the recent insight that T-odd TMDs like $h_1^\perp$ arise from the gauge link.

In order to see the parton model expectation $\kappa = \frac{\nu}{2} \to 0$ at large Q_T in the data, one has to keep in mind that the pQCD contributions (that grow as Q_T increases) will have to be subtracted. For κ perturbative corrections arise at order α_s^2, but for ν already at order α_s. To be specific, at large Q_T hard gluon radiation (to first order in α_s) gives rise to [14]

$$\nu(Q_T) = \frac{Q_T^2}{Q^2 + \frac{3}{2}Q_T^2}. \tag{7}$$

Due to this growing large-Q_T perturbative contribution the fall-off of the $h_1^\perp$ contribution will not be visible directly from the behavior of ν at large Q_T. Therefore, in order to use ν as function of Q_T to differentiate between effects, it is necessary to subtract the calculable pQCD contributions. In Fig. 3 an illustration of this point is given. The dashed curve corresponds

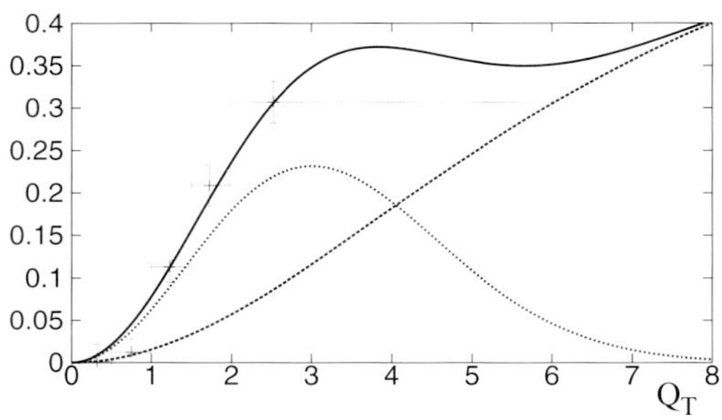

Fig. 3. *Impression* of possible contributions to ν as function of Q_T compared to DY data of NA10 (for $Q = 8$ GeV). Dashed curve: contribution from perturbative one-gluon radiation. Dotted curve: contribution from a nonzero $h_1^\perp$. Solid curve: their sum.

to the contribution of Eq. (7) at $Q = 8$ GeV. The dotted line is a possible, parton model level, contribution from $h_1^\perp$ with Gaussian transverse momentum dependence. Together these contributions yield the solid curve (although strictly speaking it is not the case that one can simply add them, since one is a noncollinear parton model contribution expected to be valid for small Q_T and the other is an order-α_s result within collinear factorization expected to be valid at large Q_T). The data are from the NA10 Collaboration for a pion beam energy of 194 GeV/c.[5]

The Q^2 dependence of the $h_1^\perp$ contribution is not known to date. Only the effect of resummation of soft gluon radiation on the $h_1^\perp$ contribution to ν (and κ) has been studied to some extent and was found to be quite important.[15] It gives rise to a considerable Sudakov suppression with increasing Q: in going from $Q = 10$ to $90\,\mathrm{GeV}$, the contribution decreases by an order of magnitude and approximately follows a $1/Q$ behavior (although it is neither a dynamical nor a kinematical higher twist effect). Interestingly, the contribution from hard gluon radiation (7) decreases more rapidly: as $1/Q^2$ at fixed Q_T. But it seems safe to conclude that using the Q^2 dependence of ν (or κ) to differentiate between effects is not feasible at present.

By assumption, nonzero $h_1^\perp$ gives rise to a factorized product of spin density matrices $\rho^{(q,\bar{q})} = \rho^{(q)} \otimes \rho^{(\bar{q})}$ with[7]

$$\rho^{(q)} = \frac{1}{2}\left\{\mathbf{1} + \frac{h_1^\perp}{f_1}\frac{x_1}{M_1}\left(\boldsymbol{e}_3 \times \boldsymbol{p}_1\right)\cdot\boldsymbol{\sigma}\right\} \equiv \frac{1}{2}\left\{\mathbf{1} + F_j\,\boldsymbol{\sigma}_j\right\}, \tag{8}$$

$$\rho^{(\bar{q})} = \frac{1}{2}\left\{\mathbf{1} - \frac{\bar{h}_1^\perp}{\bar{f}_1}\frac{x_2}{M_2}\left(\boldsymbol{e}_3 \times \boldsymbol{p}_2\right)\cdot\boldsymbol{\sigma}\right\} \equiv \frac{1}{2}\left\{\mathbf{1} + G_j\,\boldsymbol{\sigma}_j\right\}. \tag{9}$$

Therefore, $H_{ij} = F_i G_j$ with $H_{33} = 0$. Unfortunately it is hard to observe the difference between $H_{33} = 0$ and $H_{33} \neq 0$. But the factorization $H_{ij} = F_i G_j$ should shows itself via consistency among various processes, which is based on the fact that the same function $h_1^\perp$ appears in different processes. Regarding this universality, complications have recently been addressed[16] that go beyond the sign change[17] that occurs between semi-inclusive DIS ($e\,p \to e'\,\pi\,X$) and DY: $(h_1^\perp)_{\mathrm{SIDIS}} = -(h_1^\perp)_{\mathrm{DY}}$. Nevertheless, the different numerical factors with which $h_1^\perp$ arises in different processes are calculable (functions of N_c only) and can be taken into account.

4. Hadronic effect versus vacuum effect

Summarizing the features of the two approaches in a table:

Table 1. Comparison of the hadronic and the vacuum effect

	$h_1^\perp \neq 0$	QCD vacuum effect
$\rho^{(q,\bar{q})}$	$\rho^{(q)} \otimes \rho^{(\bar{q})}$	possibly entangled
Q dependence	$\kappa \overset{?}{\sim} 1/Q$	?
$Q_T \to \infty$	$\kappa \to 0$	need not disappear ($\kappa \to \kappa_0$)
flavor dependence	yes	flavor blind
x dependence	yes	yes, but flavor blind

As indicated in the table, the hadronic effect will generally be flavor dependent and have an x dependence that is flavor dependent, since there is no reason to assume that $h_1^\perp$ for the u quark should be the same as (or simply related to) that for the d quark. This is different from the QCD vacuum effect, which in this sense is flavor blind; it does not matter whether the spin correlation is between $u\,\bar{u}$ or $d\,\bar{d}$ (except for presumably small mass corrections). There will be an x dependence, since that determines the energy of the annihilation process, but this again should be flavor blind. It should be emphasized that flavor blindness in general does not imply hadron blindness or even process blindness. So the best next step would be to perform experiments with different beams ($\pi^+, p, \bar{p}, \ldots$, where π^+ and $\bar{p}$ offer the advantage of having valence anti-quarks) and in different kinematical regimes. For instance, the measurement of $\langle \cos 2\phi \rangle$ can be done at RHIC in $p\,p \to \mu^+\mu^- X$, or in $p\,\bar{p} \to \mu^+\mu^- X$ at Fermilab or GSI/FAIR.

The use of polarized beams can also help (e.g. at RHIC or GSI). In the DY process with one transversely polarized hadron, the differential cross section can namely depend on the azimuthal angle ϕ_S of the transverse hadron spin ($\boldsymbol{S}_T$) compared to the lepton plane:

$$\frac{d\sigma(p\,p^\uparrow \to \ell\,\bar{\ell}\,X)}{d\Omega\; d\phi_S} \propto 1 + \cos^2\theta + \sin^2\theta \left[\frac{\nu}{2}\cos 2\phi - \rho\,|\boldsymbol{S}_T|\sin(\phi+\phi_S)\right] + \ldots$$

Within the framework of TMDs the analyzing power ρ is proportional to the product $h_1^\perp\, h_1$,[1] which involves the transversity function h_1. A nonzero function $h_1^\perp$ will provide a relation between ν and ρ, which in case of one (dominant) flavor (usually called u-quark dominance) and Gaussian transverse momentum dependences, is approximately given by

$$\rho \approx \frac{1}{2}\sqrt{\frac{\nu}{\nu_{\text{max}}}}\,\frac{h_1}{f_1}, \tag{10}$$

where ν_{max} is the maximum value attained by $\nu(Q_T)$. This relation depends on the magnitude of h_1 compared to f_1 (see Refs. 18 and 19 for explicit examples) and this may be extracted from double transverse spin asymmetries in DY (potentially at RHIC or GSI) or from SIDIS data (from e.g. HERMES or COMPASS) by exploiting the interference fragmentation functions (which can be obtained from e^+e^- data, e.g. at BELLE).

Also semi-inclusive DIS can be used. The $\langle \cos 2\phi \rangle$ in $e\,p \to e'\,\pi\,X$ would be $\propto h_1^\perp H_1^\perp$, where $H_1^\perp$ is the Collins fragmentation function (also obtainable from BELLE). This particular SIDIS observable has been studied using model calculations.[20] All this illustrates how the consistency among processes may be used to test the $h_1^\perp$ hypothesis.

5. Conclusions

A transverse spin correlation in quark-antiquark annihilation ($q^\uparrow \bar{q}^\uparrow \to \gamma^*$) will lead to a $\cos(2\phi)$ asymmetry in the DY lepton-pair angular distribution. Such a spin correlation can arise from the chromomagnetic background field in the QCD vacuum or from noncollinear partons. If a flavor dependence is observed in future data, it would favor a hadronic effect. On the other hand, persistence of the asymmetry at large values of Q_T and Q (after subtraction of pQCD corrections if needed) would favor a vacuum effect. Several future and ongoing experiments will be able to provide crucial information on these dependences.

Acknowledgments

I thank Arnd Brandenburg, Stan Brodsky, Dae Sung Hwang, Otto Nachtmann and Andre Utermann for fruitful discussions and collaboration on this topic. The research of D.B. has been made possible by financial support from the Royal Netherlands Academy of Arts and Sciences.

References

1. D. Boer, *Phys. Rev.* **D60** (1999) 014012.
2. E. Mirkes and J. Ohnemus, *Phys. Rev.* **D51** (1995) 4891.
3. A. Brandenburg, O. Nachtmann and E. Mirkes, *Z. Phys.* **C60** (1993) 697.
4. S. Falciano *et al.* [NA10 Collaboration], *Z. Phys.* **C31** (1986) 513.
5. M. Guanziroli *et al.* [NA10 Collaboration], *Z. Phys.* **C37** (1988) 545.
6. J.S. Conway *et al.*, *Phys. Rev.* **D39** (1989) 92.
7. D. Boer, A. Brandenburg, O. Nachtmann and A. Utermann, *Eur. Phys. J.* **C40** (2005) 55.
8. O. Nachtmann and A. Reiter, *Z. Phys.* **C24** (1984) 283.
9. G.W. Botz, P. Haberl and O. Nachtmann, *Z. Phys.* **C67** (1995) 143.
10. D. Boer and P.J. Mulders, *Phys. Rev.* **D57** (1998) 5780.
11. L.P. Gamberg and G.R. Goldstein, arXiv:hep-ph/0209085.
12. D. Boer, S.J. Brodsky and D.S. Hwang, *Phys. Rev.* **D67** (2003) 054003.
13. B. Lu and B.Q. Ma, *Phys. Rev.* **D70** (2004) 094044.
14. J.C. Collins, *Phys. Rev. Lett.* **42** (1979) 291.
15. D. Boer, *Nucl. Phys.* **B603** (2001) 195.
16. A. Bacchetta, C.J. Bomhof, P.J. Mulders and F. Pijlman, *Phys. Rev.* **D72** (2005) 034030.
17. J.C. Collins, *Phys. Lett.* **B536** (2002) 43.
18. D. Boer, *Nucl. Phys. Proc. Suppl.* **79** (1999) 638.
19. D. Boer, *AIP Conf. Proc.* **675** (2003) 479.
20. L.P. Gamberg, G.R. Goldstein and K.A. Oganessyan, *Phys. Rev.* **D67** (2003) 071504; *ibid.* **D68** (2003) 051501.

"T-ODD" EFFECTS IN TRANSVERSE SPIN AND AZIMUTHAL ASYMMETRIES IN SIDIS

Leonard P. Gamberg

Physics Department, Penn State-Berks, Reading, PA 19610, USA
E-mail: lpg10@psu.edu

Gary R. Goldstein

Physics and Astronomy, Tufts University, Medford, MA 02155, USA
E-mail: gary.goldstein@tufts.edu

We consider gluon rescattering as a mechanism to calculate a leading twist T-odd pion fragmentation function, a favored candidate for filtering the transversity properties of the nucleon. We evaluate the single spin azimuthal asymmetry for a transversely polarized target in semi-inclusive deep inelastic scattering (for HERMES kinematics). Additionally, we calculate the double T-odd $\cos 2\phi$ asymmetry in this framework.

1. "T-Odd" Correlations in TSSA and Azimuthal Asymmetries

One of the persistent challenges confronting the QCD parton model is to provide a theoretical basis for the significant azimuthal and single spin asymmetries that emerge in inclusive and semi-inclusive processes.[1,2] Generally speaking, the spin dependent amplitudes for the scattering will contribute to non-zero transverse single spin asymmetries (TSSA) if there are imaginary parts of bilinear products of those amplitudes that have overall helicity change. When the transverse momentum of the produced hadron, $p_T \sim k_\perp$, where $k_\perp$ is the intrinsic transverse momentum of a quark, transverse momentum distribution (TMD) and fragmentation functions[3,4] become relevant as source for spin asymmetries in the factorized description of SIDIS.[5,6] They are indicative of a rich set of correlations among transverse momenta of quarks and/or hadrons and the transverse spin of the reacting particles. In pion production, this correlation emerges as an expectation value of $i\mathbf{s}_T \cdot (\boldsymbol{k} \times \boldsymbol{P}_\perp)$, where $\boldsymbol{s}_T$ is the fragmenting quark's spin polarization, $\boldsymbol{k}$ is fragmenting quark's momenta and $\boldsymbol{P}_\perp$ is the transverse momentum of the produced pion. The associated T-odd distributions [7–10] are of importance as they possess necessary phases and transversity properties (helicity flips) to account for SSA and azimuthal asymmetries.[8,10–12] In contrast to the SSAs generated from the interference of tree-level and one loop correction in PQCD,[13] such effects go like $\alpha_s\langle k_\perp\rangle/M$, where now M plays the role of the chiral symmetry breaking scale and $k_\perp$ is characteristic of quark intrinsic motion.[4]

We consider the leading twist T-odd contributions as the dominant source of the $\cos 2\phi$ azimuthal asymmetry and $\sin(\phi \pm \phi_s)$ TSSAs in SIDIS.[12,14] Among other interesting properties, these asymmetries contain information on the distribution of quark transverse spin in an unpolarized proton, through the Boer–Mulders function $h_1^\perp(x, k_\perp)$.[3,12] In a parton-spectator framework we estimate these asymmetries at HERMES kinematics.[2] Additionally we report on the status of Collins fragmentation function calculations within this framework and the implications for universality.

The leading twist contributions to the factorized cross-section for a transversely polarized nucleon target in lepton–nucleon scattering are [3]

$$\frac{d\sigma_{UT}^{\ell N^\uparrow\to\ell\pi X}}{dxdydzd\phi_S d^2\boldsymbol{P}_{h\perp}} = \frac{2\alpha^2}{Q^2 y}\Bigg\{|\boldsymbol{S}_T|(1-y)\sin(\phi_h+\phi_S)\sum_q e_q^2\,\mathcal{F}\left[\tfrac{\boldsymbol{p}_\perp\cdot\hat{\boldsymbol{h}}}{M_h}\,h_1^q H_1^{\perp q}\right] + |\boldsymbol{S}_T|\tfrac{(1+(1-y)^2)}{2}\sin(\phi_h-\phi_S)\textstyle\sum_q e_q^2\,\mathcal{F}\left[\tfrac{\boldsymbol{k}_\perp\cdot\hat{\boldsymbol{h}}}{M}\,f_{1T}^{\perp q} D_1^q\right]\Bigg\}, \quad (1)$$

where $\mathcal{F}\big[w(k_\perp, p_\perp) f D\big]$ is the convolution integral

$$\int d^2\boldsymbol{p}_\perp d^2\boldsymbol{k}_\perp \delta^{(2)}\Big(p_\perp - k_\perp - \frac{P_\perp}{z}\Big) w(k_\perp, p_\perp) f(x, p_\perp) D(z, k_\perp).$$

The twist two T-odd distribution and fragmentation functions appearing in Eq. (1) are projected from the correlation functions for the transverse momentum dependent distribution and fragmentation correlators, Φ and Δ

$$\begin{aligned}\Phi(x, \boldsymbol{p}_\perp) &= \tfrac{1}{2P^+}\Big\{\cdots + h_1^\perp(x, p_\perp)\tfrac{i\not{p}_\perp \not{P}}{M} + f_{1T}^\perp(x, p_\perp)\tfrac{\epsilon_T^{\mu\nu} p_{\perp\mu} S_{T\nu} \not{P}}{M} + \cdots\Big\},\\ \Delta(z, \boldsymbol{k}_\perp) &= \tfrac{1}{4P_h^-}\Big\{\cdots + H_1^\perp(z, zk_\perp)\tfrac{i\not{k}_\perp \not{P}_h}{M_h} + \cdots\Big\},\end{aligned} \quad (2)$$

where, for example, the Boer–Mulders function $h_1^\perp$ is projected

$$\int dp^- \mathrm{Tr}\left(\sigma^{\perp +}\gamma_5 \Phi\right) = \ldots \frac{2\varepsilon_{+-\perp j}\, p_{\perp j}}{M} h_1^\perp(x, p_\perp) \ldots \quad . \quad (3)$$

Formally, the path ordered gauge link entering the color gauge invariant definition of the T-odd TMD distribution [15–17] and fragmentation [17] functions generate the phases which characterize the TSSAs arising in the factorized description of SIDIS.[5,6] Considering the gauge link contributions up to order g in the so-called "tree level approximation",[17] the hadronic tensor factorizes into convolutions of distribution and fragmentation correlators that alternatively are quark–quark and quark–gluon–quark correlators which are represented by the Feynman diagrams in Fig. 1. In this approximation gluons legs connect only the hard and soft parts of diagrams. For example,

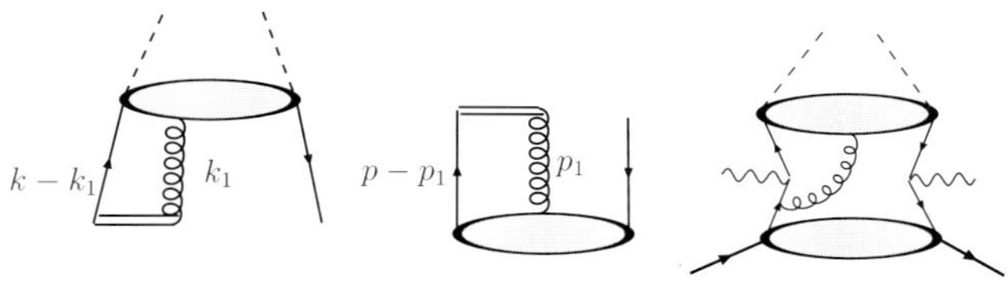

Fig. 1. Left and Center: Quark–gluon–quark correlators arising from the leading gauge link contribution in the factorized hadronic tensor. The double line represents the eikonal fermion propagator emerging from the gauge link. Right: Characteristic term appearing to first order expansion of the gauge link.

the Collins asymmetry is projected from the following contribution to the hadronic tensor

$$2MW_{\mu\nu} = \int d^2p_\perp d^2k_\perp \delta^2(p_\perp - k_\perp - \frac{P_{h\perp}}{z}) \int dk_1^+ d^2k_{1\perp} \tag{4}$$
$$\left\{ \mathrm{Tr}\left(\frac{-i\gamma_\mu \Phi(x,p_\perp)\gamma_\nu \Delta_A^-(k-k_1,k)}{\eta^+\cdot k_1 - i\epsilon}\right) + \mathrm{Tr}\left(\frac{-i\gamma_\nu \Delta_A^-(k,k-k_1)\gamma_\mu \Phi(x,p_\perp)}{\eta^+\cdot k_1^- + i\epsilon}\right)\right\},$$

where the expression for the quark–gluon–quark fragmentation correlator is given by

$$\Delta_A^-(k,k_1) = \sum_X \int \frac{d^4\zeta}{(2\pi)^4}\frac{d^4\eta}{(2\pi)^4} \mathrm{e}^{i(k-k_1)\zeta}\mathrm{e}^{i\eta k_1}\langle 0|\psi(\zeta)A^-(\eta)|P_h,X\rangle$$
$$\times \langle P_h,X|\bar\psi(0)|0\rangle. \tag{5}$$

The Sivers asymmetry is obtained from analogous terms where gauge link contribution is associated with the quark–gluon–quark correlator $\Phi_A(p,p_1)$

$$\Phi_A^+(p,p_1) = \int \frac{d^4\zeta}{(2\pi)^4}\frac{d^4\eta}{(2\pi)^4}\mathrm{e}^{i(p-p_1)\zeta}\mathrm{e}^{i\eta p_1}\langle P,S|\bar\psi(0)A^+(\eta)\bar\psi(\zeta)|P,S\rangle, \tag{6}$$

while the $\cos 2\phi$ azimuthal asymmetry arises from a convolution of $\Phi_A^+(p,p_1)$ and $\Delta_A^-(k,k_1)$, namely the Collins and Boer–Mulders functions. In the fragmentation correlator, the $[-]$ indicates that the path ordered gauge-link $\mathcal{G}_{[-\infty,\eta^+]} = \mathcal{P}\exp\big(-ig\int_{-\infty}^{\eta^+} d\xi^+ A^-(\xi)\big)$, runs from $-\infty$ to ξ^+ along the light-cone "plus" direction. In this approximation these quark–gluon–quark correlation functions are the starting points for estimating the gauge link contributions to the Boer–Mulders, Sivers, and Collins functions which enter the factorized cross sections in Eqs. (1) reported in Refs. 12, 14–18. Adopting a spectator framework we approximate the non-perturbative cor-

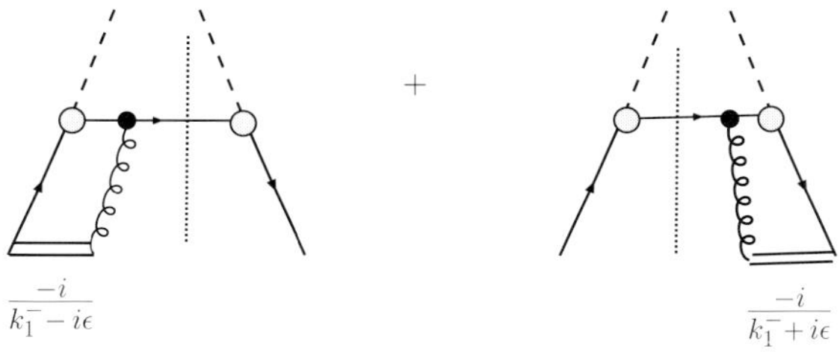

Fig. 2. Feynman diagrams depicting the gauge link contribution to T-odd contribution to quark distribution and fragmentation functions in the spectator framework. The double line represents the eikonal fermion propagator emerging from the gauge link. This propagator is associated with the gluonic pole contribution to the Collins function.

relators as a spectator diquark and quark in the TMD distribution and fragmentation functions respectively. The quark–pion correlations appearing in Eqs. (4–6) in momentum space are given by

$$\langle 0|\psi(0)|P_h, X\rangle = \left(\frac{i}{\not k - m + i\epsilon}\right)\Upsilon(k_\perp^2)\, u(k - P_h, s), \tag{7}$$

where $\Upsilon(k_\perp^2)$ is the form factor of the quark–pion vertex function. Noting that parton intrinsic transverse momentum yields a natural regularization for the moments of these distributions, we incorporate a Gaussian from factors into our model, $\Upsilon(k_\perp^2) = i\gamma_5 f_{qq\pi} e^{-b' k_\perp^2}$. The quark–gluon–pion correlation is modeled according to the Greens function

$$\begin{aligned}\langle P_h, X|A^-(k_1)\psi(0)|0\rangle = \bar{u}(k - P_h)\frac{ig(k_1^2)}{k_1^2 + i\epsilon}\gamma^- \frac{i}{\not k - \not k_1 - \not P_h + \mu + i\epsilon}\\ \times\, \Upsilon(k_\perp^2)\frac{i}{\not k - \not k_1 - m + i\epsilon}.\end{aligned} \tag{8}$$

Taking the sum of diagrams in Fig. 2 and performing the loop integral under the assumption that the virtual photon is off shell and that the form factors model the analytic structure of the non-perturbative correlation function we pick up the gluonic-pole from the gauge link. The resulting expression for the Collins function in this approach is given by

$$H_1^{\perp}(z,k_{\perp}) = \mathcal{N}'\alpha_s \frac{M_\pi}{4z}\frac{(1-z)}{z^2}\frac{(\mu - m(1-z))}{\Lambda'(k_{\perp}^2)\,k_{\perp}^2}\mathcal{R}(z,k_{\perp}^2), \tag{9}$$

where $\mathcal{R}(k_{\perp}^2;x) = \exp^{-2b(k_{\perp}^2-\Lambda(0))}\left(\Gamma(0,2b\Lambda(0)) - \Gamma(0,2b\Lambda(k_{\perp}^2))\right)$ is the regularization function. $\Lambda'(k_{\perp}^2) \equiv k_{\perp}^2 + \frac{1-z}{z^2}M_\pi^2 + \frac{\mu^2}{z} - \frac{1-z}{z}m^2$ and μ is the quark-spectator mass. $\mathcal{N}'$ is determined from the normalization on the unpolarized fragmentation function $D_1^u(z)$.[19] An analogous procedure leads to the Boer–Mulders and Sivers functions.[12,14] The scalar diquark contribution to the Boer Mulders function is

$$h_1^{\perp}(x,p_{\perp}) = \mathcal{N}\alpha_s M\frac{(1-x)(m+xM)}{\Lambda(p_{\perp}^2)\,p_{\perp}^2}\mathcal{R}(x,p_{\perp}^2) \tag{10}$$

where $\Lambda(p_{\perp}^2) = p_{\perp}^2 + (1-x)m^2 + x\lambda^2 - x(1-x)M^2$, and M, m, and λ are the nucleon, quark, and spectator diquark masses respectively. $\mathcal{N}$ is a normalization factor and $b = 1/\langle p_{\perp}^2\rangle$. Both parameters are determined with respect to the unpolarized u-quark distribution $f_1^{(u)}(x,p_{\perp})$ normalized with respect to valence distributions of,[20] where $\langle p_{\perp}^2\rangle = (0.4)^2\ \mathrm{GeV}^2$. The Collins and Sivers weighted asymmetries are projected from the differential cross section, Eq. (1): The Collins asymmetry is given by

$$\left\langle \frac{P_{h\perp}}{M_\pi}\sin(\phi+\phi_s)\right\rangle_{UT} = \frac{|S_T|2(1-y)\sum_q e_q^2 h_1(x) z H_1^{\perp(1)}(z)}{(1+(1-y)^2)\sum_q e_q^2 f_1(x) D_1(z)}\,. \tag{11}$$

In addition to TSSAs, an unpolarized nucleon target yields a double T-odd azimuthal asymmetry in SIDIS.[3] This correlation emerges as a convolution of the Boer–Mulders and Collins functions in the cross section

$$\frac{d\sigma_{UU}^{\ell N\to \ell\pi X}}{dxdydzd^2P_{h\perp}} = \frac{2\alpha^2}{Q^2 y}\left\{(1-y)\cos 2\phi_h \sum_q e_q^2\,\mathcal{F}\left[\frac{2(\boldsymbol{k}_\perp\cdot\hat{\boldsymbol{h}})(\boldsymbol{p}_\perp\cdot\hat{\boldsymbol{h}}) - \boldsymbol{k}_\perp\cdot\boldsymbol{p}_\perp}{MM_h}\,h_1^{\perp q} H_1^{\perp q}\right]\right\}.$$

An analogous azimuthal asymmetry enters the Drell–Yan process.[4,21,22] While the leading effect enters the $\cos 2\phi$ asymmetry with $h_1^{\perp}$ convoluted with the Collins function $H_1^{\perp}$,[8] an ordinary T-even kinematic sub-leading twist contribution competes at moderate Q^2. To order $1/Q^2$ it combines with the leading twist contribution as

$$\langle\cos 2\phi\rangle_{UU} = \frac{2\frac{\langle k_{\perp}^2\rangle}{Q^2}(1-y) f_1(x) D_1(z) + 8(1-y) h_1^{\perp(1)}(x) H_1^{\perp(1)}(z)}{\left[1+(1-y)^2 + 2\frac{\langle k_{\perp}^2\rangle}{Q^2}(1-y)\right] f_1(x) D_1(z)}. \tag{12}$$

The x and z-dependence of the Collins and Sivers TSSA at HERMES kinematics[2] are displayed in Fig. 3. In addition we display in Fig. 4 azimuthal

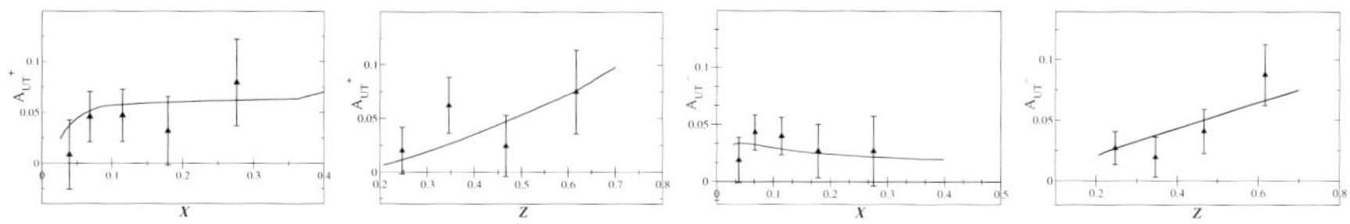

Fig. 3. Left two panels: the $\langle \sin(\phi + \phi_s) \rangle_{UT}$ asymmetry for π^+ production as a function of x and z compared to the HERMES data. Right two panels: $\langle \sin(\phi - \phi_S) \rangle_{UT}$ as a function of x and z compared with HERMES data.

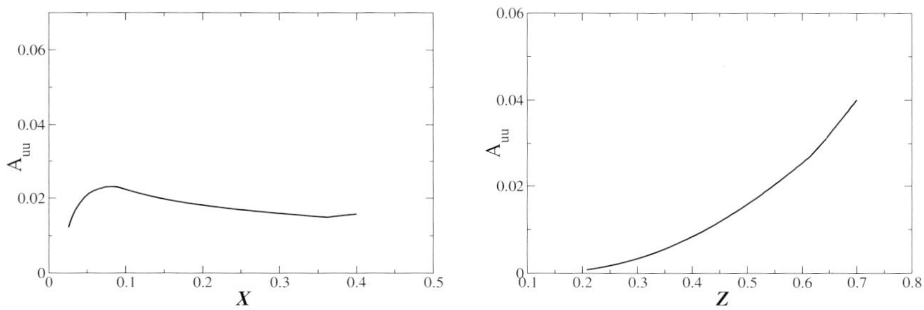

Fig. 4. Left Panel: The $\langle \cos(2\phi) \rangle_{UU}$ asymmetry for π^+ production as a function of x and z at HERMES kinematics.

asymmetry at HERMES kinematics. While this latter weighted asymmetry is fairly small (2 to 4 %), one expects unweighted asymmetry to be larger.

Now we consider the approach advocated in Refs. 23 and 24 where the gauge link contribution to the Collins function correlator was modeled from the scattering subprocess, off-shell photon–quark scattering into a pion and quark, $\gamma^\star(q) + q(p) \to \pi(P_h) + X$. $\hat{s} = (q+p)^2$ is assumed to be positive. We perform the calculation of this subprocess in the center of mass frame for the off-shell photon and incoming quark, to which we boost from the collinear nucleon–photon frame.[25] In this frame the pion momentum possesses a transverse component, $P_h = (P_h^-, P_h^+, P_\perp)$. Performing the loop integrations [25] we obtain the following contributions to the Collins function

$$H_1^\perp(z, k_\perp) = \mathcal{N}'' \alpha_s \frac{M_\pi}{4z}(1-z) \frac{\mathcal{I}_1(z, P_\perp^2) + \mathcal{I}_2(z, P_\perp^2)}{\Lambda'(P_\perp^2) P_\perp^2}, \tag{13}$$

where

$$\mathcal{I}_1 = \pi\big(\mu - m(1-z)\big) \frac{E_\pi + P\cos\theta}{P + E_\pi \cos\theta} \left[\ln \frac{(P + E_\pi \cos\theta)^2}{\mu^2} - \cos\theta \ln \frac{4P^2}{\mu^2} \right]$$

$$\mathcal{I}_2 = \pi\, z\, m \frac{P \sin^2\theta}{E_\pi - P\cos\theta} \ln \frac{4P^2}{\mu^2}, \tag{14}$$

and we denote $P \equiv |\mathbf{P}_h|$ and $P_\perp^2 = k_\perp^2/z^2$. As in the case of the gluonic pole contribution, this result survives the limit that the incoming quark mass $m \to 0$. Thus, both contributions depend on the non-perturbative spectator mass μ. It is important to point out that the contribution of Eq. (13) requires that the $\hat{s}$-channel is physical, *i.e.* $\hat{s} = k^2 = -Q^2 + p^2 + 2q \cdot p \geq (M_\pi + \mu)^2$. Connecting this to the full process kinematics for $\gamma + N(P_N) \to \pi + X$, we see that if $p = xP_N$, the collinear quark leading order assumption, then $\hat{s} = (Mx)^2 - P_h^2$, is not necessarily time-like where the photon is space-like and off mass shell. Noting, however, that the quark is not collinear with the nucleon, the relation between x and p^+/P_N^+ is altered – a small difference between the two can be magnified via Q^2 to be a large time-like $\hat{s}$. Considering the t-channel point of view, $t = (P_h - q)^2$ is restricted to be below the threshold for the t-channel to be in the physical region by the kinematics for the overall process, namely $M_X \geq M_N$. Thus, one might interpret this result as a model for the so-called final state interaction (FSI) contribution to the Collins function.[17] By contrast, allowing the M_X to be below that limit by a small amount opens the possibility that t can be above the physical threshold for a range of x values, suggesting that the calculation of the imaginary part would involve the eikonal pole and spectator propagator both on-shell. This implies the existence of the gluonic pole contribution to the Collins function which we modeled above and in Ref. 14. The narrow region of the kinematics to allow for the t-channel picture implied by the model-gluonic pole contribution[26,18,25] maybe as reasonable as the $\hat{s}$-channel picture we pursued here. In either case, there is a small "stretch" from the fully constrained SIDIS kinematics in order to get a pair of the cuts to be physical. This circumstance may reflect a problem with the simple spectator framework or a more general breakdown of universality. Said otherwise one is led to consider whether both the gluonic pole and FSI contributions contribute to the Collins function as was formally suggested by Boer Pijlman and Mulders.[17]

Finally, we note that including higher order contributions from the path-ordered Wilson line leads to a generalization of Eq. (4) that links the loop momenta of the various gluon legs[27] in these model calculations. Such contributions indicate a rich analytic structure suggesting the decomposition of T-odd fragmentation correlations into so-called "FSI" and gluonic pole contributions as suggested in Ref. 17. In this sense it is important to investigate the full analytic structure of the T-odd fragmentation functions defined from the SIDIS factorization theorems beyond one loop.

Acknowledgments

We thank Phil Ratcliffe and Enzo Barone and the organizers of *COMO 2005* for the invitation to present our work. We acknowledge fruitful discussions with D. Boer, F. Pijlman, A. Bacchetta, F. Yuan and C. Bomhof. GRG acknowledges support from the US Department of Energy, DE-FG02-29ER40702.

References

1. K. Heller *et al.*, *Phys. Rev. Lett.* **41** (1978) 607; E704 Collaboration: D.L. Adams *et al.*, *Phys. Lett.* **B261** (1991) 201; STAR Collaboration: J. Adams *et al.*, *Phys. Rev. Lett.* **92** (2004) 171801.
2. A. Airapetian *et al.*, *Phys. Rev. Lett.* **94** (2005) 012002.
3. D. Boer and P.J. Mulders, *Phys. Rev.* **D57** (1998) 5780.
4. D. Boer, *Phys. Rev.* **D60** (1999) 014012.
5. X. Ji, J. Ma and F. Yuan, *Phys. Rev.* **D71** (2005) 034005
6. J.C. Collins and A. Metz, *Phys. Rev. Lett.* **93** (2004) 252001.
7. D. Sivers, *Phys. Rev.* **D41** (1990) 83.
8. J.C. Collins, *Nucl. Phys.* **B396** (1993) 161.
9. M. Anselmino, M. Boglione and F. Murgia, *Phys. Lett.* **B362** (1995) 164.
10. S.J. Brodsky, D.S. Hwang and I. Schmidt, *Phys. Lett.* **B530** (2002) 99.
11. X. Ji and F. Yuan, *Phys. Lett.* **B543** (2002) 66.
12. G.R. Goldstein and L.P. Gamberg, arXiv:hep-ph/0209085, proc. ICHEP 2002.
13. G.L. Kane, J. Pumplin and K. Repko, *Phys. Rev. Lett.* **41** (1978) 1689.
14. L.P. Gamberg, G.R. Goldstein and K.A. Oganessyan, *Phys. Rev.* **D67** (2003) 071504.
15. J.C. Collins, *Phys. Lett.* **B536** (2002) 43.
16. A.V. Belitsky, X. Ji and F. Yuan, *Nucl. Phys.* **B656** (2003) 165.
17. Daniel Boer, P.J. Mulders, F. Pijlman, *Nucl. Phys.* **B667** (2003) 201.
18. L.P. Gamberg and G.R. Goldstein, *AIP Conf. Proc.* **792** (2005) 941.
19. S. Kretzer, *Phys. Rev.* **D62** (2000) 054001.
20. M. Glück, E. Reya and A. Vogt, *Z. Phys.* **C67** (1995) 433.
21. L.P. Gamberg and G.R. Goldstein, "T-odd effects in unpolarized Drell–Yan scattering", arXiv:hep-ph/0506127.
22. G.R. Goldstein and L.P. Gamberg "Transversity, Transversity-Odd Distributions and Asymmetries in Drell–Yan Processes", *these proceedings.*
23. A. Metz, *Phys. Lett.* **B549** (2002) 139.
24. D. Amrath, A. Bacchetta and A. Metz, *Phys. Rev.* **D71** (2005) 114018.
25. L. Gamberg and G.R. Goldstien, *in preparation.*
26. L.P. Gamberg, G.R. Goldstein and K.A. Oganessyan, *Phys. Rev.* **D68** (2003) 051501.
27. C.J. Bomhof, P.J. Mulders and F. Pijlman, *Phys. Lett.* **B596** (2004) 277.

TRANSVERSITY, TRANSVERSITY-ODD DISTRIBUTIONS AND ASYMMETRIES IN DRELL–YAN PROCESSES

Gary R. Goldstein*

Department of Physics and Astronomy,
Tufts University,
Medford, MA 02155, USA
E-mail: gary.goldstein@tufts.edu

L.P. Gamberg

Division of Science,
Penn State-Berks Lehigh Valley College,
Reading, PA 19610, USA
E-mail: lpg10@psu.edu

After a brief recap of *Transversity* it is noted that Drell–Yan unpolarized processes display large azimuthal asymmetries. One such asymmetry, $\cos(2\phi)$, is directly related to the leading twist transversity distribution $h_1^{\perp}(x, k_T)$. We use a model developed for semi-inclusive deep inelastic scattering that determines the Sivers function $f_{1T}^{\perp}(x, k_T)$ to predict the Drell–Yan asymmetry ν as a function of q^2, q_T and x. The resulting predictions include a non-leading twist contribution from spin-averaged distributions that measurably effect lower energy results.

1. Introduction

The use of *transversity* for describing exclusive scattering processes [1] was introduced at a time when spin and polarization measurements had become feasible and innovative at high energy accelerators. Transversity states are combinations of helicity states, $|\perp/\top\rangle \sim (\,|+\rangle \pm (i)|-\rangle\,)$, for a moving nucleon or spin $\frac{1}{2}$ particle. Transversity is a spin variable introduced to reveal an underlying simplicity in hadronic scattering. For nucleon–nucleon spin dependent scattering amplitudes $f_{a,b;c,d}(s,t)$, with $a \ldots d$ being transversity eigenvalues, a particular phase simplicity results [2] when analyzing polariza-

*Work partially supported by grant DE-FG02-92ER40702 of the U.S. Department of Energy.

tion data.

For single-spin asymmetry (SSA) in two-body or inclusive reactions, parity conservation requires that the spin only be oriented perpendicular to a scattering plane, $\langle \mathbf{S} \cdot \mathbf{n} \rangle$, (the eigenstates are transversity states) where $\mathbf{n}$ must be proportional to a vector product of momenta, e.g. $\mathbf{n} \sim \mathbf{p_1} \times \mathbf{p_2}$. To have SSA, then, necessitates having some transverse component of p_2 relative to p_1. Note that transversity states for spin 1 have the form $|\pm 1)_T = \left(|+1\rangle \pm \sqrt{2}|0\rangle + |-1\rangle \right)/2$ and $|0)_T = \left(|+1\rangle - |-1\rangle \right)/\sqrt{2}$. For massless photons or gluons $|+1)_T = |-1)_T$ and $|+1)_T$ or $|0)_T$ correspond to linear polarization normal to or parallel to the plane – there is no distinction between +1 and -1 transversity. This is important for the following.

A spin asymmetry for spin $\frac{1}{2}$ particle D in process A+B→C+D corresponds to

$$\langle \mathbf{S} \cdot \mathbf{n} \rangle \propto \Sigma f^*_{a,b;c,d} [\sigma \cdot \mathbf{n}]_{d,d'} f_{a,b;c,d'} \propto \Sigma \mathrm{Im}[f^*_{a,b;c,+} f_{a,b;c,-}] \tag{1}$$

in helicity basis, or

$$\propto \Sigma \left\{ |f_{a,b;c,+\perp}|^2 - |f_{a,b;c,-\perp}|^2 \right\} \tag{2}$$

in transversity basis. Hence a non-vanishing asymmetry requires both a phase difference among helicity amplitudes and helicity change. If the particles are massless quarks, there will not be helicity change. If only tree-level processes are considered, there will be no relative phases. Hence it was predicted long ago that SSAs will be small in PQCD.[3] Indeed, an exhaustive calculation was carried out that incorporates the requisite phases through interference of tree level and loop contributions *in PQCD* in an attempt to explain spin asymmetry in Λ production.[4] As expected, the single quark polarization goes like $\alpha_s m/Q$ where m represents a non-zero quark mass (to allow for helicity flip) and Q represents the hard QCD scale. For inclusive reactions, the same conclusion can be reached by taking the amplitudes as two-body helicity amplitudes for the production of a fixed hadron and a state X. Through the generalized optical theorem, SSA in inclusive reactions can be related to discontinuities in helicity flipping three-body forward scattering amplitudes.[5]

How do these concepts appear at the quark–gluon level? The structure of the nucleon, revealed through parton distribution functions, involves $f_1(x, Q^2)$, the unpolarized distribution, $g_1(x, Q^2)$, which measures the transfer of longitudinal polarization or helicity from the nucleon to the quark constituents (mediated by gluons), and $h_1(x, Q^2)$, for the transfer of transversity from nucleon to quarks. The latter is chiral odd and can not be observed in DIS – it involves chirality flip along quark lines in PQCD diagrams. An extra hadron is needed to study $h_1(x, Q^2)$ – a transverse momentum is essential for defining transversity as we suggested above – as was known earlier regarding Drell–Yan processes [6] or semi-inclusive deep inelastic scattering (SIDIS).[7] These affordances inspired the first searches for transversity properties of the nucleon.[8,9] A good, model independent determination of h_1 still remains illusive, since it must be sorted out of convoluted functions that contribute either to measurable double transversity correlations or spin and azimuthal asymmetry combinations.[10] SSAs and single azimuthal asymmetries are easier to measure and have yielded some unexpectedly large results.[11]

Focusing on SSAs and single azimuthal asymmetries (SAAs) leads us to consider "Transversity – odd" observables, distributions and fragmentation functions. Commonly referred to as "T-odd", these quantities do not violate Time Reversal Invariance, but vanish in tree-level calculations with T-conserving interactions, like leading order calculations of two-body decays in effective field theories. In that latter example it is through final state interactions that non-zero results are obtained. A similar mechanism is at work in modeling the T-odd distribution functions that we will use in predicting SAAs in Drell–Yan processes. Once transverse momentum dependence of parton distributions enters the picture of scattering processes a much larger set of distribution and fragmentation functions,[12] transverse momentum dependent(TMD) functions, become possible and relevant, particularly for spin asymmetries. Among such functions are these leading twist T-odd quark distribution [13] and fragmentation functions.[14] Their existence was suggested by Sivers to account for the significant SSAs in inclusive reactions (*e.g.* $pp^{\uparrow} \rightarrow \pi X$),[13,15] by Collins in SIDIS,[14] and by Boer [10,16] in Drell–Yan scattering. In contrast to the previously cited SSAs, such effects go like $\alpha_s\langle k_\perp\rangle/M$, where now M plays the role of the chiral symmetry breaking scale and $k_\perp$ is characteristic of quark intrinsic motion.

2. T-odd distribution functions

Considering the soft contributions to hadronic processes opens up the possibility that there are non-trivial leading-twist TMDs, with transverse single spin asymmetries at the parton level that can contribute to SSAs and SAAs.[17] These T-odd distributions [10,13,15,18] are of importance; they must possess both transversity properties and the necessary phases. One T-odd function, called the Sivers function, $f_{1T}^{\perp}(x, k_T^2)$, can be interpreted as the probability to find an unpolarized quark with transverse momentum k_T

inside a positive transversity nucleon. Another, called the Boer–Mulders function, $h_1^\perp(x, k_T^2)$, is interpreted as the probability to find a positive transversity quark with k_T inside an unpolarized nucleon.

A great deal of progress has been made in understanding the role of these functions in the last several years, following the work of Brodsky, Hwang and Schmidt (BHS) who showed that final state interactions can give rise to a polarization as expected from a Sivers function.[19] Several researchers generalized the derivation thereafter.[20,21] We have performed detailed calculations of some of the TMD functions in the quark–diquark spectator model of the quark-hadron interactions, following BHS, where absorptive effects are generated by gluon loop corrections to tree-level.[21,22] Formally, these non-zero T-odd distributions emerge from the gauge link insertion into the formal definitions of the parton distributions. These loop corrections were first applied to SIDIS, thereby giving rise to predictions for SSA's and angular asymmetries.

For the T-odd terms the basic diagram has the gluon exchange and the eikonalized struck quark line arising from the gauge link. We showed that the determination of $f_{1T}^\perp$ by BHS [19] could be extended to the other T-odd functions,[21] particularly yielding the model dependent result $h_1^\perp = f_{1T}^\perp$ for scalar diquarks. Noting that parton intrinsic transverse momentum yielded a natural regularization for the moments of these distributions, we incorporated a factor of $e^{-bk_T^2}$ into our model for the hadronic structure.[22,23] The result was

$$h_1^\perp(x, k_T) = \mathcal{N}\alpha_s M \frac{(1-x)(m+xM)}{k_T^2 \Lambda(k_T^2)} \mathcal{R}_h(k_T^2, x)\,, \tag{3}$$

where $\mathcal{R}$ is the regularization function $\mathcal{R}(k_T^2, x) = e^{-2b(k_T^2-\Lambda(0))} \times \big(\Gamma(0, 2b\Lambda(0)) - \Gamma(0, 2b\Lambda(k_T^2))\big)$ and $\mathcal{N}$ is an overall constant for our u-quark dominated model.

3. Drell–Yan and Transversity

In unpolarized Drell–Yan scattering early cross section data as a function of the transverse momentum of the muon pair indicated deviations from the Bjorken scaling prediction [24,25] . The implication was that the collinear approximation was insufficient to describe the data.[26,27] Transverse momentum of a parton arises due to hard Bremsstrahlung of gluons, which is calculable from PQCD when the momentum transfers are large.[28] On the other hand, quark confinement implies that quarks have soft, primordial or intrinsic transverse momenta $k_\perp$. This latter effect is significant at low transverse momentum, $q_T \ll Q$. q_T dependence has been incorporated into the factorized Drell–Yan model[6,17] by extending the parton probability distribution to be a function of $k_\perp$. Once the parton distributions within the incoming hadrons have transverse momentum dependence, there will be a continuum of values of their $k_\perp$ for which a time-like photon of fixed 4-momentum will be formed. This complicates the convolution of the quark-antiquark distributions.

The angular dependence for the Drell–Yan process can be expressed as

$$\frac{dN}{d\Omega} = \frac{3}{4\pi}\frac{1}{\lambda+3}\left(1 + \lambda\cos^2\theta + \mu\sin 2\theta\cos\phi + \frac{\nu}{2}\sin^2\theta\cos 2\phi\right). \tag{4}$$

The solid angle Ω refers to the lepton pair orientation in the pair rest frame relative to the boost direction and the incoming hadrons' plane.[29] The functions λ, μ, ν depend on $s, x, m^2_{\mu\mu}, q_T$, the center of mass energy, the fraction of quark momentum in the hadron, the invariant mass of the produced lepton pair, and the transverse momentum of the dimuon pair, respectively.

In an early attempt to explain experimental disagreement with collinear PQCD predictions Collins and Soper[29] used transverse momentum dependent distributions to obtain, among the other asymmetry functions, a T-even non-leading twist contribution to the $\cos 2\phi$ asymmetry

$$\nu_4 = \frac{\frac{1}{Q^2}\sum_a e_a^2 \mathcal{F}\left[w_4\, f_1(x,k_\perp)\bar{f}_1(\bar{x},p_\perp)\right]}{\sum_a e_a^2 \mathcal{F}\left(f_1(x,k_\perp)\bar{f}_1(\bar{x},p_\perp)\right)}, \tag{5}$$

where $w_4 = 2\big(\hat{\boldsymbol{h}}\cdot(k_\perp - p_\perp)\big)^2 - (k_\perp - p_\perp)^2$. Experimental measurements of $\pi + p \to \mu^+ + \mu^- + X$[30,31] discovered unexpectedly large values of the asymmetries compared to naive quark model expectations resulting in a serious violation of the Lam–Tung relation,[32] $1 - \lambda - 2\nu = 0$. Recently D. Boer[16] proposed that there is a dominant leading twist contribution to ν coming from the T-odd distributions $h_1^\perp(x,k_T)$ for both hadrons which dominates in the kinematic range, $\boldsymbol{q_T} \ll Q$. The $\cos 2\phi$ azimuthal asymmetry in unpolarized $p\,\bar{p} \to \mu^+\,\mu^-\,X$ would involve the convolution[16,33–35]

$$\nu_2 = \frac{\sum_a e_a^2 \mathcal{F}\left[w_2\, h_1^\perp(x,k_\perp)\bar{h}_1^\perp(\bar{x},p_\perp)/(M_1 M_2)\right]}{\sum_a e_a^2 \mathcal{F}\left[f_1(x,k_\perp)\bar{f}_1(\bar{x},p_\perp)\right]}, \tag{6}$$

where $w_2 = (2\hat{\boldsymbol{h}}\cdot k_\perp \cdot \hat{\boldsymbol{h}}\cdot p_\perp - p_\perp \cdot k_\perp)$ is the weight in the convolution integral, $\mathcal{F}$.

Boer first used a simple model for these distributions, inspired by Collins' *ansatz* for the transversity fragmentation functions. That assumption led to a simple Q_T dependent ν which could be fit to the large values of the πp data. Further work along those lines [33] incorporated a more realistic model for the T-odd functions, as first developed in SIDIS [19] for the functions $f_{1T}^{\perp}(x, k_T)$ which were related to the $h_1^{\perp}(x, k_T)$ in Ref. 21.

Herein we extend our calculations for T-odd contributions to the unpolarized Drell–Yan $p + \bar{p}$ scattering. first reported in Ref. 34. We perform a detailed analysis of this effect and compare the double T-odd contribution to the conventional subleading twist T-even contribution.[29] That is, we combine both Eq. (6) and Eq. (5).

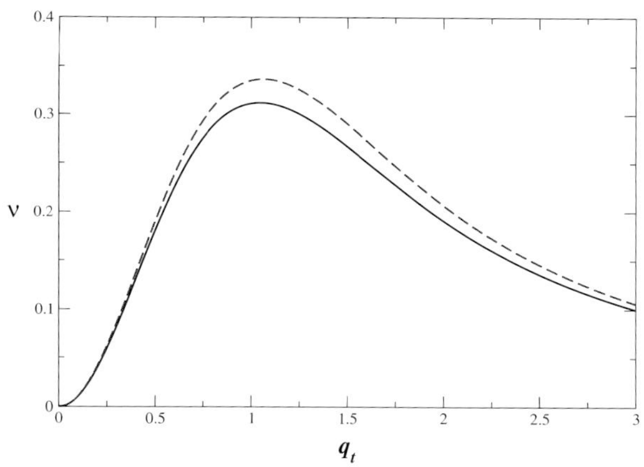

Fig. 1. The variable ν plotted as a function of q_T for $s = 50\,\text{GeV}^2$.

In evaluating the convolution, with x and $\bar{x}$ being the fractional longitudinal momenta of the quark and antiquark, there are constraints, $x\bar{x} = \tau = Q^2/s$ and $x - \bar{x} = x_F$. We have obtained values of the asymmetry ν as a function of three variables, x, q_T, and q (or $m_{\mu\mu}$). Due to the constraints the allowed range of x is restricted for each q value, from $x_{\text{min}} = q^2/s$ to 1. Furthermore, evaluating the convolutions of $h_1^{\perp}\bar{h}_1^{\perp}$ and $f_1\bar{f}_1$ for a sampling of x will not treat the $\bar{x}$ and the corresponding antiparticle structure functions symmetrically. So it is more appropriate to use the symmetrical variable x_F. At $s = 50\,\text{GeV}^2$ we take $q_T \leq 3\,\text{GeV}/c$ and $3\,\text{GeV}/c \leq q \leq 6\,\text{GeV}/c$. $\nu(q_T)$ and $\nu(x)$ are shown in Figs. 1 and 2.

The additional contribution of the twist 4, Eq. (6), for $s = 50\,\text{GeV}^2$ to each of the partially integrated functions ν is shown in Figs. 1 and 2 as slightly higher curves. There are corresponding $\nu(q)$, $\nu(\zeta)$ and $\nu(x_F)$ shown in our recent paper.[36] At most the additional contribution is around 10%.

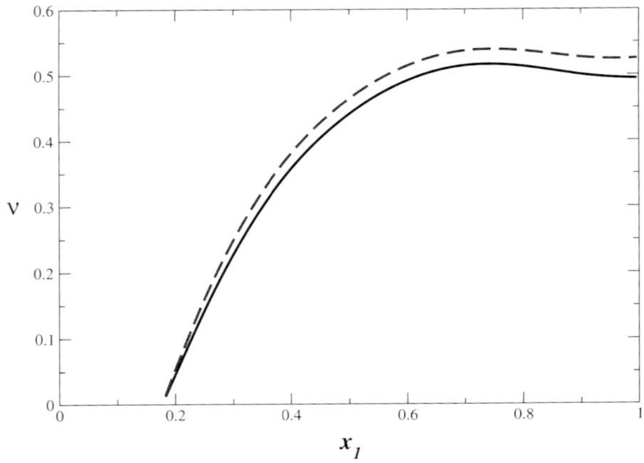

Fig. 2. The variable ν plotted as a function of x for $s = 50\,\mathrm{GeV}^2$.

For higher s values the effect is even smaller, as expected.

A perusal of the figures shows that the $\cos 2\phi$ azimuthal asymmetry ν is not small at center of mass energies of $50\,\mathrm{GeV}^2$. The distinction between the leading order T-odd and sub-leading order T-even contributions diminishes at center of mass energy of $s = 500\,\mathrm{GeV}^2$. Thus, aside from the competing T-even effect, the experimental observation of a strong x-dependence would indicate the presence of T-odd structures in *unpolarized* Drell–Yan scattering, implying that novel transversity properties of the nucleon can be accessed without invoking beam or target polarization.

Acknowledgments

We wish to thank the organizers of this very productive workshop set on beautiful Lake Como, as well as the many participants who advanced our understanding of transversity.

References

1. G.R. Goldstein and M.J. Moravcsik, *Ann. Phys.* **98** (1976) 128.
2. F. Arash, G.R. Goldstein and M.J. Moravcsik, *Phys. Rev.* **D31** (1985) 2360.
3. G.L. Kane, J. Pumplin, and K. Repko, *Phys. Rev. Lett.* **41** (1978) 1689.
4. W.G.D. Dharmaratna and G.R. Goldstein, *Phys. Rev.* **D41** (1990) 1731; *Phys. Rev.* **D53** (1996) 1073.
5. G.R. Goldstein and J.F. Owens, *Nucl. Phys.* **B103** (1976) 145.
6. J. Ralston and D.E. Soper, *Nucl. Phys.* **B152** (1979) 109.
7. J.C. Collins, *Nucl. Phys.* **B396** (1993) 161.

8. G. Bunce, N. Saito, J. Soffer, W. Vogelsang, *Ann. Rev. Nucl. Part. Sci.* **50** (2000) 525.
9. A. Airapetian *et al.*, *Phys. Rev. Lett.* **84** (2000) 4047.
10. D. Boer and P.J. Mulders, *Phys. Rev.* **D57** (1998) 5780.
11. A. Airapetian *et al.*, *Phys. Rev. Lett.* **94** (2005) 012002.
12. R.D. Tangerman and P.J. Mulders, *Nucl. Phys.* **B461** (1996) 197.
13. D. Sivers, *Phys. Rev.* **D41** (1990) 83.
14. J.C. Collins, *Nucl. Phys.* **B396** (1993) 161.
15. M. Anselmino, M. Boglione and F. Murgia, *Phys. Lett.* **B362** (1995) 164.
16. D. Boer, *Phys. Rev.* **D60** (1999) 014012.
17. D.E. Soper, *Phys. Rev. Lett.* **43** (1979) 1847.
18. J.C. Collins, *Phys. Lett.* **B536** (2002) 43.
19. S.J. Brodsky, D.S. Hwang, and I. Schmidt, *Phys. Lett.* **B530** (2002) 99.
20. X. Ji and F. Yuan, *Phys. Lett.* **B543** (2002) 66; A.V. Belitsky, X. Ji and F. Yuan, *Nucl. Phys.* **B656** (2003) 165.
21. G.R. Goldstein and L.P. Gamberg, arXiv:hep-ph/0209085, proc. ICHEP 2002, eds. S. Bentvelsen *et al.*, Amsterdam, The Netherlands (North-Holland 2003), p. 452.
22. L.P. Gamberg, G.R. Goldstein and K.A. Oganessyan, *Phys. Rev.* **D67** (2003) 071504.
23. L.P. Gamberg, G.R. Goldstein and K.A. Oganessyan, *Phys. Rev.* **D68** (2003) 051501.
24. A.S. Ito *et al.*, *Phys. Rev.* **D23** (1981) 604.
25. D. Antreasyan *et al.*, *Phys. Rev. Lett.* **48** (1982) 302.
26. G. Altarelli, R.K. Ellis, M. Greco, and G. Martinelli, *Nucl. Phys.* **B246** (1984) 12.
27. J. Collins, D. Soper, and G. Sterman, *Nucl. Phys.* **B250** (1985) 199.
28. J.C. Collins, *Phys. Rev. Lett.* **42** (1979) 291.
29. J.C. Collins and D.E. Soper, *Phys. Rev.* **D16** (1977) 2219.
30. E615 Collaboration: J.S. Conway *et al.*, *Phys. Rev.* **D39** (1989) 92.
31. S. Falciano *et al.* [NA10 Collaboration], *Z. Phys.* **C31** (1986) 513.
32. C.S. Lam and W.K. Tung, *Phys. Rev.* **D21** (1980) 2712.
33. D. Boer, S.J. Brodsky, and D.S. Hwang, *Phys. Rev.* **D67** (2003) 054003.
34. L.P. Gamberg, G.R. Goldstein, and K.A. Oganessyan, *AIP Conf. Proc.* **747** (2005) 159 [arXiv:hep-ph/0411220].
35. Z. Lu and B.-Q. Ma, *Phys. Lett.* **B615** (2005) 200 [arXiv:hep-ph/0504184].
36. L.P. Gamberg and G.R. Goldstein, arXiv:hep-ph/0506127.

ALTERNATIVE APPROACHES TO TRANSVERSITY: HOW CONVENIENT AND FEASIBLE ARE THEY?

M. Radici

Dipartimento di Fisica Nucleare e Teorica, Università di Pavia, and
Istituto Nazionale di Fisica Nucleare, Sezione di Pavia
via Bassi 6, I-27100 Pavia, Italy
E-mail: radici@pv.infn.it

The complete knowledge of the nucleon spin structure at leading twist requires also addressing the transverse spin distribution of quarks, or transversity, which is yet unexplored because of its chiral-odd nature. Elaborating strategies to extract it from (spin) asymmetry data represents a unique opportunity to explore more generally the transverse spin structure of the nucleon and the transverse momentum dynamics of partons inside it. Here, we critically review some of the most promising approaches.

1. Introduction

The predictions that the nucleon tensor charge is much larger than its helicity and that the evolution of transversity should be weaker than the helicity one, are counterintuitive and they represent a basic test of QCD in the nonperturbative domain (for a review, see Refs. 1 and 2).

The pioneering suggestion of extracting the transversity from the Drell–Yan process with transversely polarized protons [3] opened the way to a deeper insight of the spin structure of the proton. In fact, if the cross section depends explicitly upon the transverse momentum of the lepton pair, interesting information can be inferred by using unpolarized and single-polarized Drell–Yan with antiproton beams. Two leading-twist convolutions of novel distribution functions open the door on studies of the orbital motion of partons inside hadrons.[4] In Sec. 2, we will simulate the corresponding spin asymmetries in order to explore the feasibility of such measurements at the future HESR facility at GSI. We will also compare the results with what can be expected if the antiproton beam is replaced by a pion beam in the kinematic conditions reachable at COMPASS, such that the center-of-mass (cm) energy of the reaction is the same.

The growing interest in the transversity reflected in a rich experimental program at several laboratories. In particular, single-spin asymmetries have been measured in SIDIS with transversely polarized targets.[5] A possible interpretation in terms of the Collins effect requires the cross section to depend explicitly upon the transverse momentum of the detected pion with respect to the jet axis. This fact brings in several complications, including the overlap with other competing mechanisms and more complicated factorization proofs and evolution equations.

It seems more convenient to consider more exclusive final states, where, e.g., 2 pions are inclusively detected.[6] The chiral-odd partner of transversity can be represented by the so-called Interference Fragmentation Function (IFF) $H_1^{\sphericalangle}$[7] and it can be extracted by looking for asymmetric orientations of the plane containing the pion pair with respect to the scattering plane.[8] In Sec. 3, we will briefly recall the advantages of this strategy and present new results in comparison with upcoming data from the HERMES collaboration.[9]

2. Single-polarized Drell–Yan at GSI and COMPASS

The polarized part of the cross section for the process $\bar{p}p^\uparrow \to l^+l^-X$ contains at leading twist the terms[4]

$$\frac{d\Delta\sigma^\uparrow}{d\Omega dx_1 dx_2 d\mathbf{q}_T} \propto \sum_f e_f^2 \, |\mathbf{S}_{2_T}| \left\{ - B(y) \sin(\phi + \phi_{S_2}) \, F \left[\hat{\mathbf{h}} \cdot \mathbf{p}_{1_T} \, \frac{\bar{h}_1^{\perp\, f} \, h_1^f}{M_1} \right] \right.$$
$$\left. + A(y) \sin(\phi - \phi_{S_2}) \, F \left[\hat{\mathbf{h}} \cdot \mathbf{p}_{2_T} \, \frac{\bar{f}_1^f \, f_{1T}^{\perp\, f}}{M_2} \right] \; \cdots \right\}, \quad (1)$$

where the annihilating partons with charge e_f carry transverse momenta $\mathbf{p}_{1,2_T}$ and longitudinal fractions $x_{1,2}$ of the proton momentum with mass M and transverse polarization $\mathbf{S}_{2_T}$. The functions $A(y)$ and $B(y)$ depend only on the leptonic scattering angle via $y = (1 + \cos\theta)/2$. The convolution F is defined as

$$F[\bar{f}_1 f_1] = \int d\mathbf{p}_{1_T} d\mathbf{p}_{2_T} \, \delta(\mathbf{p}_{1_T} + \mathbf{p}_{2_T} - \mathbf{q}_T) \, [\bar{f}_1(x_1, \mathbf{p}_{1_T}) f_1(x_2, \mathbf{p}_{2_T}) + 1 \leftrightarrow 2]. \quad (2)$$

2.1. *The Boer–Mulders effect*

The first term in Eq. (1) involves the transversity h_1 convoluted with the chiral-odd distribution $h_1^\perp$, which describes the influence of the quark transverse polarization on its momentum distribution inside an unpolarized parent hadron. Extraction of the latter is of great importance, because $h_1^\perp$ is believed to be responsible for the well known violation of the Lam–Tung sum rule,[10] an anomalous big azimuthal asymmetry of the corresponding unpolarized Drell–Yan cross section that still awaits for a justification.

This contribution is simulated in a Monte Carlo along the lines described in Ref. 11. The spin asymmetry is produced by dividing the events into two groups, one for positive (U) and one for negative (D) values of $\sin(\phi + \phi_{S_2})$ for each bin x_2, and then constructing the ratio $(U - D)/(U + D)$ for each bin x_2 after integrating upon x_1, θ, and $\mathbf{q}_T$. Two different test functions (ascending and descending) are used to probe the x_2 dependence. The goal is to explore under which conditions such different behaviours can be recognized also in the corresponding asymmetry A_T. In fact, in that case the measurement of A_T would allow for the extraction of unambiguous information on the analytical form of both $h_1^\perp(x)$ and $h_1(x)$.

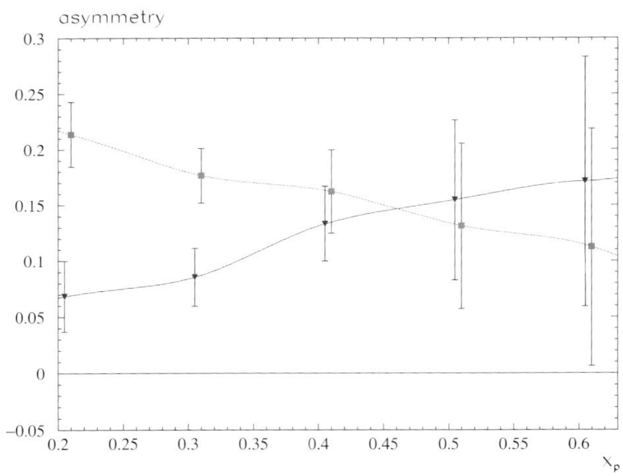

Fig. 1. Asymmetry $(U - D)/(U + D)$ from Boer–Mulders effect (see text) as function of x_2 for the process $\bar{p}p^\uparrow \to \mu^+\mu^- X$. Full squares for the descending input test function, downward triangles for the ascending one (see text). Continuous lines are drawn to guide the eye. Error bars due to statistical errors only, obtained by 20 independent repetitions of the simulation.

In Fig. 1, the asymmetry is shown for the $\bar{p}p^\uparrow \to \mu^+\mu^- X$ process that could be realized at GSI in so-called asymmetric collider mode, i.e. with $E_{\bar{p}} = 15\,\text{GeV}$ and $E_p = 3.3\,\text{GeV}$ such that the cm energy squared is

$s \approx 200\,(\text{GeV})^2$. The significant sample is made of 8000 selected events and statistical error bars are obtained by repeating 20 times the simulation for each bin x_2. For an hypothetical luminosity of $10^{31}\,\text{cm}^{-2}\,\text{s}^{-1}$, this corresponds to a running time of approximately three months (for further details see Ref. 11). From the figure, we deduce that in the range $0.1 < x_2 < 0.4$ it seems possible to extract information on the x dependence of both the transversity and $h_1^{\perp}$.

This statement can be reinforced if antiprotons are replaced by pions. The abundance of such particles allows for realistically increasing the significant sample by an order of magnitude. We selected the COMPASS setup with a π^- beam of energy $E_\pi \sim 100\,\text{GeV}$, impinging on a transversely polarized NH_3 fixed target. The corresponding $\pi^- p^\uparrow \rightarrow \mu^+\mu^- X$ process at $s \sim 200\,(\text{GeV})^2$ can be directly compared with the asymmetric collider setup at GSI. Combining the increased statistics with the stronger dilution factor, for a sample of 125000 events we can considerably shrink the statistical error bars of the asymmetry, as it is evident from Fig. 2.

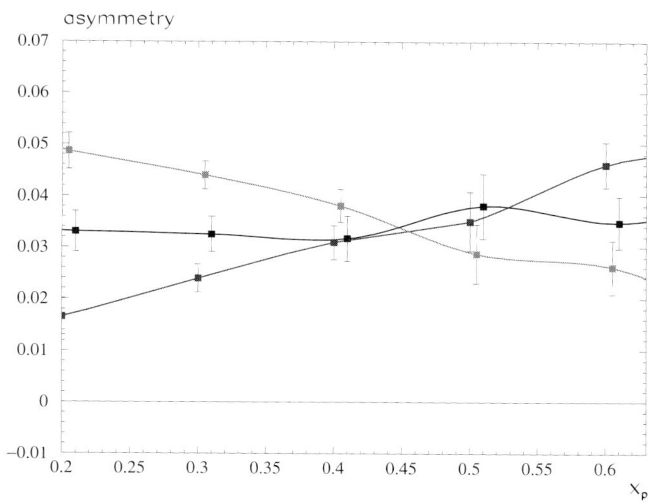

Fig. 2. Asymmetry $(U-D)/(U+D)$ (see text) as function of x_2 for the process $\pi^- p^\uparrow \rightarrow \mu^+\mu^- X$, using the same conditions as in the previous figure. The case of constant input test function is also included. Continuous lines are drawn to guide the eye. Error bars due to statistical errors only, obtained by 10 independent repetitions of the simulation.

2.2. *The Sivers effect*

The second term in Eq. (1) has a different azimuthal dependence and it involves the standard unpolarized distribution f_1 and the Sivers function $f_{1T}^{\perp}$, which describes how the distribution of unpolarized quarks is affected by the transverse polarization of the parent proton. A measurement of a nonvanishing asymmetry would be a direct evidence of the orbital angular momentum of quarks.

In order to perform the Monte Carlo simulation, we adopted for $f_{1T}^{\perp}(x, \mathbf{p}_T)$ the parametrization of Ref. 12, which was determined by fitting the HERMES data for one-pion inclusive production in DIS regime [5] assuming that such asymmetry is produced by the Sivers effect only. We have further simplified the expression by neglecting the contribution of antipartons. Similarly to the Boer–Mulders case, the asymmetry is generated in the Monte Carlo by dividing the events into two groups, one for positive (U) and one for negative (D) values of $\sin(\phi - \phi_{S_2})$ for each bin x_2, and then constructing the ratio $(U - D)/(U + D)$ for each bin x_2 after integrating upon x_1, θ, and $\mathbf{q}_T$.

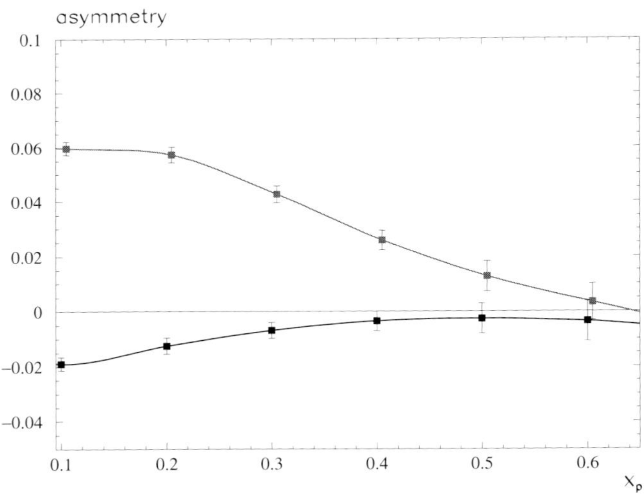

Fig. 3. Asymmetry $(U - D)/(U + D)$ from Sivers effect (see text) as function of x_2 for the process $\pi^{\pm} p^{\uparrow} \to \mu^+ \mu^- X$. Upper curve is the π^+ case, lower curve the π^- one. Continuous lines are drawn to guide the eye. Error bars due to statistical errors only, obtained by 10 independent repetitions of the simulation.

In Fig. 3, the mechanism is considered for the process $\pi^{\pm} p^{\uparrow} \to \mu^+ \mu^- X$

in the COMPASS setup, as in Fig. 2. The resulting statistical error bars are very small because of the high statistics, and selection of the beam charge can even determine the sign and size of the spin asymmetry.

3. Two-hadron inclusive production at HERMES

As already anticipated in Sec. 1, looking for more exclusive final states in SIDIS can represent a competitive alternative to the Collins mechanism. In fact, for hadron pairs collinear with the jet axis the cross section at leading twist for two-hadron inclusive SIDIS production has a very simple structure: the unpolarized contribution and the factorized product of h_1 and $H_1^{\sphericalangle}$, which describes the fragmentation of transversely polarized quarks.[8] There is no overlap with other mechanisms; collinear factorization holds and evolution properties of such objects should be determined straightforwardly.[13] The unknown IFF can be extracted from the $e^+e^- \to (h_1h_2)_{\mathrm{jet}1}(h_1h_2)_{\mathrm{jet}2}X$ process by looking for an azimuthal asymmetry in the position of the hadron pair planes with respect to the lab frame.[14]

Since IFF are built on T-odd structures, it is possible to study in detail the mechanisms involved in the residual interactions of the outgoing hadrons by expanding the amplitudes in relative partial waves. If the hadrons are two pions, the main partial-wave contributions are the s and p waves.[15,16] Residual interactions come from interference of amplitudes for different channels with different phases. Each interference component ($s-p$ or $p-p$) can be disentangled by a suitable selection of the integration phase space.[16] In particular, the following weighted asymmetry

$$A_{UT}^{\sin(\phi_R+\phi_S)\,\sin\theta} = \frac{\displaystyle\int d\cos\theta\, d\phi_R\, d\phi_S\, \sin(\phi_R+\phi_S)\, d\sigma_{UT}}{\displaystyle\int d\cos\theta\, d\phi_R\, d\phi_S\, d\sigma_{UU}}$$

$$\propto \frac{|\mathbf{S}_T||\mathbf{R}|}{M_h}\, \frac{\sum_f e_f^2\, h_1^f\, H_{1,sp}^{\sphericalangle\, f}}{\sum_f e_f^2\, f_1^f\, (\frac{3}{4}\, D_1^{pp\, f} + \frac{1}{4}\, D_1^{ss\, f})} \qquad (3)$$

has been measured at HERMES for the $ep^{\uparrow} \to e'(\pi\pi)X$ process, where the proton polarization $\mathbf{S}_T$ forms an azimuthal angle ϕ_S with the scattering plane, and the final pion pair with invariant mass M_h has a relative momentum $\mathbf{R}$ oriented with the azimuthal angle ϕ_R. The angle θ defines the direction of the back-to-back emission in the pair cm frame with respect to the jet axis.

In the literature, there are two predictions for the asymmetry of Eq. (3).

The first one[15] is based on the guess that the asymmetry should depend on the properties of the $\pi - \pi$ phase shifts for the considered s and p channels. The second one[8] is based on the calculation in the spectator model of the interference diagram where the two pions are emitted directly or through the decay of the ρ resonance. The results have strikingly different features, because the first model predicts a marked M_h dependence with a sign change around $M_h \approx m_\rho$, which is not observed in the second model.

Therefore, we have considered a more refined version of the spectator model.[17] The amplitude for the p channel contains the coherent sum of the resonant decays $\rho, \omega \to \pi^+\pi^-$, and the incoherent sum of the channel $\omega \to \pi^+\pi^-\pi^0$, properly integrated upon the third pion π^0. In the s channel, the amplitude is the incoherent sum of the $K_s^0 \to \pi^+\pi^-$ decay and of a background, represented by the direct production of the charged pion pair. The diagonal s and p contributions enter D_1^{ss} and D_1^{pp} in Eq. (3), respectively, while $H_{1\,sp}^{\sphericalangle}$ contains the $s - p$ interference.

The parameters of the resonances are taken from the Particle Data Group,[18] while the free parameters of the model are then fixed by reproducing the invariant mass distribution $D_1(M_h)$ as it is output by the HERMES Monte Carlo program with no corrections for acceptance and with proper kinematical cuts in order to remove elastic, single- and double-diffractive events.[19]

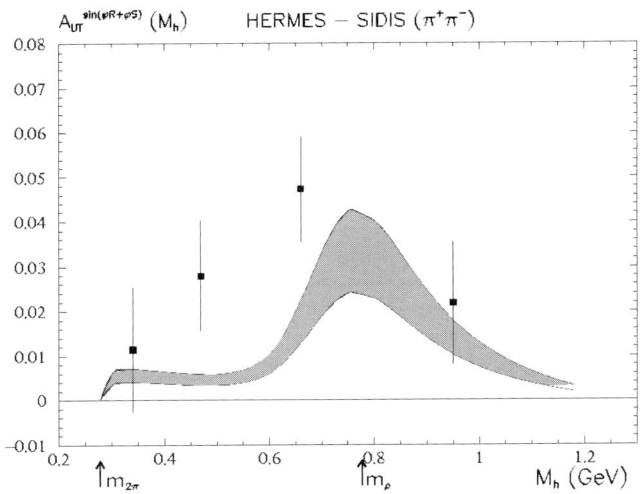

Fig. 4. The single-spin asymmetry from the interference of 2 pion production channels in relative s and p waves, for the process $ep^\uparrow \to e'(\pi\pi)X$ in the HERMES kinematics.

Once all the free parameters have been fixed to the invariant mass spectrum, we predict the single-spin asymmetry $A_{UT}^{\sin(\phi_R+\phi_S)\sin\theta}$ of Eq. (3). We further integrate it upon all variables but M_h, following the experimental cuts. The net result is compared in Fig. 4 with available experimental data;[9] in particular, no evidence for a sign change of the asymmetry is displayed. The uncertainty band is given only by various possible choices for the distribution functions f_1, h_1. All the other ingredients of the calculations are fixed and, therefore, we can interpret the result as a true prediction.

Acknowledgments

A fruitful collaboration is warmly acknowledged with A. Bacchetta and A. Bianconi, who coauthored most of the results here presented.

References

1. R.L. Jaffe, in Proc. of the *Ettore Majorana International School on the Spin Structure of the Nucleon*, Erice (Italy) 1995, `hep-ph/9602236`.
2. V. Barone and P.G. Ratcliffe, *Transverse Spin Physics*, ed. World Scientific (River Edge, USA, 2003).
3. J.P. Ralston and D.E. Soper, *Nucl. Phys.* **B152** (1979) 109.
4. D. Boer, *Phys. Rev.* **D60** (1999) 014012.
5. A. Airapetian *et al.*, *Phys. Rev. Lett.* **94** (2005) 012002.
6. J.C. Collins, S.F. Heppelmann, and G.A. Ladinsky, *Nucl. Phys.* **B420** (1994) 565.
7. A. Bianconi *et al.*, *Phys. Rev.* **D62** (2000) 034008.
8. M. Radici, R. Jakob, and A. Bianconi, *Phys. Rev.* **D65** (2002) 074031.
9. P. van der Nat, in proc. of the *13th International Workshop on Deep Inelastic Scattering (DIS05)*, Madison (Michigan-USA) 2005, ed. A.I.P., in press.
10. J.S. Conway *et al.*, *Phys. Rev.* **D39** (1989) 92.
11. A. Bianconi and M. Radici, *Phys. Rev.* **D71** (2005) 074014.
12. M. Anselmino *et al.*, *Phys. Rev.* **D71** (2005) 074006.
13. D. Boer, *Nucl. Phys. Proc. Suppl.* **105** (2002) 76.
14. D. Boer, R. Jakob, and M. Radici, *Phys. Rev.* **D67** (2003) 094003.
15. R.L. Jaffe, X. Jin, and J. Tang, *Phys. Rev. Lett.* **80** (1998) 1166.
16. A. Bacchetta and M. Radici, *Phys. Rev.* **D67** (2003) 094002.
17. A. Bacchetta and M. Radici, in preparation.
18. S. Eidelman *et al.*, *Phys. Lett.* **B592** (2004) 1.
19. E. Aschenhauer, private communication.

RELATIONS BETWEEN SINGLE AND DOUBLE TRANSVERSE ASYMMETRIES*

O.V. Teryaev

Joint Institute for Nuclear Research, Dubna, 141980 Russia

The effective (non-universal) nature of Sivers function reflects the process dependence of the imaginary phase required for T-odd Single Spin asymmetry. It is supported by recent calculation of asymmetry in Semi-Inclusive Deeply Virtual Compton Scattering. The explicit account for the phase allows to relate T-odd (single) and T-even (double) Spin asymmetries. The emerging sum rules relating Sivers function to the twist-3 part of transverse spin structure function $\bar{g}_2$ are discussed

1. Introduction

Single and double spin asymmetries (SSA's and DSA's) are quite different. While the former ones are more simple for experimental studies, as they require only one polarized particle, the underlying theory is rather complicated. The main reason is the necessity for the imaginary phase, whose appearance is easy to understand because of the factor i from γ_5 Dirac matrix entering the definition of fermion density matrix. As the observables are real, this factor is compensated by the phase difference of interfering amplitudes.

SSA's represent T-odd observables and imaginary phases allow to get them in T-conserving theory. To understand that, recall, that the sign of the imaginary phase is defined by the sign of $i\varepsilon$ in the causal propagators which, in turn, is defined by the time direction. To some extent, such effects are similar to the phenomenon of spontaneous symmetry breaking, when lagrangian does respect some symmetry, but the matrix elements do not.

Contrary to that, DSA's are not related to such a complications. At the same time, transverse SSA's and DSA's (the latter being described, in particular, by structure function $g_2(x)$) have something in common: as massless

*This work is partially supported by grant RFBR 03-02-16816.

particles are always polarized longitudinally, both SSA's and DSA's include some mass parameter and correspond in that sense to twist 3. However, as it will be seen later, this does not necessary mean the suppression like $1/Q$. This resolve the seeming contradiction: it is quite common now to speak about such effects as being of leading twist, although they correspond to the objects being of twist 3 in field theoretical sense.

In this paper the particular source of such unsuppressed SSA in QCD, the T-odd Sivers [1] function is explored. It is "effective" [2] or "non-universal" [3,4] function in the sense that the required imaginary phase comes from the hard subprocess [5] and the result is therefore process dependent. Moreover, the assumption of its universality would lead [6] to the appearance of SSA in DIS where it is forbidden by T-invariance (cf. 7). Although this property of Sivers function was established few years ago,[2,3] yet another derivation of that is presented here. This includes the calculation of the SSA in the new hard process, Semi-Inclusive Deeply Virtual Compton Scattering (SIDVCS).[8,9] The twist 3 asymmetry in the specific kinematical domain ($p_T \ll Q$) shows the appearance of leading ($\sim Q^0$) asymmetry manifesting the effective Sivers function. The answer is proportional to the particular contribution to the twist-3 large distance matrix element (gluonic pole). The latter, in turn, may be related to the twist 3 contribution to structure function g_2. This establishes the relation between SSA and DSA and, consequently, between seemingly rather distant branches of transverse spin physics.

2. Unsuppressed twist-3 asymmetries

The twist-3 SSA's are known for many years. Their key ingredient [10] is the generation of the imaginary phase by the interaction between the hard and soft parts of amplitudes mediated by the physical gluons. Such physical gluons are known to be the crucial element of twist 3 factorization for polarized [11–13] case. The imaginary phase emerges when one of the incoming partons is soft, and only the case of soft quarks (fermionic poles) was originally [10] considered. Further development [14] had led to the identification of another possibility, when gluon is soft(gluonic poles), and this contribution will be of special interest here, as it gives rise to effective Sivers function.

The model calculation of SSA [15] is qualitatively similar to these twist 3 calculations. At the same time, the asymmetry is not suppressed as $1/Q$. For this reason it was considered to be the manifestation of twist-2 Sivers function.[4,16]

Let us discuss whether these approaches are really different. To do so one may consider [8] the new hard process, the semi-inclusive photon production in DIS (SIDVCS) which appears when the produced hadron (usually pion) in SIDIS is substituted by photon. The result of calculations [8,9] may be summarized in the following way.

There are various contribution to SSA having the common factor Mp_T/Q^2. The special role is played by one of them, proportional to quark–gluon correlator $b_V(x_1, x_2)$. Here $x_{1,2}$ are the momentum fractions carried by quarks, so that the gluon momentum fraction $x_g = x_1 - x_2$. Moreover, in SIDVCS x_g is fixed by kinematics like x_B in DIS, so that $x_g \sim p_T^2/Q^2$. The existence of the gluonic pole implies [17] the behaviour of correlator $b_V \sim 1/x_g$. As a result, the asymmetry in the presence of gluonic pole behaves like $Mp_T/Q^2 \times Q^2/p_T^2 \sim Mp_T/p_T^2$. This is the typical behaviour of SSA found also in Ref. 15. The violation of positivity at small p_T is cured by the higher twists contributions whose partial resummation should lead [10] to the behaviour Mp_T/m_T^2.

For this term the spin-dependent partonic cross-section coincides with the spin-independent one, and the azimuthal dependence is just $\sin \phi_{\rm Sivers}$, both these facts signaling the appearance of effective Sivers function in agreement with Refs. 2, 3.

Note that the manifestation of Sivers function in Ref. 2 at M/Q level only is entirely related to the only possible observable SSA studied in this (transverse-momentum integrated Drell–Yan) process. Let us stress, that the same twist-3 effects lead to the $1/Q$ asymmetry in that case and to the leading order asymmetry in SIDVCS.

As a result, we get the unsuppressed twist-3 contribution. This conclusion is in fact complementary to the derivation using gluonic exponentials.[4,16] In the latter case one starts with unsuppressed contribution of unphysical gluons which acquires the physical piece due to transverse part of the link at large distances. At the same time, the suggested approach starts with suppressed contribution of physical gluons, getting the unsuppressed piece due to the existence of the large distance contribution of gluonic poles.

3. Relating effective Sivers function and $\bar{g}_2$

As soon as the effective character of Sivers function is established and the (process-dependent) source of imaginary phase is identified, one may look for other manifestations of gluonic pole twist-3 matrix element. We start with the relation first found in Ref. 2

$$x f_T^1(x) = \frac{1}{2M} T(x,x) = \frac{1}{4}\varphi_V(x). \tag{1}$$

Here $T(x,x)$ is the standard matrix element defining the gluonic pole contribution [14] related by the equations of motion [17] to the residue $\varphi_V(x)$ [18] of the pole in the quark–gluon correlator [10,11] b_V.

$$b_V(x_1,x_2) = \frac{\varphi_V(x_1)}{x_1 - x_2} + b_V^r(x_1,x_2). \tag{2}$$

Taking into account the sum rules [18,19] resulting from the equation of motion, gauge and rotational invariance one comes to the following relation:

$$\int_0^1 x^2 \hat{g}_2(x)\, dx = \tag{3}$$

$$-\frac{1}{3\pi}\int_{|x_1,x_2,x_1-x_2|\leq 1} dx_1 dx_2 \sum_f e_f^2\, b_V(x_1,x_2)\,(x_1 - x_2).$$

One needs some additional assumption to separate the gluonic pole contribution and the simplest one is its domination over the regular piece leading to the relation [18]

$$\int_0^1 x^2 \bar{g}_2(x)\, dx = -\frac{1}{3\pi}\int_{|x_1,x_2,x_1-x_2|\leq 1} dx_1 dx_2 \sum_f e_f^2 \varphi_V(x_1). \tag{4}$$

Performing the integration over x_2 one gets:

$$\int_0^1 x^2 \bar{g}_2(x)\, dx = -\frac{1}{3\pi}\int_{-1}^1 dx_1 \sum_f e_f^2 \varphi_V(x_1)(2 - |x_1|). \tag{5}$$

The assumption of the dominance of singular piece is physically equivalent to the smooth variation of gluonic field strength which is supposed to be almost constant in the integration range. The similar assumption was also adopted in the model derivation of Ref. 20 and its "unintegrated" modification.[21] Note that the physical picture of almost classical slow gluon field is quite popular in the consideration of small-x phenomena.

From the formal point of view, let us note that the representation [18] (2) does not provide the correct symmetry properties for the singular part itself. Although the violation is of order of neglected regular part, one may use instead the following definition

$$b_V(x_1,x_2) = \frac{\varphi_V\left(\frac{x_1+x_2}{2}\right)}{x_1 - x_2} + b_V^r(x_1,x_2). \tag{6}$$

As a result the integration in (4) should be performed in $x_1 - x_2$ leading to

$$\int_0^1 x^2 \bar{g}_2(x)\, dx = -\frac{2}{3\pi} \int_{-1}^1 dz \sum_f e_f^2 \varphi_V(z) \big(1 - \theta(|z| - 1/2)(2|z| - 1)\big). \quad (7)$$

Finally, separating contributions of various flavours and recalling (1) one may get the sum rule for the Sivers function

$$\int_0^1 x^2 \bar{g}_2^f(x)\, dx = -\frac{8}{3\pi} \int_{-1}^1 dz z f_T^{1,f}(z) \big(1 - \theta(|z| - 1/2)(2|z| - 1)\big). \quad (8)$$

Note that the integration over negative z corresponds to antiquarks contribution.

4. Numerical comparison of Sivers function and $\bar{g}_2$

The first numerical comparison of SSA and $\bar{g}_2$ was performed ten years ago[20] As soon as only large hadronic SSA's were available at that time while $\bar{g}_2$ is small, their compatibility was questioned at that time.

At the same, the current situation using the Sivers function, extracted from SIDIS and extensively described[22] at this conference, is different.

Say, in the parametrization of Ref. 23 the integral[a]

$$\int_0^1 dx\, x f_T^{1,u,d}(x) = \pm 0.0072(0.0042 - 0.014). \quad (9)$$

So, its scale is compatible to the scale of $\bar{g}_2^{u,d}$.[24] At the same time, the specific (mirror) flavour structure of Sivers function was never mentioned in the case of $\bar{g}_2$. So one may conclude that the scale of SSA and DSA is compatible but the flavour structure is not.

Let us note that the qualitative reason for numerical compatibility is quite clear. HERMES and COMPASS asymmetries are sensitive to the region of low $x < 0.4$ providing the small contribution to the second moment of $\bar{g}_2$. In the approach of Ref. 23 the strong suppression of large x region due to similarity of Sivers function and GPD's[25] was suggested. At the same time, the large x contribution should be manifested in the large hadronic SSA.

Two comments are in order here. First, as Sivers function represents some particular part of twist 3 contribution, it should not be taken together with the latter in order to avoid double counting.

---

[a]I am grateful to P. Schweitzer for providing this number.

Second, although the sign of effective Sivers function in hadronic SSA deserves special investigation, it may be conjectured using the following reasons.[5] Let us first compare SIDIS and DY process. The imaginary phases in these processes are coming from the different sources. The only positive variable in SIDIS $lH(p) \to l^{'} h(p_f)X$ is the c.m. energy squared $s = (p + q)^2$, which does not coincide with the mass of unobserved particles $M_X^2 = (p + q - p_f^2)^2$. For Drell–Yan process $H(p_1)H(p_2) \to l^+l^-(Q)X$ besides the variable $s = (p_1 + p_2)^2$ there is also variable Q^2 which may produce the imaginary cuts. Moreover,[5] it is just this variable which is responsible for the appearance of the twist-3 SSA (both $1/Q$ suppressed [26] and unsuppressed) at Born level. As soon as the dependence on the kinematical variables enters through the ratio Q^2/s, the cuts is s and Q^2 provide the opposite signs for the imaginary parts. As soon as there are no analogs of cuts in Q^2 for hadronic SSA, one may expect to get the same sign of effective Sivers function as in SIDIS.

This may explain the compatibility of Sivers function from SIDIS [27] and previous analysis of hadronic SSA.[b] Note that the strong suppression of large x was not assumed in that analysis, so their second moment of f_T^1, although being compatible with Ref. 23 is larger. One may still wonder whether it is possible to describe simultaneously SIDIS and hadronic SSA quantitatively. Note that $\bar{g}_2$ may strongly oscillate [18] due to zero sum rules,[28,29] implying also the similar oscillating behaviour for f_T^1. As a result, one may have effectively two Sivers functions at large and small x, the first responsible for SIDIS, and the second for hadronic SSA, while the oscillations in-between may provide the small moment of $\bar{g}_2$.

Still, the different flavour dependence of Sivers function and $\bar{g}_2$ needs to be explained. This may come from the regular part of (6), as well as from the contribution [18] of another, axial, quark–gluon correlator, responsible for fermionic poles. For DIS, the regular part of vector correlator and axial correlator (which does not contribute to the third moment) contribute at the same order as gluonic pole part. The same is also true for the contributions of fermionic poles to hadronic SSA,[30] where up to now only gluonic poles were included [31] into numerical analysis.

The appearance of new more accurate data for SIDIS, Drell–Yan and hadronic SSA, as well as for the twist-3 contribution to g_2 may allow the new, global, fits of the data, and this would mark the new stage of the analysis of T-odd spin effects, which was already achieved for spin-independent

[b] I am indebted to M. Anselmino for this comment.

and spin-dependent DIS. In this analysis taking into account the QCD evolution of Sivers function will be necessary. To do so the coordinate space analysis [32] may be useful.

5. Conclusions

The effective nature of Sivers function is explicitly manifested in the behaviour of SSA in the new hard process, SIDVCS. If there is a gluonic pole contribution to quark–gluon correlator, corresponding to slowly varying gluonic field strength, the generic $1/Q$ suppression of twist 3 SSA is disappearing. This may be considered as a compensation of perturbative (coming from coefficient function) suppression by nonperturbative enhancement. The relation of Sivers function to gluonic poles opens a possibility to relate SSA's and DSA's, the latter represented by twist 3 contribution to structure function $\bar{g}_2$. Its scale is compatible with the scale of Sivers function extracted from SIDIS, although the flavour dependence seems to disagree. The resolution of this problem may require to take into account the contributions of fermionic poles to DIS and hadronic SSA, and performing the global fit of $\bar{g}_2$ and SSA (hadronic and SIDIS).

Finally, I would like to express my deep gratitude to Enzo Barone and Phil Ratcliffe for warm hospitality in Como and useful discussions related to the subject of this paper. I benefited also from many enlightening discussions with the participants of this very interesting workshop, in particular, with M. Anselmino, U. d'Alesio, A. Bacchetta, D. Boer, S. Brodsky, M. Burkardt, A.V. Efremov, A. Kotzinian, F. Murgia, A. Prokudin, D. Sivers, I. Schmidt, and P. Schweitzer.

References

1. D.W. Sivers, *Phys. Rev.* **D43** (1991) 261.
2. D. Boer, P.J. Mulders and O.V. Teryaev, *Phys. Rev.* **D57** (1998) 3057 [arXiv:hep-ph/9710223].
3. D. Boer, P.J. Mulders and F. Pijlman, *Nucl. Phys.* **B667** (2003) 201 [arXiv:hep-ph/0303034].
4. J.C. Collins, *Phys. Lett.* **B536** (2002) 43 [arXiv:hep-ph/0204004].
5. O.V. Teryaev, "T odd effects in QCD", *RIKEN Rev.* **28** (2000) 101.
6. O.V. Teryaev, *Czech. J. Phys.* **53** (2003) 47 [arXiv:hep-ph/0306301].
7. A. Bacchetta, "Single-spin asymmetries and Qiu–Sterman effect(s)", these proceedings, arXiv:hep-ph/0511085.
8. O. Teryaev, in the proc. of Advanced Study Institute "Spin and Symmetry", Prague, July 2005, *Czech. J. Phys.*, to appear
9. O. Teryaev and S. Srednyak, in the proc. of Advanced Study Institute "Spin and Symmetry", Prague, July 2005, *Czech. J. Phys.*, to appear

10. A.V. Efremov and O.V. Teryaev, *Phys. Lett.* **B150** (1985) 383.
11. A.V. Efremov and O.V. Teryaev, *Yad. Fiz.* **39** (1984) 1517.
12. A.P. Bukhvostov, E.A. Kuraev and L.N. Lipatov, *JETP Lett.* **37** (1983) 482 [*Pisma Zh. Eksp. Teor. Fiz.* **37** (1983) 406 *Sov. Phys. JETP*, **60** (1984) 22 *Zh. Eksp. Teor. Fiz.*, **87** (1984) 37].
13. P.G. Ratcliffe, *Nucl. Phys.* **B264** (1986) 493.
14. J.w. Qiu and G. Sterman, *Phys. Rev. Lett.* **67** (1991) 2264.
15. S.J. Brodsky, D.S. Hwang and I. Schmidt, *Phys. Lett.* **B530** (2002) 99 [arXiv:hep-ph/0201296].
16. A.V. Belitsky, X. Ji and F. Yuan, *Nucl. Phys.* **B656** (2003) 165 [arXiv:hep-ph/0208038].
17. V.M. Korotkiian and O.V. Teryaev, JINR-E2-94-200; *Phys. Rev.* **D52** (1995) 4775.
18. O.V. Teryaev, proc., Prospects of spin physics at HERA, Hamburg, DESY-95-200, pp. 132, arXiv:hep-ph/0102296.
19. A.V. Efremov and O.V. Teryaev, *Phys. Lett.* **B200** (1988) 363.
20. B. Ehrnsperger, A. Schafer, W. Greiner and L. Mankiewicz, *Phys. Lett.* **B321** (1994) 121 [arXiv:hep-ph/9312264].
21. D. Boer, Talk at the Workshop at BNL on "Hadron Structure From Lattice QCD", March 18–22, 2002 (`http://quark.phy.bnl.gov/www/riken/HadronWorkshop/proceedings/boer.ps.gz/`).
22. M. Anselmino *et al.*, these proceedings, arXiv:hep-ph/0511017.
23. J.C. Collins *et al.*, these proceedings, arXiv:hep-ph/0510342.
24. P.L. Anthony *et al.* [E155 Collaboration], *Phys. Lett.* **B553** (2003) 18 [arXiv:hep-ex/0204028].
25. M. Burkardt, these proceedings.
26. N. Hammon, O. Teryaev and A. Schafer, *Phys. Lett.* **B390** (1997) 409 [arXiv:hep-ph/9611359].
27. M. Anselmino, M. Boglione, U. D'Alesio, A. Kotzinian, F. Murgia and A. Prokudin, arXiv:hep-ph/0511249.
28. H. Burkhardt and W.N. Cottingham, *Ann. Phys. (N.Y.)* **16** (1970) 543.
29. A.V. Efremov, O.V. Teryaev and E. Leader, *Phys. Rev.* **D55** (1997) 4307 [arXiv:hep-ph/9607217].
30. A. Efremov, V. Korotkiian and O. Teryaev, *Phys. Lett.* **B348** (1995) 577.
31. J.-W. Qiu and G. Sterman, *Phys. Rev.* **D59** (1999) 014004 [arXiv:hep-ph/9806356].
32. O.V. Teryaev, arXiv:hep-ph/0310069, proc. of DIS-03, p. 827; proc. of SPIN-03, p. 200.

CROSS SECTIONS, ERROR BARS AND EVENT DISTRIBUTIONS IN SIMULATED DRELL–YAN AZIMUTHAL ASYMMETRY MEASUREMENTS

A. Bianconi

Dip. di Chimica e Fisica per l'Ingegneria e per i Materiali,
Via Valotti 9, 25123 Brescia, Italy
E-mail: bianconi@bs.infn.it

A short summary of results of recent simulations of (un)polarized Drell–Yan experiments is presented here. Dilepton production in pp, $\bar{p}p$, π^-p and π^+p scattering is considered, for several kinematics corresponding to interesting regions for experiments at GSI, CERN–Compass and RHIC. A table of integrated cross sections, and a set of estimated error bars on measurements of azimuthal asymmetries (associated with collection of 5, 20 or 80 Kevents) are reported.

1. Introduction

The aim of this work is to give some useful reference numbers for planning Drell–Yan experiments aimed at the measurement of transverse spin/momentum related azimuthal asymmetries.

Table 1. Total cross sections (nb) for several colliding particle combinations, mass ranges and S-values.

M (GeV/c^2)	1.5–2.5	4–9	12–40	1.5–2.5	4–9	12–40
S (GeV^2)		pp			$\bar{p}p$	
30	0.03	$\ll$ pb	$\ll$ pb	1.3	0.3 pb	$\ll$ pb
200	0.7	0.01	$\ll$ pb	4.4	0.35	$\ll$ pb
$(200)^2$	3.5–17	1.2–5.5	0.06–0.26	5–18	2.6–7	0.24–0.4
		π^-p			π^+p	
30	0.9	1 pb	$\ll$ pb	0.25	0.1 pb	$\ll$ pb
200	1.9	0.25	$\ll$ pb	0.7	0.07	$\ll$ pb
$(200)^2$	1.8–5.6	0.75–2.1	0.1–0.2	1.5–4.8	0.5–1.4	0.04–0.1

Note: For the high-energy case, two different parametrizations have been used, leading to pairs of σ values. See text for details.

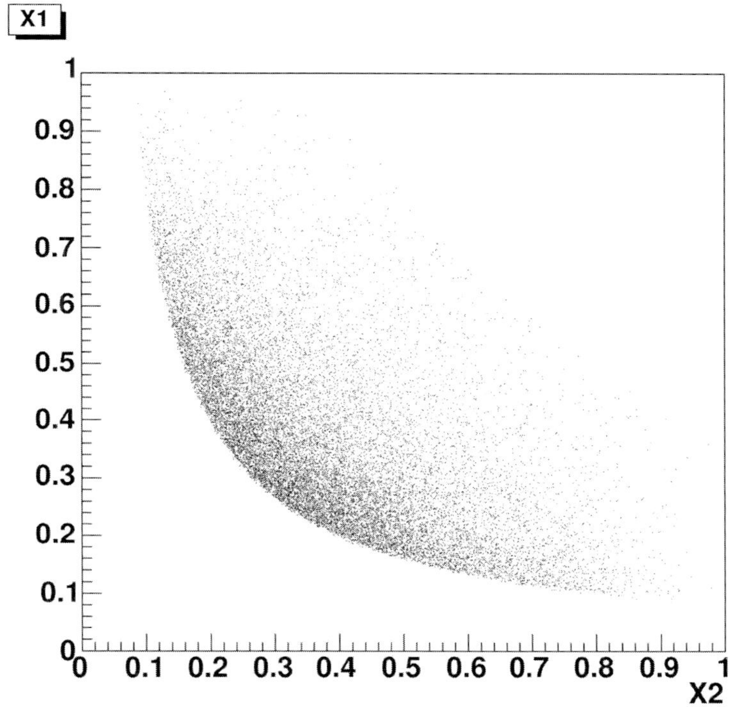

Fig. 1. The $\bar{p}p$ event scatter plot for $S = 200\,\mathrm{GeV}^2$, and dilepton mass in the range 4–9 GeV/c^2.

This includes total cross sections for several kinematical options for pp, $\bar{p}p$ and $\pi_{\pm}p$ Drell–Yan dilepton production: squared hadron CM energy $S = 30\,\mathrm{GeV}^2$, $200\,\mathrm{GeV}^2$, $(200)^2\,\mathrm{GeV}^2$, and dilepton masses in the ranges 1.5–2.5 GeV/c^2, 4–9 GeV/c^2, 12–40 GeV/c^2. For some relevant situations, estimates of the asymmetry error bars are reported, for sets of 5, 20, 80 Kevents divided into 10 bins of the longitudinal fraction x. Previous Drell–Yan data and fitting relations [1,2] are the basis of the initial core of the used simulation code. Several details on the formalism, together with the most recent examples of simulated asymmetries, are presented elsewhere in this workshop,[3] and in published work by myself and M. Radici.[4,5]

2. Total Cross Sections.

Total cross sections are shown in Table 1. For the two lower S-values they have been evaluated with the differential cross-section fit relations [1,2]

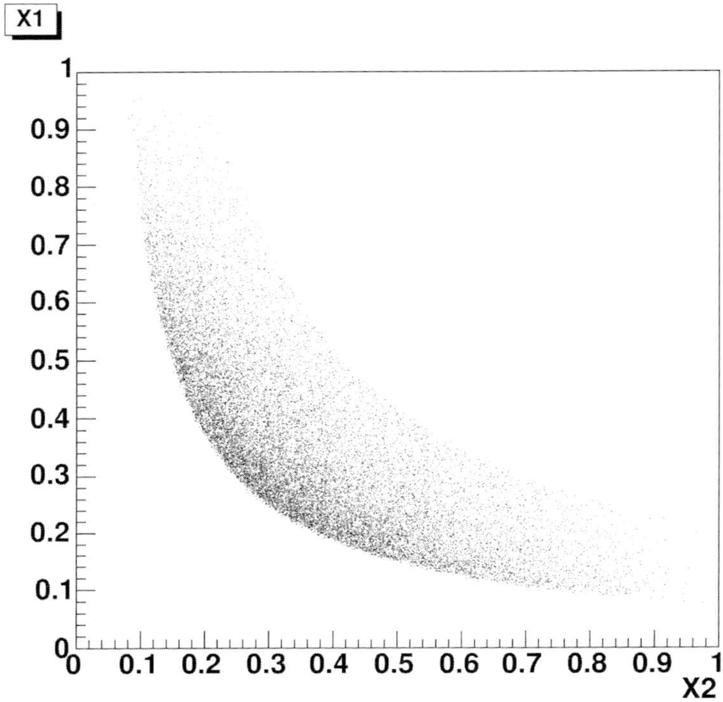

Fig. 2. The $\bar{p}p$ event scatter plot for $S = 30\,\mathrm{GeV}^2$, and dilepton mass in the range 1.5–2.5 GeV/c^2.

coming from measurements of $\pi^- A$ and $\bar{p}A$ at $S = 250$–$400\,\mathrm{GeV}^2$. The two cases $\pi^- p$ and $\bar{p}p$ correspond to these cross sections for $Z/A = 1$. For $\pi^+ p$ we have assumed $\pi^- p \equiv \pi^+ p$ for the pion sea contribution, $\pi^- p \equiv (1/4)\pi^+ n$ for the pion valence contribution, and calculated $\pi^- A$ for $Z/A = 0$. Cross sections for pp have been evaluated by substituting the $(sea + valence)(sea + valence)$ structure of $\bar{p}p$ with the structure $1/2[sea(sea + valence) + (sea + valence)sea]$. For the largest S case the calculation based on the previous parametrization (with sea $\sim$ const for small x) has been sided by an alternative calculation using the (LO-intermediate gluon) MRST distributions,[6] with sea $\sim x^{-\lambda}$ ($\lambda \approx 0.2$–0.3 for mass $< 9\,\mathrm{GeV}/c$), and based on data sets including several recent Drell–Yan measurements.[7] No evolution was applied, which is improper for $M \gg 10\,\mathrm{GeV}/c^2$. The K-factors have been assumed as constant but tuned to reproduce with both methods the measured cross sections at

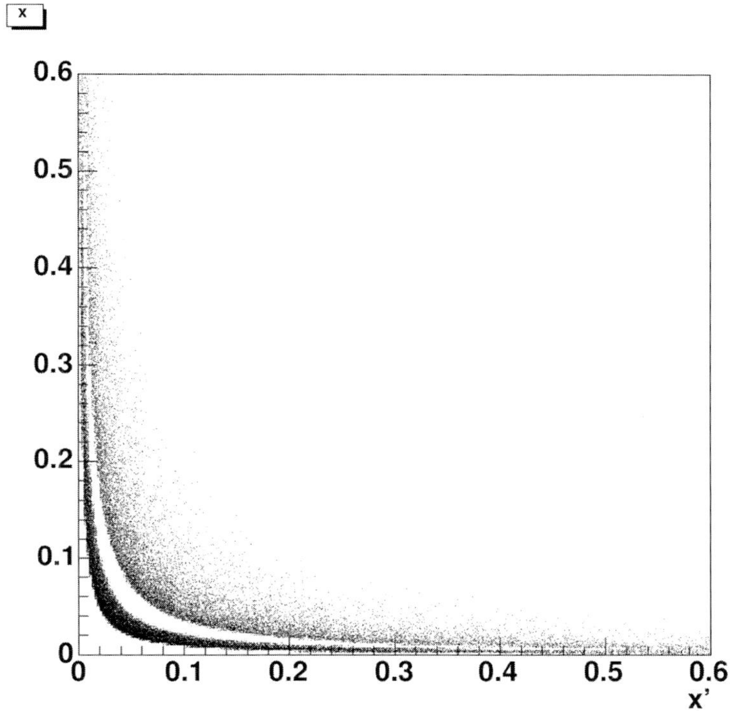

Fig. 3. The pp event scatter plot for $S = (200)^2\,\mathrm{GeV}^2$, and dilepton mass in the double range 4–9 and 12–40 GeV/c^2.

$S = 250\,\mathrm{GeV}^2$, masses 4–9 GeV/c^2.[2] The former strategy leads to the smaller reported cross section values, the latter to the bigger ones. For the $\pi^{\pm}p$ high-energy case double values refer to different parametrizations for the distribution functions of the proton only, pion distribution functions have not been changed. For the lowest mass range 1.5–2.5 the smaller number of contributing quarks introduces a reduction factor $\approx 1/2$. The difference between the other two mass ranges is not essential.

3. Event distributions

The event distribution has the general form

$$N(S, x, x', P_T, \xi) = F(S, x, x', P_T) \cdot [1 + A(x, x', P_T, \xi)]. \tag{1}$$

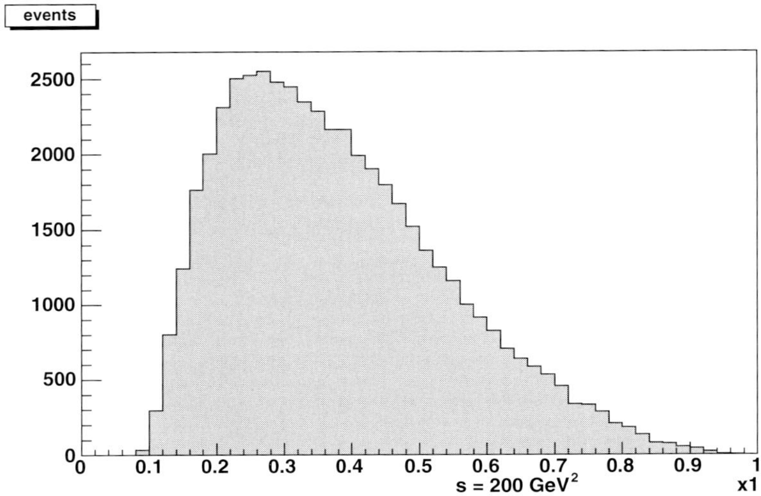

Fig. 4. X2-integrated distribution of the scatter plot of Fig. 1. Events are divided into 50 X1-bins.

where x, x', P_T describe the virtual photon kinematics in the hadron center of mass ($P_L/(S/2) \equiv x - x'$, $M^2/S \equiv xx'$), while ξ represents compactly the set of variables describing the angular distribution of the leptons in the Collins–Soper frame ($\equiv$ the photon polarization). $F(S, x, x', P_T)$ alone gives the virtual photon event distribution. $A(x, x', P_T, \xi)$ averages to zero over all the solid angle, and describes the asymmetry properties of the lepton distribution in unpolarized, single or double polarized DY.

The scatter plots of Figs. 1, 2 and 3 reproduce the event distribution

$$N(S, x, x') = \int d^2\vec{P}_T F(S, x, x', P_T) \tag{2}$$

for some relevant $\bar{p}p$ and pp cases. Fig. 4 reports

$$N(S, x) = \int dx' N(S, x, x') \tag{3}$$

where the integrated distribution is the one of Fig. 1. Events in the xx' scatter plots concentrate near the hyperbole $xx' = M^2{}_{\min}/S$, because $\sigma \propto q(x)\bar{q}(x')/M^2$. For the same reason, in the case of Fig. 3 the lower event band (M in the range 4–9 GeV/c^2) contains 95 % of all the events reported in the figure.

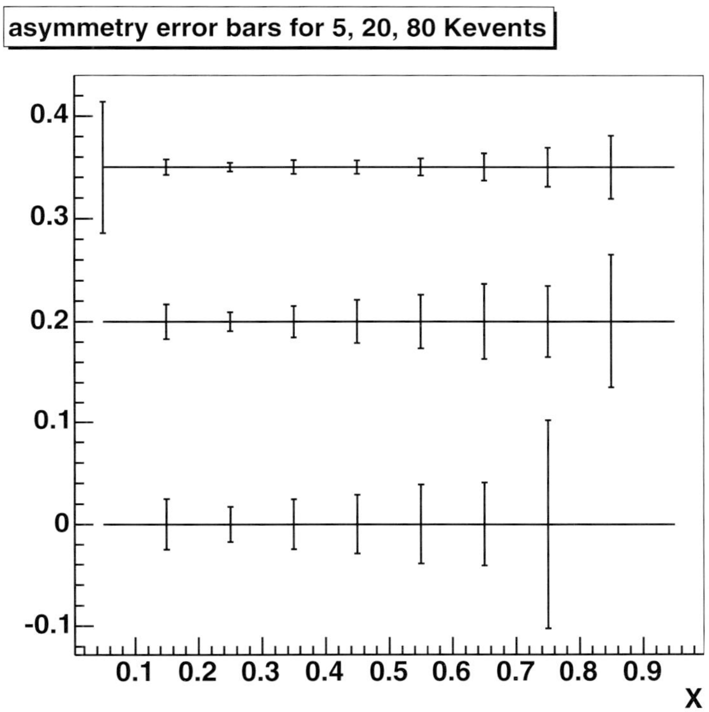

Fig. 5. Error bars on the azimuthal asymmetry for $\bar{p}p$, $S = 30\,\text{GeV}^2$, dilepton mass in the range 1.5–2.5 GeV/c^2.

4. Asymmetry Error Bars

The integrated $N(S, x)$ distribution has its peak at $x \approx 3M^2{}_{\text{min}}/S$. To the right of the peak, $N(S, x) \sim 1/x^n$ with $n > 1$. This selects for each S, M_{min} the x-range where error bars are smaller. The shown error bars in Figs. 5, 6 and 7 refer to the special case of Sivers asymmetry, however they are the same for any kind of left/right asymmetry with respect to a Collins–Soper azimuthal angle ϕ. The asymmetry is defined as $(A - B)/(A + B)$, where A and B are the event numbers with positive or negative $\sin(\phi - \phi_S)$. Error bars have been calculated by assuming constant 0.05 asymmetry everywhere, and repeating the simulation 10 times to calculate fluctuations. Despite the size of the error bars is reasonably stable for $N_{\text{repetitions}} > 5$, with $N_{\text{repetitions}} = 10$ there are still small but evident fluctuations in the error bar size. This is due to most x-bins being filled

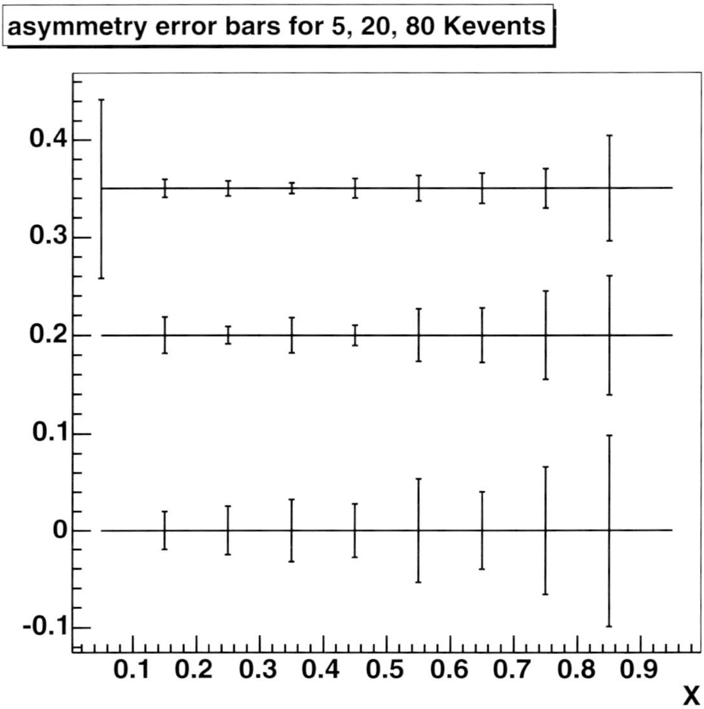

Fig. 6. Error bars on the azimuthal asymmetry for $\pi^- p$, pion beam energy 100 GeV, fixed target, dilepton mass in the range 4–9 GeV/c^2.

with event numbers $N(+) + N(-) < 1000$. For $N(+) + N(-) < 50$ error bars are not shown. Typically the largest shown error bars refer to event numbers $N(+) + N(-) \sim 100$. Examination of the error bar fluctuations suggests that the reported error bars are reliable within a factor $1/\sqrt{2} \div \sqrt{2}$. If one repeats the calculation by assuming asymmetry 0 or 0.1, systematic changes of these (purely statistic) error bars are smaller than the above fluctuations. So, unless they damp asymmetries by orders, all those reducing coefficient like polarization dilution etc are not influent on the error bar sizes. For asymmetry $> 33\,\%$, $N(-) < N(+)/2$. So, in the case of really large asymmetries, the consequent small value of $N(-)$ in those bins whose population is $\ll 1000$ introduces errors whose size may be uncontrollably larger than the estimated ones. In this case the error simulation must be more specific to be appropriate.

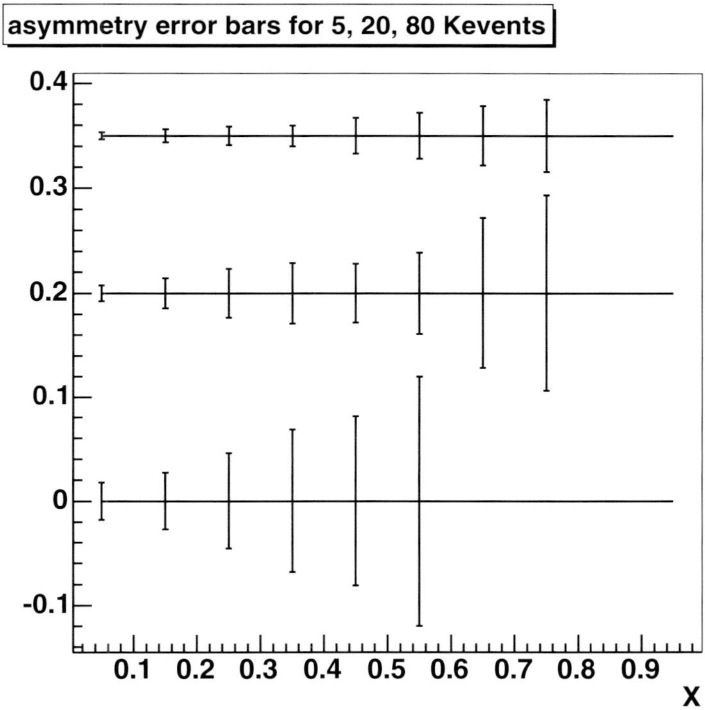

Fig. 7. Error bars on the azimuthal asymmetry for pp, $S = (200)^2$ GeV^2, dilepton mass in the range 12–40 GeV/c^2.

References

1. J.S. Conway *et al.*, *Phys. Rev.* **D39** (1989) 92.
2. E. Anassontzis *et al.*, *Phys. Rev.* **D38** (1988) 1377.
3. M. Radici, this workshop.
4. A. Bianconi and M. Radici, *Phys. Rev.* **D34** (1980) 1729.
5. A. Bianconi and M. Radici *Phys. Rev.* **D25** (1992) L527.
6. A.D. Martin *et al.*, *Eur. Phys. J.* **C4** (1998) 463; and *Phys. Lett.* **B443** (1998) 301.
7. E605 collab., G. Moreno *et al.*, *Phys. Rev.* **D43** (1991) 2815; E772 collab., P.L. McGaughey *et al.*, *Phys. Rev.* **D50** (1994) 3038; NA51 collab., A. Baldit *et al.*, *Phys. Lett.* **B332** (1994) 244. E866 collab., E.A. Hawker *et al.*, *Phys. Rev. Lett.* **80** (1998) 3715.

NEXT-TO-LEADING ORDER QCD CORRECTIONS FOR TRANSVERSELY POLARIZED PP AND $\bar{P}P$ COLLISIONS

A. Mukherjee

Physics Department, Indian Institute of Technology Bombay, Powai, Mumbai 400076, India

M. Stratmann

Institute für Theoretische Physik, Universität Regensburg M D 93040 Regensburg, Germany

W. Vogelsang

Physics Department, Brookhaven National Laboratory Upton, New York 11973, USA
and
RIKEN-BNL Research Center, Bldg. 510a, Brookhaven National Laboratory, Upton, New York 11973 – 5000, USA

We present a calculation of the next-to-leading order QCD corrections to the partonic cross sections contributing to single-inclusive high-p_T hadron production in collisions of transversely polarized hadrons. We use a recently proposed projection technique and give some predictions for the double-spin asymmetry A^{π}_{TT} for the proposed experiments at RHIC and at the GSI.

1. Introduction

The leading-twist partonic structure of a spin-$\frac{1}{2}$ hadron is given in terms of the unpolarized parton distribution functions $f(x, Q^2)$, the helicity distributions $\Delta f(x, Q^2)$, and the transversity distributions $\delta f(x, Q^2)$. Transversity describes the number density of a parton with the same transverse polarization as the nucleon, minus the number density for opposite polarization. Among the various parton distributions, the $\delta f(x, Q^2)$ are the ones about which we have the least knowledge. They are at present the focus of much experimental activity. Transversity will be probed by double-transverse spin asymmetries in transversely polarized pp collisions at the BNL Relativistic Heavy Ion Collider (RHIC).[1] The most promising reaction is the Drell–

Yan process, which offers the largest spin asymmetries, but whose main drawback is the rather moderate event rate.[2] Other relevant processes include high p_T prompt photon[3,4] and jet production.[3] However, for these the asymmetry is expected to be much smaller because of the absence of gluon-initiated subprocesses in the transversely polarized cross section. Recently, it has also been investigated whether one could extract transversity from measurements of $A_{\rm TT}$ for the Drell–Yan process in transversely polarized $\bar{p}p$ collisions at the planned GSI-FAIR facility.[5–10] Here (for details, see Ref. 11), we consider the spin asymmetry in single-inclusive production of pions at large transverse momentum p_T as a possible means for determining transversity at RHIC or the GSI. We report on the calculation of the next-to-leading order (NLO) corrections for this reaction, and also present some phenomenological results.

2. Projection Technique

It is known that the NLO QCD corrections are required in order to have firm theoretical predictions for hadronic scattering. Only with their knowledge can one reliably extract information on the partonic (spin) structure of nucleons. Apart from this motivation, also interesting new technical questions arise beyond leading order (LO) in the calculations of cross sections with transverse polarization. Unlike the longitudinally polarized case, where the spin vectors are aligned with the momenta, the transverse spin vectors specify extra spatial directions and, as a result, the cross section has non-trivial dependence on the azimuthal angle of the observed particle. For A_{TT} this dependence is of the form[4]

$$\frac{d^3\delta\sigma}{dp_T d\eta d\Phi} \equiv \cos(2\Phi) \left\langle \frac{d^2\delta\sigma}{dp_T d\eta} \right\rangle , \tag{1}$$

for a parity conserving theory with vector coupling. Here, the spin vectors are taken to point in the $\pm x$ direction. We furthermore consider the scattering in the center-of-mass frame of the initial hadrons and use their momenta to define the z axis. Because of the $\cos(2\Phi)$ dependence, integration over all azimuthal angles is not appropriate. This makes it difficult to use the standard techniques developed for NLO calculations of unpolarized and longitudinally polarized processes, because these usually rely on the integration over the full azimuthal phase space and also on the choice of particular reference frames that are related in complicated ways to the center-of-mass frame of the initial hadrons. In Ref. 4 a general technique was introduced that facilitates NLO calculations for transverse polarization by conveniently projecting on the azimuthal dependence of the cross section in a covariant way. The projector

$$F(p, s_a, s_b) = \frac{s}{\pi t u} \left[2\,(p \cdot s_a)\,(p \cdot s_b) \, + \, \frac{tu}{s}\,(s_a \cdot s_b) \right] \tag{2}$$

reduces to $\cos(2\Phi)/\pi$ in the center-of-mass frame of the initial hadrons. Here p is the momentum of the observed particle in the final state and the s_i are the initial transverse spin vectors. The squared matrix element for the partonic process is multiplied with this projector and integrated over the full azimuthal phase space. Integrations of terms involving the product of the transverse spin vectors with the final-state momenta can be performed using a tensor decomposition. After this step, no scalar products involving the s_i are left in the squared matrix element. For the remainder of the phase space integrations, one can now use techniques familiar from the unpolarized and longitudinally polarized cases. This method is particularly convenient at NLO, where one uses dimensional regularization and the phase space integrations are performed in $n \neq 4$ dimensions.

3. Application to single-inclusive hadroproduction: phenomenological results

Next, we give some phenomenological results for transversely polarized pp collisions at RHIC ($\sqrt{S} = 200$ and $500\,\text{GeV}$) and $\bar{p}p$ collisions at the GSI-FAIR facility in an asymmetric collider mode with proton and antiproton energies of 3.5 GeV and 15 GeV, respectively.[11] For our numerical predictions, we model the transversity distributions by saturating the Soffer inequality [12] at some low input scale $\mu_0 \approx 0.6\,\text{GeV}$, using the NLO (LO) GRV [13] and GRSV ('standard scenario')[14] densities $q(x, \mu_0)$ and $\Delta q(x, \mu_0)$, respectively.

Figure 1 (left) shows our predictions for the transversely polarized single-inclusive pion production cross sections at LO and NLO for the two different c.m.s. energies at RHIC. We also display the scale uncertainty. For the K-factor shown in the lower panel the choice is $\mu = 2p_T$. A significant decrease of scale dependence is observed when going from LO to NLO. Figure 1 (right) shows the asymmetry $A_{\rm TT}^{\pi^0}$ defined as the ratio of the polarized and unpolarized cross sections. Here, the scale is set to $\mu = p_T$. We also display the statistical errors that may be achievable in experiment. We have calculated these using

$$\delta A_{\rm TT} \simeq \frac{1}{P_1 P_2 \sqrt{L\,\sigma_{\rm bin}}}\,, \tag{3}$$

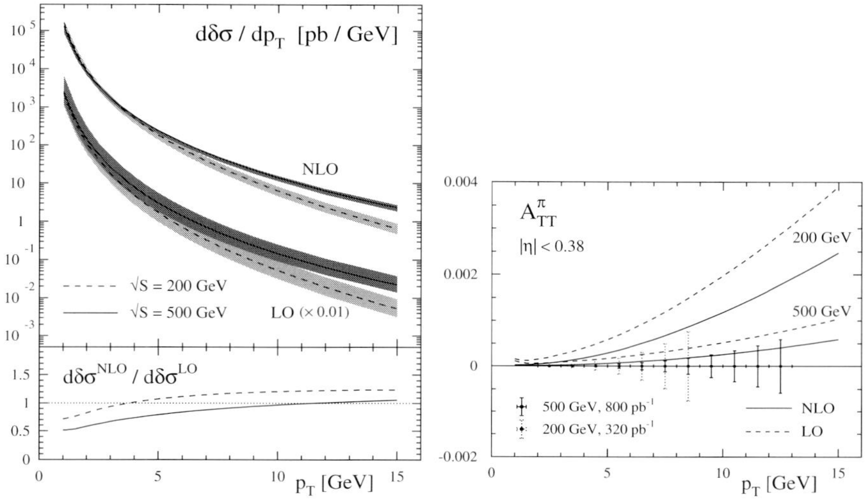

Fig. 1. Predictions for the transversely polarized single-inclusive pion production cross sections at LO and NLO at RHIC (left), and for the transverse double-spin asymmetry $A_{\rm TT}^{\pi^0}$ (right). The shaded bands represent the range of predictions when the scale μ is varied in the range $p_T \leq \mu \leq 4p_T$. The lower panel on the left shows the ratios of the NLO and LO results (the 'K-factors').

where $P_1 = P_2 = 0.7$ are the transverse polarizations of the proton beams, L the integrated luminosity of the collisions indicated in the figure, and $\sigma_{\rm bin}$ the unpolarized cross section integrated over the p_T-bin for which the error is to be determined. The asymmetry is very small.

Figure 2 (left) shows the corresponding cross sections in transversely polarized $\bar{p}p$ collisions at $\sqrt{S} = 14.5\,\text{GeV}$ in an asymmetric collider mode at the GSI-FAIR facility We have assumed beam polarizations of 30% and 50 % for the antiprotons and protons, respectively. At GSI energies, the scale dependence does not really improve from LO to NLO. A resummation of the double-logarithmic corrections to the partonic cross sections [15] would be very desirable for the future, along with a study of power corrections. Figure 2 (right) shows the predicted transverse double-spin asymmetry at the GSI, which is much larger than at RHIC.

Acknowledgments

A.M. thanks the organizers of Transversity 2005 for the kind invitation and support. W.V. is grateful to RIKEN, Brookhaven National Laboratory and the Department of Energy (contract number DE-AC02-98CH10886) for

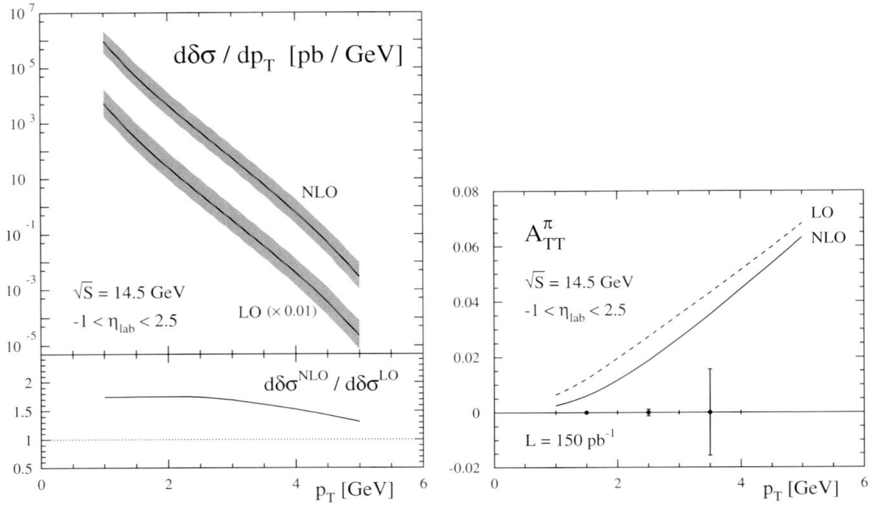

Fig. 2. LO and NLO predictions for the cross section (left) and the transverse spin asymmetry $A_{\mathrm{TT}}^{\pi^0}$ (right) for single-inclusive pion production in $\bar{p}p$ collisions at the GSI. The shaded bands represent the range of predictions if the scale μ is varied in the range $p_T \leq \mu \leq 4p_T$. The lower panel (left) shows the ratios of the NLO and LO results (the 'K-factors').

providing the facilities essential for the completion of this work. This work is supported in part by the "Bundesministerium für Bildung und Forschung (BMBF)" and by FOM, The Netherlands.

References

1. See, for example: G. Bunce, N. Saito, J. Soffer, and W. Vogelsang, *Ann. Rev. Nucl. Part. Sci.* **50** (2000) 525.
2. O. Martin, A. Schäfer, M. Stratmann, and W. Vogelsang, *Phys. Rev.* **D57** (1998) 3090; **D60** (1999) 117502.
3. J. Soffer, M. Stratmann, and W. Vogelsang, *Phys. Rev.* **D65** (2002) 114024.
4. A. Mukherjee, M. Stratmann, and W. Vogelsang, *Phys. Rev.* **D67** (2003) 114006.
5. P. Lenisa and F. Rathmann [the PAX Collaboration], `hep-ex/0505054` and `http://www.fz-juelich.de/ikp/pax/`
6. M. Maggiora [the ASSIA Collaboration], `hep-ex/0504011`; GSI-ASSIA Technical Proposal, Spokesperson: R. Bertini, `http://www.gsi.de/documents/DOC-2004-Jan-152-1.ps`
7. M. Anselmino, V. Barone, A. Drago, and N.N. Nikolaev, *Phys. Lett.* **B594** (2004) 97.
8. A.V. Efremov, K. Goeke, and P. Schweitzer, *Eur. Phys. J.* **C35** (2004) 207.

9. H. Shimizu, G. Sterman, W. Vogelsang, and H. Yokoya, *Phys. Rev.* **D71** (2005) 114007.
10. A. Bianconi and M. Radici, `hep-ph/0504261`.
11. A. Mukherjee, M. Stratmann, and W. Vogelsang, *Phys. Rev.* **D72** (2005) 034011.
12. J. Soffer, *Phys. Rev. Lett.* **74** (1995) 1292; D. Sivers, *Phys. Rev.* **D51** (1995) 4880.
13. M. Glück, E. Reya, and A. Vogt, *Eur. Phys. J.* **C5** (1998) 461.
14. M. Glück, E. Reya, M. Stratmann, and W. Vogelsang, *Phys. Rev.* **D63** (2001) 094005.
15. D. de Florian and W. Vogelsang, *Phys. Rev.* **D71** (2005) 114004.

DOUBLE TRANSVERSE-SPIN ASYMMETRIES IN DRELL–YAN AND J/ψ PRODUCTION FROM PROTON–ANTIPROTON COLLISIONS

M. Guzzi[1,2,3], V. Barone[4,5], A. Cafarella[6],
C. Corianò[1,2] and P.G. Ratcliffe[3,7]

[1]*Dip. di Fisica, Università di Lecce, 73100 Lecce, Italy*
[2]*INFN, Sezione di Lecce, 73100 Lecce, Italy*
[3]*Dip. di Fisica e Matematica, Università dell'Insubria, 22100 Como, Italy*
[4]*Di.S.T.A., Università del Piemonte Orientale "A. Avogadro" 15100 Alessandria, Italy*
[5]*INFN, Gruppo Collegato di Alessandria, 15100 Alessandria, Italy*
[6]*Dept. of Physics, University of Crete, 71003 Heraklion, Greece*
[7]*INFN, Sezione di Milano, 20133 Milano, Italy*

We perform a NLO numerical study of the double transverse-spin asymmetries in the J/ψ resonance region for proton–antiproton collisions. We analyze the large x kinematic region, relevant for the proposed PAX experiment at GSI, and discuss the implication of the results for the extraction of the transversity densities.

1. Introduction

The purpose of this talk is to illustrate a numerical analysis of the double transverse-spin asymmetries in Drell–Yan processes in the J/ψ resonance region and to discuss the results with regard to the proposed PAX experiment and the possibility of accessing the transversity densities in proton–antiproton collisions.

2. Access to transversity densities

The missing leading-twist piece in the QCD perturbative description of the nucleon is the transversity density,[1] which is defined as the difference of probabilities for finding a parton of flavour q at energy scale Q^2 and light-cone momentum fraction x with its spin aligned ($\uparrow\uparrow$) or anti-aligned ($\uparrow\downarrow$) with that of transversely polarized parent nucleon

$$\Delta_T q(x, Q^2) = q_{\uparrow\uparrow}(x, Q^2) - q_{\uparrow\downarrow}(x, Q^2)\,. \tag{1}$$

Given its chirally odd nature, transversity may be accessed in collisions of two transversely polarized nucleons (Drell–Yan) via the double transverse-spin asymmetries which are defined as the ratio

$$A_{TT} = \frac{\sigma_{\uparrow\uparrow} - \sigma_{\uparrow\downarrow}}{\sigma_{\uparrow\uparrow} + \sigma_{\uparrow\downarrow}} = \frac{\Delta_T \sigma}{\sigma_{\text{unp}}} \tag{2}$$

between the transversely polarized and unpolarized cross-sections.

Doubly polarized Drell–Yan production (illustrated in Fig. 1) is the cleanest process for probing transversity distributions. It has recently been suggested that collisions of transversely polarized protons and antiprotons should provide a very good opportunity to determine the nucleon transversity via measurement of A_{TT}.

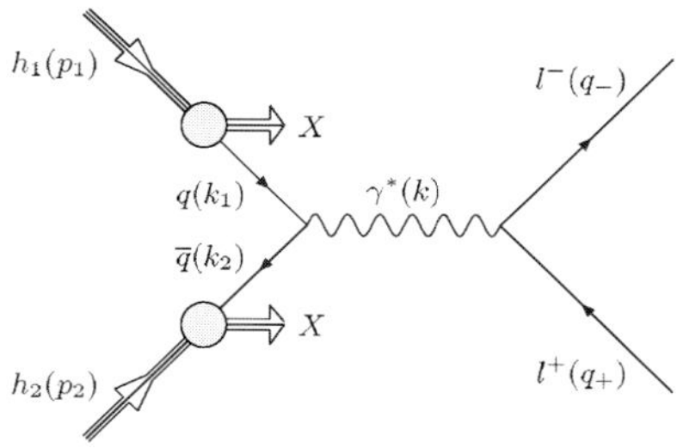

Fig. 1. Drell–Yan process

Double transverse-spin asymmetries depend may only on quark and antiquark transversity distributions

$$A_{TT} = \frac{\sigma_{\uparrow\uparrow} - \sigma_{\uparrow\downarrow}}{\sigma_{\uparrow\uparrow} + \sigma_{\uparrow\downarrow}} = \hat{a}_{TT}(\varphi) \frac{\sum_q e_q^2 \, \Delta_T q(x_1, M^2) \, \Delta_T \bar{q}(x_2, M^2) + (1 \leftrightarrow 2)}{\sum_q e_q^2 \, q(x_1, M^2) \, \bar{q}(x_2, M^2) + (1 \leftrightarrow 2)}, \tag{3}$$

where $\hat{a}_{TT}(\varphi)$ contains the azimuthal angular dependence

$$\hat{a}_{TT}(\varphi) = \tfrac{1}{2} \cos 2\varphi \tag{4}$$

and M is the dilepton invariant mass.

Measurement of A_{TT}^{pp} in the case of proton–proton collisions is planned at RHIC but the asymmetry is expected to be small (2–3%).[2,3] In fact, A_{TT}^{pp} contains antiquark distributions and the RHIC kinematics ($\sqrt{s} = 200\,\text{GeV}$, $M < 10\,\text{GeV}$, $x_1 x_2 = M^2/s < 3 \times 10^{-3}$) probes the low-$x$ region where, compared to $q(x)$, $\Delta_T q(x)$ is suppressed by QCD evolution. Such problems may be avoided by measuring $A_{TT}^{p\bar{p}}$ in proton–antiproton collisions at lower centre-of-mass energies;[2,4] this is the program of the PAX experiment at

GSI.[5] For the GSI kinematics we have $s = 30$ or $45\,\mathrm{GeV}^2$ in fixed-target, and $s = 200\,\mathrm{GeV}^2$ in collider mode, $M > 2\,\mathrm{GeV}$ and $\tau = x_1 x_2 = M^2/s > 0.1$.

The GSI kinematics is such that the asymmetries for double transverse Drell–Yan proton–antiproton processes are dominated by valence distributions and thus probe the product $\Delta_T q \times \Delta_T q$. At LO we can write

$$A_{TT}^{p\bar{p}} = \hat{a}_{TT} \frac{\sum_q e_q^2 [\Delta_T q(x_1, M^2) \Delta_T q(x_2, M^2) + \Delta_T \bar{q}(x_1, M^2) \Delta_T \bar{q}(x_2, M^2)]}{\sum_q e_q^2 [q(x_1, M^2) q(x_2, M^2) + \bar{q}(x_1, M^2) \bar{q}(x_2, M^2)]} \tag{5}$$

and $A_{TT}^{p\bar{p}}/\hat{a}_{TT}$ is found to be of order of 30%.[4,6]

At NLO the factorization formula of the cross-section for dilepton production in transversely polarized proton–antiproton scattering is[7,8]

$$\begin{aligned} \frac{\mathrm{d}\Delta_T \sigma}{\mathrm{d}M\,\mathrm{d}y\,\mathrm{d}\varphi} = \sum_q e_q^2 \int_{\xi_1}^1 \mathrm{d}x_1 \int_{\xi_2}^1 \mathrm{d}x_2 \, \big[\Delta_T q(x_1, \mu^2) \Delta_T q(x_2, \mu^2) \\ + \Delta_T \bar{q}(x_1, \mu^2) \Delta_T \bar{q}(x_2, \mu^2)\big] \frac{\mathrm{d}\Delta_T \hat{\sigma}}{\mathrm{d}M\,\mathrm{d}y\,\mathrm{d}\varphi}, \end{aligned} \tag{6}$$

where μ^2 is the factorization scale, y is the rapidity of the dilepton pair and the momentum fractions ξ_1 and ξ_2 are defined as

$$\xi_1 = \sqrt{\tau}\, e^{y}, \qquad \xi_2 = \sqrt{\tau}\, e^{-y}, \qquad y = \frac{1}{2} \ln \frac{\xi_1}{\xi_2}. \tag{7}$$

The NLO hard-scattering cross-section is

$$\begin{aligned} \frac{\mathrm{d}\Delta_T \hat{\sigma}^{(1),\overline{\mathrm{MS}}}}{\mathrm{d}M\,\mathrm{d}y\,\mathrm{d}\varphi} &= \frac{2\alpha^2}{9sM} C_F \frac{\alpha_s(\mu^2)}{2\pi} \frac{4\tau(x_1 x_2 + \tau)}{x_1 x_2 (x_1 + \xi_1)(x_2 + \xi_2)} \cos(2\varphi) \\ &\times \Bigg\{ \delta(x_1 - \xi_1)\delta(x_2 - \xi_2) \left[\frac{1}{4} \ln^2 \frac{(1-\xi_1)(1-\xi_2)}{\tau} + \frac{\pi^2}{4} - 2\right] \\ &+ \delta(x_1 - \xi_1) \left[\frac{1}{(x_2 - \xi_2)_+} \ln \frac{2x_2(1-\xi_1)}{\tau(x_2 + \xi_2)} + \left(\frac{\ln(x_2 - \xi_2)}{x_2 - \xi_2}\right)_+ + \frac{\ln(\xi_2/x_2)}{x_2 - \xi_2}\right] \\ &+ \frac{1}{2[(x_1 - \xi_1)(x_2 - \xi_2)]_+} + \frac{(x_1 + \xi_1)(x_2 + \xi_2)}{(x_1 \xi_2 + x_2 \xi_1)^2} - \frac{3 \ln\left(\frac{x_1 x_2 + \tau}{x_1 \xi_2 + x_2 \xi_1}\right)}{(x_1 - \xi_1)(x_2 - \xi_2)} \\ &+ \ln \frac{M^2}{\mu^2} \left[\delta(x_1 - \xi_1)\delta(x_2 - \xi_2) \left(\frac{3}{4} + \frac{1}{2} \ln \frac{(1-\xi_1)(1-\xi_2)}{\tau}\right) \right. \\ &\left. + \delta(x_1 - \xi_1) \frac{1}{(x_2 - \xi_2)_+}\right] \Bigg\} + \Big[1 \longleftrightarrow 2\Big]. \end{aligned} \tag{8}$$

In order to predict asymmetries, some assumption for the transversity distributions is needed. For instance, we may take transversity equal to helicity at some low scale (as suggested by certain models)

$$\Delta_T f(x,\mu) = \Delta f(x,\mu) \quad \text{(minimal bound)} \tag{9}$$

or, alternatively, saturation of the Soffer inequality [9]

$$2\,|\Delta_T f(x,\mu)| = f(x,\mu) + \Delta f(x,\mu)\,. \tag{10}$$

We use NLO GRV input densities,[10] with starting scale $\mu = 0.63\,\text{GeV}$. The relation between transversity and the GRV distributions is set at this scale. QCD evolution is performed via the appropriate NLO DGLAP equations.[11,12] In Fig. 2 we see that in the energy range relevant for the PAX experiment the asymmetries are around 35%. From Fig. 3 we see

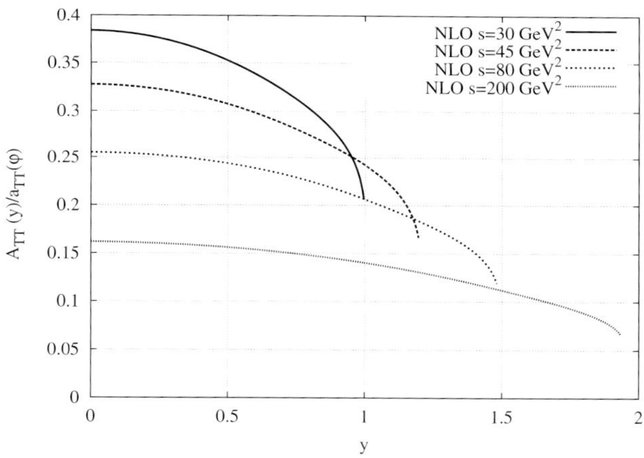

Fig. 2. $A_{TT}(y)/\hat{a}_{TT}(\varphi)$ at NLO, with M integrated from 2 to 3 GeV using GRV input with the minimal bound $\Delta_T q(x,\mu) = \Delta q(x,\mu)$.

that in the case of the Soffer bound the asymmetries are systematically larger than the asymmetries obtained in the case of the minimal bound. In Fig. 4 we display the asymmetry at larger M, where it grows up to 45% (but one should recall that the cross-section falls rapidly as M increases). Applying the constraint $\Delta_T f(x,\mu) = \Delta f(x,\mu)$ at, say, 1 GeV instead of 0.63 GeV would produce slightly larger asymmetries; this is due to QCD evolution effects since $\Delta_T f(x,\mu)$ is less suppressed by evolving from 1 GeV than from 0.63 GeV. The comparison between LO and NLO results is shown in Fig. 5, where one sees that NLO corrections have very little affect on the asymmetries.

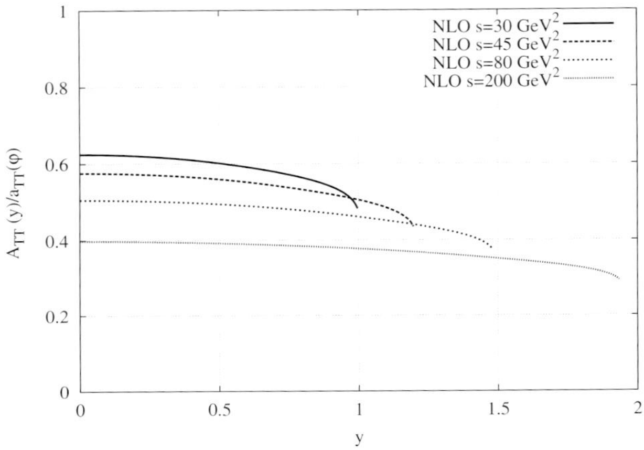

Fig. 3. $A_{TT}(y)/\hat{a}_{TT}(\varphi)$ at NLO, with M integrated from 2 to 3 GeV using GRV input and saturating the Soffer bound.

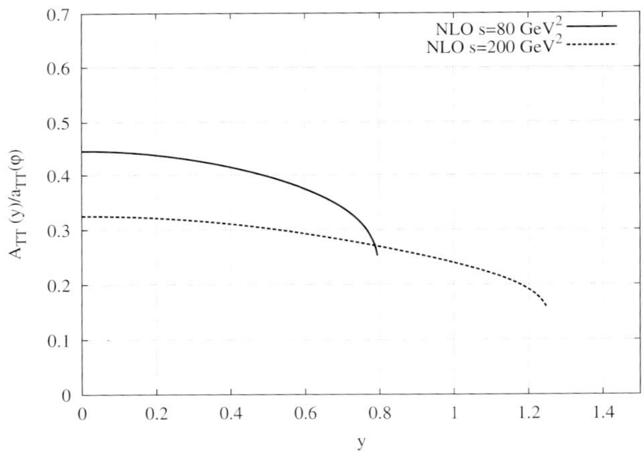

Fig. 4. $A_{TT}(y)/\hat{a}_{TT}(\varphi)$ at NLO with M integrated from 4 to 7 GeV using GRV input with the minimal bound.

3. Dilepton production via the J/ψ resonance in the GSI regime

To achieve a higher counting rate, one may exploit the J/ψ peak, where the cross-section is two orders of magnitude larger. If J/ψ production is dominated by $q\bar{q}$ annihilation channel, the corresponding asymmetry has the same structure as in the continuum region, since the J/ψ is a vector

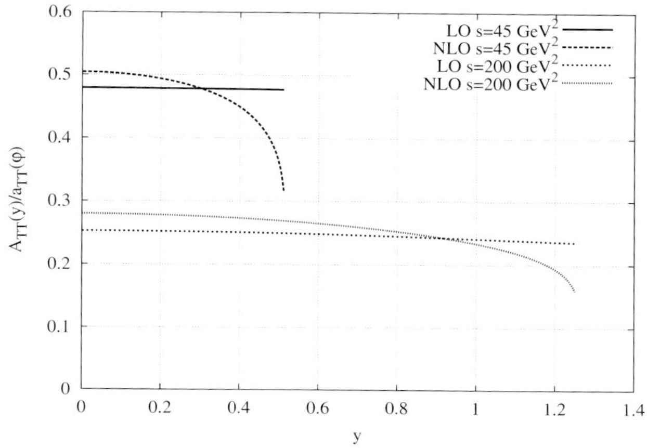

Fig. 5. $A_{TT}(y)/\hat{a}_{TT}(\varphi)$ at LO (solid curve) *vs.* NLO (dashed curve) at $M = 4\,\text{GeV}$ and $s = 45\,\text{GeV}^2$ and $A_{TT}(y)/\hat{a}_{TT}(\varphi)$ at LO (dotted curve) *vs.* NLO (dot-dashed curve) at $M = 4\,\text{GeV}$ and $s = 200\,\text{GeV}^2$ using GRV input with the minimal bound.

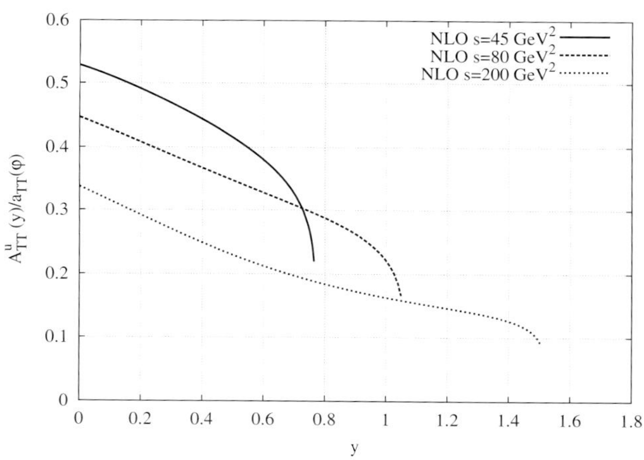

Fig. 6. The double transverse-spin asymmetry in the J/ψ resonance region for various c.m. energies. As usual, the minimal bound is used for the input distributions.

particle and the $q\bar{q} - J/\psi$ couplings are similar to $q\bar{q} - \gamma^*$.[4] SPS data[13] show the $p\bar{p}$ cross-section for J/ψ production at $s = 80\,\text{GeV}^2$ to be about 10 times larger than the corresponding pp cross-section, indicating the dominance of the $q\bar{q}$ annihilation mechanism. Thus, the helicity structure of the asymmetries is preserved and, replacing the couplings in Eq. (5), we can write

$$A_{TT}^{J/\psi} = \hat{a}_{TT} \times \frac{\sum_q (g_q^V)^2 \left[\Delta_T q(x_1, M^2)\Delta_T q(x_2, M^2) + \Delta_T \bar{q}(x_1, M^2)\Delta_T \bar{q}(x_2, M^2)\right]}{\sum_q (g_q^V)^2 \left[q(x_1, M^2)q(x_2, M^2) + \bar{q}(x_1, M^2)\bar{q}(x_2, M^2)\right]} \tag{11}$$

In the large x_1, x_2 region the u and d valence quarks dominate and, since the $q\bar{q} - J/\psi$ coupling is the same for u and d quarks, the asymmetry becomes

$$A_{TT}^{J/\psi} \simeq \hat{a}_{TT} \frac{\Delta_T u(x_1, M^2)\Delta_T u(x_2, M^2) + \Delta_T d(x_1, M^2)\Delta_T d(x_2, M^2)}{u(x_1, M^2)u(x_2, M^2) + d(x_1, M^2)d(x_2, M^2)}. \tag{12}$$

The condition $\Delta_T u(x) \gg \Delta_T d(x)$, satisfied by all models at large x, permits a further simplification and one obtains

$$A_{TT}^{J/\psi} \simeq \hat{a}_{TT} \frac{\Delta_T u(x_1, M^2)\Delta_T u(x_2, M^2)}{u(x_1, M^2)u(x_2, M^2)}. \tag{13}$$

The J/ψ asymmetry is then essentially the DY asymmetry evaluated at $M_{J/\psi}$ and, for $s = 80\,\text{GeV}^2$, lies in the range 0.25–0.45 (see Fig. 6). Inasmuch as the gg fusion diagram may be neglected, as old $p\bar{p}$ data suggest, this remains true at NLO (i.e. considering gluon radiation).

4. Threshold resummation

The kinematic region corresponding to $M \approx 1 - 4\,\text{GeV}$ and with a centre-of-mass energy $s \approx 30\,\text{GeV}^2$ is not properly contained in the domain of perturbative QCD, (i.e. factorization, parton model etc.). Thus, depending on kinematics, higher-order corrections to the cross-sections may be important and must be well understood.

For the sake of simplicity, we shall merely sketch what occurs, with little quantitative detail. The factorization theorem for the hadronic cross-section in terms of twist-2 parton densities is not exact, but holds only to the leading power of M, and the corrections generally increase as τ increases. In the region $z = \tau/(x_1 x_2) \simeq 1$ the kinematics is such that virtual and real-emission diagrams become strongly unbalanced (real-gluon emission is suppressed) and in these conditions there are large higher-order logarithmic corrections to the partonic cross-section of the form

$$\alpha_s^k \frac{[\ln(1-z)]^{2k-1}}{(1-z)} \tag{14}$$

The region $z \approx 1$ is dominant in the kinematic regime relevant for GSI, hence the large logarithmic contributions need to be resummed to all orders in α_s. NLL-resummed perturbation theory has been extensively studied[14] and resummation corrections for A_{TT} are found to be less than 10% and rather dependent on the infrared cut-off for the soft gluon emission.

5. Conclusions

In the GSI regime Drell–Yan double transverse-spin asymmetries are sizable, of the order of 30%, and are not spoiled by NLO (and resummation) effects. Transverse asymmetries for J/ψ production at moderate energies are expected to be similar (with the advantage of much higher counting rates). Transversely polarized antiproton experiments at GSI will thus provide an excellent window onto nucleon transversity.

Acknowledgments

We would like to thank M. Anselmino, N.N. Nikolaev and our colleagues of the PAX collaboration for prompting the study reported here and for various useful discussions. This work is supported in part by the Italian Ministry of Education, University and Research (PRIN 2003).

References

1. V. Barone, A. Drago and P.G. Ratcliffe, *Phys. Rep.* **359** (2002) 1.
2. V. Barone, T. Calarco and A. Drago, *Phys. Rev.* **D56** (1997) 527.
3. O. Martin, A. Schäfer, M. Stratmann and W. Vogelsang, *Phys. Rev.* **D57** (1998) 3084.
4. M. Anselmino, V. Barone, A. Drago and N.N. Nikolaev *Phys. Lett.* **B5** (2004) 97.
5. PAX Collaboration, V. Barone *et al.*, "Antiproton–Proton Scattering Experiments with Polarization", Technical Proposal, hep-ex/0505054 (2005).
6. A. Efremov, K. Goeke and P. Schweitzer, *Eur. Phys. J.* **C35** (2004) 207.
7. O. Martin, A. Schäfer, M. Stratmann and W. Vogelsang, *Phys. Rev.* **D60** (1999) 117502.
8. A. Mukherjee, M. Stratmann and W. Vogelsang, *Phys. Rev.* **D67** (2003) 114006.
9. J. Soffer, *Phys. Rev. Lett.* **74** (1995) 1292.
10. M. Glück, E. Reya and A. Vogt, *Eur. Phys. J.* **C5** (1998) 461; M. Glück, E. Reya, M. Stratmann and W. Vogelsang, *Phys. Rev.* **D63** (2001) 094005.
11. A. Cafarella and C. Corianò, *Comput. Phys. Commun.* **160** (2004) 213.
12. A. Cafarella, C. Corianò and M. Guzzi, *JHEP* **0311** (2003) 059.
13. M.J. Corden, *Phys. Lett.* **B98** (1981) 220.
14. H. Shimizu, G. Sterman, W. Vogelsang and H. Yokoya, *Phys. Rev.* **D71** (2005) 11407.

THE QUARK–QUARK CORRELATOR: THEORY AND PHENOMENOLOGY

Elvio Di Salvo

Dipartimento di Fisica, Università di Genova – INFN, sez. Genova
Via Dodecaneso 33, 16146 Genova – Italy
E-mail: disalvo@ge.infn.it

New properties of the quark correlator are found via equations of motion. As a first result, approximate relations can be established among the "soft" distribution functions; one such relation may help in determining the quark transversity in a nucleon. Secondly, the Q^2 dependence of the T-odd functions is deduced; the result is compared to unpolarized Drell–Yan data. Lastly, important remarks are made about the contributions to $g_2(x)$.

1. Introduction

The correlator, a very important theoretical tool encoding the "soft" functions involved in high energy reactions, was originally introduced by Ralston and Soper in 1979 [1] and successively adopted by other authors,[2–4] who studied some properties of this matrix. However, as I shall show, further results can be established in the sector, thanks to the equations of motion. These new results have important phenomenological consequences. In particular, I point out that progress can be done about three problems which arise in high energy physics, *i.e.*, determining transversity, interpreting azimuthal asymmetries and disentangling the contributions to $g_2(x)$.

a) It is not easy to determine experimentally the transversity in a spinning hadron, because this density is chiral odd and has to be coupled with another chiral odd function, for example the transversity of an antiquark in doubly polarized Drell–Yan, or the Collins function (or the Jaffe interference function) in singly polarized semi-inclusive deep inelastic scattering (SIDIS). The various methods proposed for determining this function present more or less serious drawbacks.

b) Azimuthal asymmetries of surprisingly large size have been observed in unpolarized Drell–Yan, in singly polarized SIDIS and in inclusive hadronic

reactions. Such asymmetries do not find an explanation in ordinary perturbative QCD. Among possible interpretations, we recall quark–quark–gluon correlations [5] and, more recently, the T-odd functions,[3,6] which take into account the intrinsic transverse momentum of a quark inside a hadron. These functions provide a discrete description of the azimuthal asymmetries, however, as regards the Q^2 dependence of such asymmetries, they do not agree neither with the predictions of quark–quark–gluon correlations, nor with unpolarized Drell–Yan data.[7]
c) The function $g_2(x)$ has been studied by several authors [8,9] from various viewpoints, but there is no agreement about the contributions it involves.

In my talk I shall show that the equations of motion [10] allow to establish for the correlator some new properties, which lead to partial answers to the three problems just illustrated. First of all, I shall define the correlator. Secondly, I shall split it into a T-even and a T-odd part. Thirdly, I shall expand it in powers of the coupling and I shall apply the equations of motion, getting some conditions on the first three terms of the expansion. I shall discuss each such term in some detail. Lastly, I shall draw a short conclusion.

2. The Correlator

The correlator is defined as [2]

$$\Phi_{ij}(p; P_0, S) = \int \frac{d^4x}{(2\pi)^4} e^{ipx} \langle P_0, S|\bar{\psi}_j(0)L(x)\psi_i(x)|P_0, S\rangle. \qquad (1)$$

Here ψ is the quark field, p the quark four-momentum and $|P_0, S\rangle$ denotes a nucleon state with a given four-momentum P_0 and Pauli–Lubanski four-vector S. Moreover $L(x)$ is the gauge link operator,[2] *i.e.*,

$$L(x) = \mathrm{P}\exp\left[-ig\Lambda_P(x)\right], \qquad \Lambda_P(x) = \int_0^x \lambda_a A^a_\mu(z)dz^\mu. \qquad (2)$$

Here "P" denotes the path-ordered product along the integration contour P; g, λ_a and A^a_μ are respectively the strong coupling constant, the Gell-Mann matrices and the gluon fields. The link operator depends on the choice of P, which has to be fixed so as to make a physical sense.[11] According to previous treatments,[11,2] I define two different contours, $P_\pm$, as sets of three pieces of straight lines, from the origin to $x_{1\infty} \equiv (\pm\infty, 0, \mathbf{0}_\perp)$, from $x_{1\infty}$ to $x_{2\infty} \equiv (\pm\infty, x^+, \mathbf{x}_\perp)$ and from $x_{2\infty}$ to $x \equiv (x^-, x^+, \mathbf{x}_\perp)$; here the + or − sign has to be chosen, according as to whether final or initial state interactions [11] are involved in the reaction. I have adopted a frame – to be used throughout this talk – whose z-axis is taken along the nucleon momentum, with $x^{\pm} = 1/\sqrt{2}(t \pm z)$.

The correlator enjoys two important properties, due to the hermiticity condition and parity conservation:

$$\Phi^{\dagger} = \gamma_0 \Phi \gamma_0, \qquad \Phi(p, P_0, S) = \gamma_0 \Phi(\bar{p}, \bar{P}_0, -\bar{S})\gamma_0. \tag{3}$$

Here $\bar{p} \equiv (p_0, -\mathbf{p})$, having set $p \equiv (p_0, \mathbf{p})$; $\bar{P}_0$ and $\bar{S}$ are defined analogously. Time reversal invariance does not give rise to any condition on Φ. Indeed, we may have T-even and T-odd functions, the latter ones being generated by interference between two amplitudes which behave differently under time reversal.

3. Splitting

I set

$$\Phi_{E(O)} = \tfrac{1}{2}[\Phi_+ \pm \Phi_-], \tag{4}$$

where $\Phi_{\pm}$ corresponds to the contour $P_{\pm}$ in Eqs. (2), while Φ_E and Φ_O select respectively the T-even and the T-odd "soft" functions. These two correlators contain respectively the link operators $L_E(x)$ and $L_O(x)$, where

$$L_{E(O)}(x) = \tfrac{1}{2}\mathrm{P}\left\{\exp\left[-ig\Lambda_{P_+}(x)\right] \pm \exp\left[-ig\Lambda_{P_-}(x)\right]\right\} \tag{5}$$

and $\Lambda_{P_{\pm}}(x)$ are defined by the second Eq. (2). Notice that, for T-even functions, the result is independent of the contour (P_+ or P_-), while T-odd functions change sign according as to whether they are generated by initial or final state interactions.[11] In this sense, such functions are not strictly universal.[11] Eq. (5) has important consequences on Φ for small values of g. Indeed, the zero order term is just T-even, while the first order correction contains T-odd contributions. This confirms that no T-odd terms occur without interactions among partons, as claimed by other authors.[6,11]

An immediate advantage of the splitting (4) is that we can define separately T-even and T-odd functions, by projecting $\Phi_E(x)$ and $\Phi_O(x)$ over the various Dirac components. These projections are defined as $\Phi^{\Gamma} = \frac{1}{2}\int dp^- tr(\Phi\Gamma)$, where Γ is a Dirac operator. In particular, among the T-even functions, I consider the three main densities of quarks in the nucleon, *i.e.*, the unpolarized density $f_1(x, \mathbf{p}_{\perp}^2)$, the longitudinally polarized density $g_{1L}(x, \mathbf{p}_{\perp}^2)$ and the transversity $h_{1T}(x, \mathbf{p}_{\perp}^2)$. They are related, respectively, to $\Phi_E^{\gamma^+}$, to $\Phi_E^{\gamma_5\gamma^+}$ and to $\Phi_E^{\gamma_5\gamma^+\gamma^i}$, $i = 1, 2$. Among the T-odd functions, it is worth mentioning the unpolarized quark density $f_{1T}^{\perp}(x, \mathbf{p}_{\perp}^2)$

in a transversely polarized nucleon (the Sivers[12] function) and the transversity $h_1^\perp(x, \mathbf{p}_\perp^2)$ in an unpolarized nucleon (analogous to the Collins[13] fragmentation function), which may be derived, respectively, from $\Phi_O^{\gamma^+}$ and $\Phi_O^{\gamma_5\gamma^+\gamma^i}$, $i = 1, 2$.

Now I consider a particular gauge, that is, *an axial gauge with antisymmetric boundary conditions.*[2] In this case one has $\Lambda_{P_-}(x) = -\Lambda_{P_+}(x)$ and therefore

$$L_E(x) = \mathrm{P}\cos\left[g\Lambda_{P_+}(x)\right], \qquad L_O(x) = -i\mathrm{P}\sin\left[g\Lambda_{P_+}(x)\right]. \tag{6}$$

In this particular gauge – to be referred to as G-gauge in the following – the T-even (T-odd) part of the correlator consists of a series of even (odd) powers of g, each term having an even (odd) number of gluon legs.

4. Equations of motion

Now I invoke the Politzer theorem on equations of motion,[10] *i.e.*,

$$\langle P_0, S|F(\psi)(i\not{D} - m_q)\psi(x)|P_0, S\rangle = 0. \tag{7}$$

Here m_q is the quark rest mass, $F(\psi)$ a functional of the quark field and $D_\mu = \partial_\mu - ig\lambda_a A_\mu^a$ the covariant derivative. The result (7) survives renormalization. I adopt the G-gauge and expand the correlator in powers of g, *i.e.*,

$$\Phi_E(p) = \Phi_E^{(0)}(p) + g^2\Phi_E^{(2)}(p) + O(g^4), \tag{8}$$

$$\Phi_O(p) = g\Phi_O^{(1)}(p) + O(g^3). \tag{9}$$

Setting $F(\psi) = \bar{\psi}(0)L(x)$, Eq. (7) yields, after some steps,[14]

$$(\not{p} - m_q)\Phi_E^{(0)}(p) = 0, \tag{10}$$

$$(\not{p} - m_q)\Phi_O^{(1)}(p) = \Psi_O(p), \tag{11}$$

$$(\not{p} - m_q)\Phi_E^{(2)}(p) = \Psi_E(p). \tag{12}$$

Here

$$\Psi_O = \Psi[\not{A}_2 - \not{A}] \tag{13}$$

$$\Psi_E = \Psi[A_2^2] - (\not{p} - m_q)\Psi[\not{A}_1\not{A}](\not{p} - m_q)^{-1}, \tag{14}$$

$\not{A} = \lambda_a \not{A}^a(x)$, $\not{A}_i = \lambda_a \not{A}^a(x_{i\infty})$ (i = 1, 2), $A_2^2 = \not{A}_2\not{A}_2$ and

$$\{\Psi[N]\}_{ij} = \int \frac{d^4x}{(2\pi)^4} e^{ipx}\langle P_0, S|\bar{\psi}_j(0)N_{ik}\psi_k(x)|P_0, S\rangle \tag{15}$$

is a functional of the generic Dirac operator N. As a consequence of Eqs. (8) to (14), the expansions of $\Phi_E(p)$ and $\Phi_O(p)$ in powers of g read [14]

$$\Phi_E(p) = p^+ \rho\delta(p^2 - m_q^2) + g^2\kappa_\perp^2 M_E(p) + O(g^4), \tag{16}$$

$$\Phi_O(p) = g\kappa_\perp M_O(p) + O(g^3), \tag{17}$$

where

$$\rho = \tfrac{1}{2}(\not p + m_q)(f_1 + \gamma_5 \not S_\parallel^q g_{1L} + \gamma_5 \not S_\perp^q h_{1T}), \tag{18}$$

$$M_E\kappa_\perp^2 = (\not p - m_q)^{-2}\Psi_E, \qquad M_O\kappa_\perp = (\not p - m_q)^{-1}\Psi_O \tag{19}$$

and

$$\kappa_\perp = \frac{|\mathbf{p}_\perp|}{p^+}. \tag{20}$$

$S_\parallel^q$ and $S_\perp^q$ are respectively the longitudinal and transverse components of the Pauli–Lubanski four-vector of the quark. Notice that four-momentum conservation in the correlators (13) and (14) avoids singularities in Eqs. (19).

5. Consequences and comments

A) At zero order in g the correlator reduces essentially to the density matrix ρ, Eq. (18) of a free, on-shell quark. Since ρ depends just on three functions, one deduces several approximate relations among the functions involved in the parametrization of the correlator.[2,3] In particular, we have, in the G-gauge,

$$h_{1T}(x, \mathbf{p}_\perp^2) = g_{1T}(x, \mathbf{p}_\perp^2) + O(\alpha_s\kappa_\perp^2), \tag{21}$$

$$h_{1L}(x, \mathbf{p}_\perp^2) = g_{1L}(x, \mathbf{p}_\perp^2) + O(\alpha_s\kappa_\perp^2). \tag{22}$$

Here g_{1T} is the helicity density of a quark in a transversely polarized nucleon, while h_{1L} is the transversity of a quark in a longitudinally polarized nucleon. Relations (21) and (22) are especially important, because they relate chiral odd functions to chiral even ones, which may be determined much more easily.[15] Such relations are not altered by QCD perturbative evolution, since the Politzer theorem survives renormalization. Therefore Eq. (21) is useful in determining transversity. However, such a relation, as well as Eq. (22), may be slightly modified by terms of order $O(\alpha_s\kappa_\perp^2)$ and also by other contributions, as we shall see in a moment.

B) Equation (17) implies that first order corrections in the correlator are of order $O(g\kappa_\perp)$; this conclusion turns out to be gauge independent.[14] Moreover the asymmetries in SIDIS and Drell–Yan which involve T-odd functions are suppressed at least as Q^{-n}, where Q is the hard QCD scale and n is the number of T-odd functions involved. This finds an experimental confirmation in the azimuthal asymmetry of unpolarized Drell–Yan.[7] Indeed, in the formalism of T-odd functions, this asymmetry is proportional to the convolutive product $h_1^\perp \otimes \bar{h}_1^\perp$,[16] therefore, according to our predictions, it decreases as Q^{-2}. This result is confirmed by the best fits to data,[7] see Figs. 1 and 2. Lastly, Eq. (13) shows that T-odd terms are produced by quark–quark–gluon correlations, which may be approximated by "effective" quark distributions only under some particular conditions.[11]

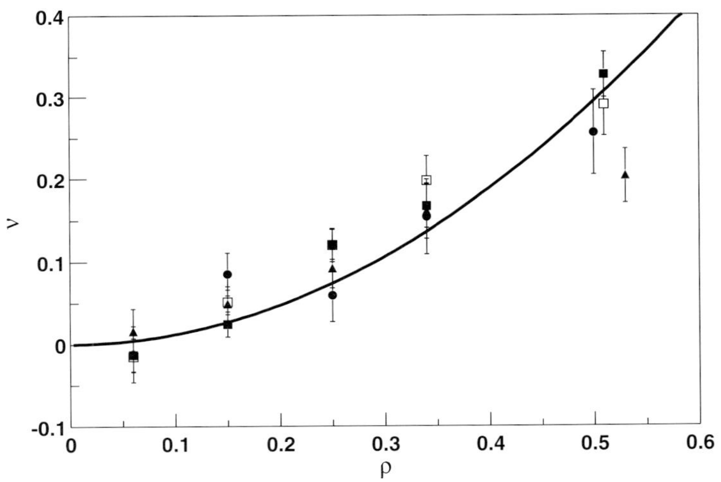

Fig. 1. Unpolarized Drell–Yan (see Refs. 7): the asymmetry parameter ν *vs.* $\rho = |\mathbf{q}_\perp|/Q$, where $\mathbf{q}_\perp$ and Q are respectively the transverse momentum and the effective mass of the muon pair.

C) Other important consequences can be drawn from Eq. (16) in the G-gauge and in the limit of small g. In particular, as regards the functions g_1 and g_2, I derive

$$g_T = g_1 + g_2 = (m_q/xM)h_1 + O[\alpha_s(\pi_\perp/p^+)^2], \tag{23}$$

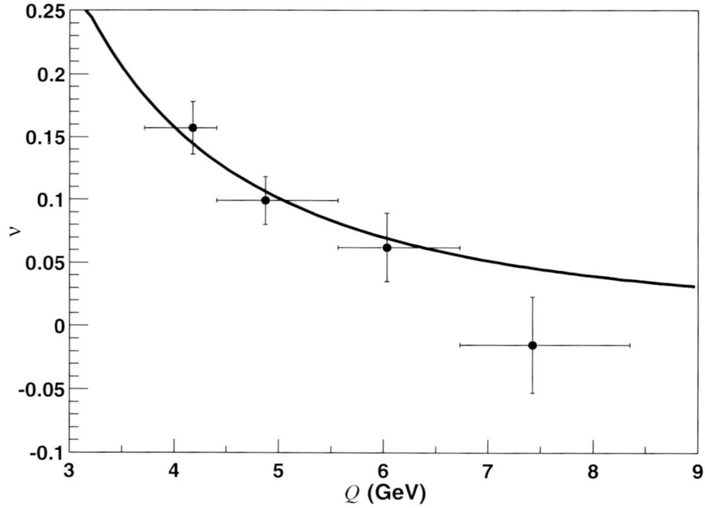

Fig. 2. Unpolarized Drell–Yan (see Refs. 7): the asymmetry parameter ν *vs.* Q at fixed $|\mathbf{q}_\perp|$. Same notations as in Fig. 1.

where M and $\pi_\perp$ are, respectively, the nucleon rest mass and the mean value of $|\mathbf{p}_\perp|$. This relation is gauge invariant. It does not include the contribution of the anomalous coupling of the singlet axial current with gluons, nor of nonperturbative fluctuations of the nucleon, like, *e.g.*, $N \to \Delta\pi$:[17] indeed, the latter term demands a large value of g, making approximation (8) meaningless. Therefore nontrivial twist-2 and twist-3 contributions to the combination $g_1 + g_2$, which are allowed by the operator product expansion,[8] may arise only from these two terms. Incidentally, contributions from nonperturbative fluctuations are of order $0.1g_1$[17] and may, in principle, affect by a few percent all the Dirac components of the correlator (8), and therefore also Eqs. (21) and (22).

Moreover, one has $\sqrt{2}p^+ \simeq Q$ in processes like Drell–Yan and SIDIS; therefore Eqs. (23) and (20) imply that the main power correction to g_T is of order $\alpha_s(\pi_\perp/Q)^2$. This correction comes from terms of the type (14), with 2 gluon legs. Therefore the $\gamma_5\gamma^i$ projection ($i = 1, 2$) of the quark–quark–gluon correlator, Eq. (13), trivially fulfills the Burkhardt–Cottingham sum rule,[18] in accord with the result of Ref. 19.

6. Conclusions

Here I shortly recall the main results exposed in my talk, which cast a new light on the problems illustrated in the introduction.

- I establish approximate relations between chiral-even and chiral-odd functions, which, in particular, may help in determining transversity.

- T-odd functions are of order $g\pi_\perp/Q$. This implies a suppression by at least a factor of Q^{-n} in the asymmetries involving n such functions, in accord with data of unpolarized Drell–Yan and with the quark–quark–gluon correlations. Our prediction could also be compared with data of incoming HERMES, COMPASS and CLAS experiments.

- Nontrivial twist-2 and twist-3 contributions to the combination $g_1 + g_2$ come only from the quark–gluon anomalous coupling in the singlet axial current or from nonperturbative fluctuations of the nucleon.

References

1. J. Ralston and D.E. Soper, *Nucl. Phys.* **B152** (1979) 109.
2. P.J. Mulders and R.D. Tangerman, *Nucl. Phys.* **B461** (1996) 197.
3. D. Boer, R. Jakob and P.J. Mulders, *Nucl. Phys.* **B564** (2000) 471.
4. K. Goeke, A. Metz and M. Schlegel, *Phys. Lett.* **B618** (2005) 90.
5. J. Qiu and G. Sterman, *Phys. Rev. Lett.* **67** (1991) 2264; *Nucl. Phys.* **B378** (1992) 52; *Phys. Rev.* **D59** (1998) 014004.
6. S.J. Brodsky, D.S. Hwang and I. Schmidt, *Phys. Lett.* **B530** (2002) 99; *Nucl. Phys.* **B642** (2002) 344.
7. NA10 Coll., S. Falciano *et al.*, *Z. Phys.* **C31** (1986) 513; NA10 Collab., M. Guanziroli *et al.*, *Z. Phys.* **C37** (1988) 545.
8. M. Anselmino, A. Efremov and E. Leader, *Phys. Rep.* **261** (1995) 1.
9. R.L. Jaffe and X. Ji, *Phys. Rev.* **D43** (1991) 724.
10. H.D. Politzer, *Nucl. Phys.* **B172** (1980) 349.
11. J.C. Collins, *Phys. Lett.* **B536** (2002) 43.
12. D.W. Sivers, *Phys. Rev.* **D41** (1990) 83; *Phys. Rev.* **D43** (1991) 261.
13. J.C. Collins, *Nucl. Phys.* **B396** (1993) 161.
14. E. Di Salvo, in preparation.
15. E. Di Salvo, *Int. J. Mod. Phys.* **A18** (2003) 5277.
16. D. Boer, S. Brodsky and D.-S. Huang, *Phys. Rev.* **D67** (2003) 054003.
17. A.I. Signal, DIS05, Madison, Wisconsin (hep-ph/0507327).
18. H. Burkhardt and W. Cottingham, *Ann. Phys.* **56** (1970) 453.
19. N.L. Ter-Isaakyan, *Phys. Lett.* **B340** (1994) 189.

CHIRAL QUARK MODEL SPIN FILTERING MECHANISM AND HYPERON POLARIZATION

S.M. Troshin and N.E. Tyurin

Institute for High Energy Physics, Protvino, Moscow Region, 142281 Russia

The model combined with unitarity and impact parameter picture provides simple mechanism for generation of hyperon polarization in collision of unpolarized hadrons. We concentrate on a particular problem of Λ-hyperon polarization and derive its linear x_F-dependence as well as its energy and transverse momentum independence at large $p_\perp$ values. Mechanism responsible for the single-spin asymmetries in pion production is also discussed.

One of the most interesting and persistent for a long time spin phenomena was observed in inclusive hyperon production in collisions of unpolarized hadron beams. A very significant polarization of Λ-hyperons has been discovered almost three decades ago.[1] Experimentally the process of Λ-production has been studied more extensively than other hyperon production processes. Therefore we will emphasize on the particular riddle of Λ-polarization because spin structure of this particle is most simple and is determined by strange quark only. This mechanism can also be used for the explanation of single-spin asymmetries in the inclusive pion production.

It should be noted that understanding of transverse single-spin asymmetries in DIS (in contrast to the hyperon polarization) has observed significant progress during last years; this progress is related to an account of final-state interactions from gluon exchange[2,3] – coherent effect not suppressed in the Bjorken limit.

Experimental situation with hyperon polarization is widely known and stable for a long time. Polarization of Λ produced in the unpolarized inclusive pp-interactions is negative and energy independent. It increases linearly with x_F at large transverse momenta ($p_\perp \geq 1\,\mathrm{GeV}/c$), and for such values of transverse momenta is almost $p_\perp$-independent.[1]

On the theoretical side, perturbative QCD with a straightforward collinear factorization scheme leads to small values of Λ-polarization[4,5] which are far below of the corresponding experimental data. Modifications of this scheme and account for higher twists contributions allows to obtain higher magnitudes of polarization but do not change a decreasing dependence proportional to $p_\perp^{-1}$ at large transverse momenta.[6–8] It is difficult to reconcile this behavior with the flat experimental data dependence on the transverse momenta. Inclusion of the internal transverse momentum of partons ($k_\perp$-effects) into the polarizing fragmentation functions leads to similarly decreasing polarization.[9] In addition it should be noted that the perturbative QCD has also problems in the description of the unpolarized scattering, e.g. in inclusive cross-section for π^0-production, at the energies lower than the RHIC energies.[10]

The essential point of the approaches mentioned above is that the vacuum at short distances is taken to be a perturbative one. There is an another possibility. It might happen that the polarization dynamics in strangeness production originates from the genuine nonperturbative sector of QCD (cf. e.g. Ref. 11). In the nonperturbative sector of QCD the two important phenomena, confinement and spontaneous breaking of chiral symmetry (χSB)[12] should be reproduced. The relevant scales are characterized by the parameters $\Lambda_{\rm QCD}$ and Λ_χ, respectively. Chiral $SU(3)_L \times SU(3)_R$ symmetry is spontaneously broken at the distances in the range between these two scales. The χSB mechanism leads to generation of quark masses and appearance of quark condensates. It describes transition of current into constituent quarks. Constituent quarks are the quasiparticles, i.e. they are a coherent superposition of bare quarks, their masses have a magnitude comparable to a hadron mass scale. Therefore hadron is often represented as a loosely bounded system of the constituent quarks. These observations on the hadron structure lead to understanding of several regularities observed in hadron interactions at large distances. It is well known that such picture provides reasonable values for the static characteristics of hadrons, for instance, their magnetic moments. The other well known direct result is appearance of the Goldstone bosons.

The most recent approach to single-spin asymmetries (SSA) based on nonperturbative QCD has been developed in Ref. 13 where, in particular, Λ-polarization has been related to the large magnitude of the transverse flavor dipole moment of the transversely polarized baryons in the infinite momentum frame. It is based on the parton picture in the impact parameter space and assumed specific helicity-flip generalized parton distribution.

The instanton-induced mechanism of SSA generation was considered in Refs. 14, 15 and relates those asymmetries with a genuine nonperturbative QCD interaction. It should be noted that the physics of instantons (cf. e.g. Ref. 16) can provide microscopic explanation for the χSB mechanism.

We discuss here mechanism for hyperon polarization based on chiral quark model[a,12] and the filtering spin states related to unitarity in the s-channel. This mechanism connects polarization with asymmetry in the position (impact parameter) space.

As it was already mentioned constituent quarks and Goldstone bosons are the effective degrees of freedom in the chiral quark model. We consider a hadron consisting of the valence constituent quarks located in the central core which is embedded into a quark condensate. Collective excitations of the condensate are the Goldstone bosons and the constituent quarks interact via exchange of Goldstone bosons; this interaction is mainly due to a pion field which is of the flavor- and spin-exchange nature. Thus, quarks generate a strong field which binds them.[18]

At the first stage of hadron interaction common effective self-consistent field is appeared. Valence constituent quarks are scattered simultaneously (due to strong coupling with Goldstone bosons) and in a quasi-independent way by this effective strong field. Such ideas were already used in the model[19] which has been applied to description of elastic scattering and hadron production.[20]

The initial state particles (protons) are unpolarized. It means that states with spin up and spin down have equal probabilities. The main idea of the proposed mechanism is the filtering of the two initial spin states of equal probability due to different strength of interactions. The particular mechanism of such filtering can be developed on the basis of chiral quark model, formulas for inclusive cross section (with account for the unitarity)[21] and notion on the quasi-independent nature of valence quark scattering in the effective field.

We will exploit the feature of chiral quark model that constituent quark $Q_\uparrow$ with transverse spin in up-direction can fluctuate into Goldstone boson and another constituent quark $Q'_\downarrow$ with opposite spin direction, i.e. perform a spin-flip transition:[17]

$$Q_\uparrow \to GB + Q'_\downarrow \to Q + \bar{Q}' + Q'_\downarrow. \qquad (1)$$

An absence of arrows means that the corresponding quark is unpolarized.

---

[a]It has been successfully applied for the explanation of the nucleon spin structure.[17]

To compensate quark spin flip $\delta\mathbf{S}$ an orbital angular momentum $\delta\mathbf{L} = -\delta\mathbf{S}$ should be generated in final state of reaction (1). The presence of this orbital momentum $\delta\mathbf{L}$ in its turn means shift in the impact parameter value of the final quark $Q'_\downarrow$ (which is transmitted to the shift in the impact parameter of Λ)

$$\delta\mathbf{S} \Rightarrow \delta\mathbf{L} \Rightarrow \delta\tilde{\mathbf{b}}.$$

Due to different strengths of interaction at the different values of the impact parameter, the processes of transition to the spin up and down states will have different probabilities which leads eventually to polarization of Λ.

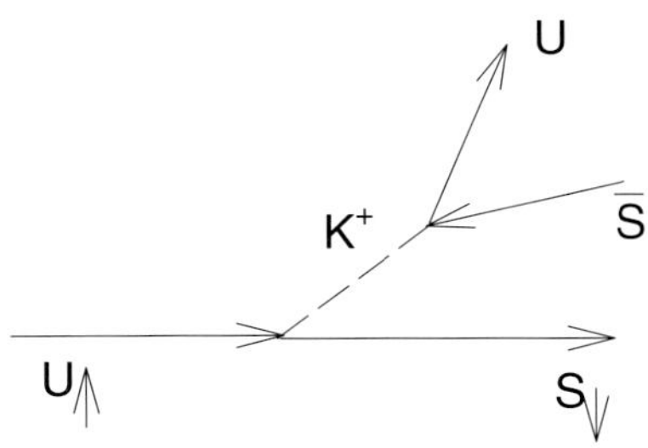

Fig. 1. Transition of the spin-up constituent quark U to the spin-down strange quark.

In a particular case of Λ-polarization the relevant transitions of constituent quark U (cf. Fig. 1) will be correlated with the shifts $\delta\tilde{b}$ in impact parameter $\tilde{b}$ of the final Λ-hyperon, i.e.:

$$\begin{aligned} U_\uparrow &\to K^+ + S_\downarrow \Rightarrow \quad -\delta\tilde{\mathbf{b}} \\ U_\downarrow &\to K^+ + S_\uparrow \Rightarrow \quad +\delta\tilde{\mathbf{b}}. \end{aligned} \tag{2}$$

Eqs. (2) clarify mechanism of the filtering of spin states: when shift in impact parameter is $-\delta\tilde{\mathbf{b}}$ the interaction is stronger compared to the case when shift is $+\delta\tilde{\mathbf{b}}$, and the final S-quark (and Λ-hyperon) is polarized negatively. Thus, the particular mechanism of filtering of spin states is related to the emission of Goldstone bosons by constituent quarks.

It is important to note here that the shift of $\tilde{\mathbf{b}}$ (the impact parameter of final hyperon) is translated to the shift of the impact parameter of the initial particles according to the relation between impact parameters in the multiparticle production:[22]

$$\mathbf{b} = \sum_i x_i \tilde{\mathbf{b}}_i. \tag{3}$$

The variable $\tilde{b}$ is conjugated to the transverse momentum of Λ, but relations between functions depending on the impact parameters $\tilde{b}_i$ are nonlinear. We consider production of Λ in the fragmentation region, i.e. at large x_F and therefore use approximate relation

$$b \simeq x_F \tilde{b}, \tag{4}$$

which results from Eq. (3).[b]

The explicit formulas for inclusive cross-sections of the process

$$h_1 + h_2 \to h_3^{\uparrow} + X,$$

where hadron h_3 is a hyperon whose transverse polarization is measured were obtained in Ref. 21. The main feature of this formalism is an account of unitarity in the direct channel of reaction. The corresponding formulas have the form

$$d\sigma^{\uparrow,\downarrow}/d\xi = 8\pi \int_0^\infty bdb\, I^{\uparrow,\downarrow}(s,b,\xi)/|1 - iU(s,b)|^2, \tag{5}$$

where b is the impact parameter of the initial particles. Here the function $U(s,b)$ is the generalized reaction matrix (for unpolarized scattering) which is determined by the basic dynamics of elastic scattering.

The functions $I^{\uparrow,\downarrow}$ in Eq. (5) are related to the functions $|U_n|^2$, where U_n are the multiparticle analogs of the U (cf. Ref. 21). The kinematical variables ξ (x_F and $p_\perp$) describe the state of the produced particle h_3. Arrows $\uparrow$ and $\downarrow$ denote transverse spin directions of the final hyperon h_3.

Polarization can be expressed in terms of the functions I_-, I_0 and U:

$$P(s,\xi) = \frac{\int_0^\infty bdb\, I_-(s,b,\xi)/|1 - iU(s,b)|^2}{2\int_0^\infty bdb\, I_0(s,b,\xi)/|1 - iU(s,b)|^2}, \tag{6}$$

where $I_0 = 1/2(I^\uparrow + I^\downarrow)$ and $I_- = (I^\uparrow - I^\downarrow)$.

On the basis of the described chiral quark filtering mechanism we can assume that the functions $I^\uparrow(s,b,\xi)$ and $I^\downarrow(s,b,\xi)$ are related to the functions $I_0(s,b,\xi)|_{\tilde{b}+\delta\tilde{b}}$ and $I_0(s,b,\xi)|_{\tilde{b}-\delta\tilde{b}}$, respectively, i.e.

$$I_-(s,b,\xi) = I_0(s,b,\xi)|_{\tilde{b}+\delta\tilde{b}} - I_0(s,b,\xi)|_{\tilde{b}-\delta\tilde{b}} = 2\frac{\delta I_0(s,b,\xi)}{\delta\tilde{b}}\delta\tilde{b}. \tag{7}$$

We can connect $\delta\tilde{b}$ with the radius of quark interaction $r^{\rm flip}_{U\to S}$ responsible for the transition $U_\uparrow \to S_\downarrow$ changing quark spin and flavor:

$$\delta\tilde{b} \simeq r^{\rm flip}_{U\to S}.$$

[b]We make here an additional assumption on the small values of Feynman x for other particles.

The following expression for polarization $P_\Lambda(s,\xi)$ can be obtained

$$P_\Lambda(s,\xi) \simeq x_F r^{\text{flip}}_{U\to S} \frac{\int_0^\infty bdb\, I_0'(s,b,\xi)db/|1-iU(s,b)|^2}{\int_0^\infty bdb\, I_0(s,b,\xi)/|1-iU(s,b)|^2}, \tag{8}$$

where $I_0'(s,b,\xi) = dI_0(s,b,\xi)/db$.

It is clear that polarization of Λ - hyperon (8) should be negative because $I_0'(s,b,\xi) < 0$.

The generalized reaction matrix $U(s,b)$ (in a pure imaginary case) is the following

$$U(s,b) = i\tilde{U}(s,b) = ig(s)\exp(-Mb/\zeta),\ g(s) \equiv g_0\left[1+\alpha\frac{\sqrt{s}}{m_Q}\right]^N, \tag{9}$$

M is the total mass of N constituent quarks with mass m_Q in the initial hadrons; α and g_0 are the parameters of model. Parameter ζ is the one which determines a universal scale for the quark interaction radius, i.e. $r_Q = \zeta/m_Q$. To evaluate polarization dependence on x_F and $p_\perp$ we use semiclassical correspondence between transverse momentum and impact parameter values. Choosing the region of small $p_\perp$ we select the large values of impact parameter and therefore we have

$$P_\Lambda(s,\xi) \propto -x_F r^{\text{flip}}_{U\to S}\frac{M}{\zeta}\frac{\int_{b>R(s)} bdb\, I_0(s,b,\xi)\tilde{U}(s,b)}{\int_{b>R(s)} bdb\, I_0(s,b,\xi)}, \tag{10}$$

where $R(s) \propto \ln s$ is the hadron interaction radius, which serve as a scale of large and small impact parameter values. At large values of impact parameter b: $\tilde{U}(s,b) \ll 1$ for $b \gg R(s)$ and therefore we will have small polarization $P_\Lambda \simeq 0$ in the region of small and moderate $p_\perp \le 1\,\text{GeV}/c$. At small values of b (and large $p_\perp$): $\tilde{U}(s,b) \gg 1$ and the following approximate relations are valid

$$\int_{b<R(s)} bdb\, \frac{I_0(s,b,\xi)\tilde{U}(s,b)}{[1+\tilde{U}(s,b)]^3} \simeq \int_{b<R(s)} bdb\, I_0(s,b,\xi)\tilde{U}(s,b)^{-2}, \tag{11}$$

since we can neglect unity in the denominators of the integrands. Thus, the energy and $p_\perp$-independent behavior of polarization P_Λ takes place at large values of $p_\perp$:

$$P_\Lambda(s,\xi) \propto -x_F r^{\text{flip}}_{U\to S} M/\zeta. \tag{12}$$

This flat transverse momentum dependence results from the similar rescattering effects for the different spin states. The numeric value of polarization P_Λ can be large: there are no small factors in (12). In (12) M is proportional to two nucleon masses, the value of parameter $\zeta \simeq 2$. We expect that

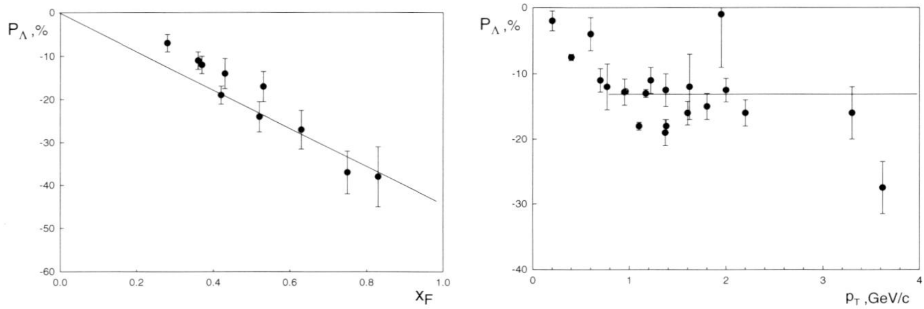

Fig. 2. x_F (left panel) and p_T (right panel) dependencies of the Λ-hyperon polarization

$r^{\rm flip}_{U\to S} \simeq 0.1 - 0.2\,{\rm fm}$ on the basis of the model,[19,21] however, this is a crude estimate. The above qualitative features of polarization dependence on x_F, $p_\perp$ and energy are in a good agreement with the experimentally observed trends.[1] For example, Fig. 2 demonstrates that the linear x_F dependence is in a good agreement with the experimental data in the fragmentation region ($x_F \geq 0.4$) where the model should work. Of course, the conclusion on the $p_\perp$-independence of polarization is a rather approximate one and deviation from such behavior cannot be excluded.

The proposed mechanism deals with effective degrees of freedom and takes into account collective aspects of QCD dynamics. Together with unitarity, which is an essential ingredient of this approach, it allows to obtained results for polarization dependence on kinematical variables in agreement with the experimental behavior of Λ-hyperon polarization, i.e. linear dependence on x_F and flat dependence on $p_\perp$ at large $p_\perp$ in the fragmentation region are reproduced. Those dependencies together with the energy independent behavior of polarization at large transverse momenta are the straightforward consequences of this model. We discussed here particle production in the fragmentation region. In the central region where correlations between impact parameter of the initial and impact parameters of the final particles being weakened, the polarization cannot be generated due to chiral quark filtering mechanism. Moreover, it is clear that since antiquarks are produced through spin-zero Goldstone bosons we should expect $P_{\bar\Lambda} \simeq 0$. The chiral quark filtering is also relatively suppressed when compared to direct elastic scattering of quarks in effective field and therefore should not play a role in the reaction $pp \to pX$ in the fragmentation region, i.e. protons should be produced unpolarized. These features take place in the experimental data set. The application of this mechanism to description of polarization of other hyperons is more complicated problem, since they could have two or three strange quarks and spins of U and D quarks can also make contributions into their polarizations. Finally, it was shown that the mechanism reversed to chiral quark filtering can provide description of the SSA in π^0 production measured at FNAL and recently at RHIC in the fragmentation region and it leads to the energy independence of the asymmetry.

One of the authors (S.T.) is very grateful to the Organizing Committee of "Transversity 2005" for the warm hospitality in Como during this very interesting workshop.

References

1. G. Bunce at al., *Phys. Rev. Lett.* **36** (1976) 1113.
2. S.J. Brodsky, *Acta Phys. Polon.* **B36** (2005) 635.
3. A. Metz, M. Schlegel, *Annalen Phys.* **13** (2004) 699.
4. G.L. Kane, J. Pumplin, W. Repko, *Phys. Rev. Lett.* **41** (1978) 1689.
5. W.G.D. Dharmaratna, G.R. Goldstein, *Phys. Rev.* **D53** (1996) 1073
6. A.V. Efremov, O.V. Teryaev, *Sov. J. Nucl. Phys.* **36** (1982) 140.
7. J. Qiu, G. Sterman, *Phys. Rev.* **D59** (1999) 014004.
8. Y. Kanazawa, Y. Koike, *Phys. Rev.* **D64** (2001) 034019.
9. M. Anselmino, D. Boer, U. D'Alesio, F. Murgia, *Phys. Rev.* **D63** (2001) 054029.
10. C. Bourrely, J. Soffer, *Eur. Phys. J.* **C36** (2004) 371.
11. S.M. Troshin, N.E. Tyurin, *Sov. J. Nucl. Phys.* **38** (1983) 639; *Phys. Lett.* **B355** (1995) 543; *AIP Conf. Proc.* **675** (2003) 579.
12. H. Georgi, A. Manohar, *Nucl. Phys.* **B234** (1984) 189;
D. Diakonov, V. Petrov, *Nucl. Phys.* **B245** (1984) 259;
E.V. Shuryak, *Phys. Rep.* **115** (1984) 151.
13. M. Burkardt, *Phys. Rev.* **D69** (2004) 057501; hep-ph/0505189.
14. N.I. Kochelev, JETP Lett. **72** (2000) 481.
15. D. Ostrovsky, E. Shuryak, hep-ph/0409253.
16. D. Diakonov, *Prog. Part. Nucl. Phys.* **51** (2003) 173.
17. J.D. Bjorken, Report No. SLAC-PUB-5608, 1991 (unpublished);
E.J. Eichten, I. Hinchliffe, C. Quigg, *Phys. Rev.* **D45** (1992) 2269;
T.P. Cheng, L.-F. Li, *Phys. Rev. Lett.* **80** (1998) 2789; Invited talk at 13th. International Symposium on High-Energy Spin Physics (SPIN 98), Protvino, Russia, 8–12 Sep 1998, p. 192.
18. D. Diakonov, hep-ph/0406043.
19. S.M. Troshin, N.E. Tyurin, *Nuovo Cim.* **A106** (1993) 327; *Phys. Rev.* **D49** (1994) 4427.
20. S.M. Troshin, N.E. Tyurin, *J. Phys.* **G29** (2003) 1061.
21. S.M. Troshin, N.E. Tyurin, *Teor. Mat. Fiz.* **28** (1976) 139; *Z. Phys.* **C45** (1989) 171.
22. B.R. Webber, *Nucl. Phys.* **B87** (1975) 269.

CLOSING LECTURE

WHERE WE'VE BEEN ....
.... AND WHERE WE'RE GOING

G. Bunce

Brookhaven National Laboratory and RIKEN BNL Research Center
Physics Department, Upton, NY 11973, USA
E-mail: bunce@bnl.gov

Large transverse spin effects have been measured for the past 30 years, at high energy. However, only recently are we beginning to recognize that these effects appear in the hard scattering domain, with the observation of similarly large effects at RHIC and Belle, and with the separation of initial state and final state asymmetries at Hermes, where also large effects are reported. Indeed, large spin effects may be a feature of hard scattering.

1. Introduction

The observation of very large spin effects at high energy goes back to 30 years ago, with the discovery that Λ hyperons are produced polarized in unpolarized collisions of a 300 GeV proton beam on various targets (Be, then p), Fermilab experiment E8.[1] Also in this time period, the Argonne ZGS was polarized, and very large spin asymmetries were measured for the elastic scattering of a 12 GeV polarized proton beam from a polarized proton target.[2] Neither was expected, with the prevailing view that the observed large spin effects had to be from a soft physics effect.[3] It is interesting that this view was somewhat self-contradictory even then. "Soft physics" was (and is) seen as a complicated domain with many mechanisms contributing to the cross section. However, large spin effects necessarily must have a simple explanation. As an aside, those of us who have been fortunate enough to see these asymmetries coming from our otherwise symmetric experiments were beguiled by the beauty of spin.

The modern era of large spin phenomena also was not anticipated. The field of the spin structure of the nucleon began with the development of a polarized electron source in a small room at Yale by Vernon Hughes and colleagues in the late 1950s. This led to measurements at SLAC that showed *expected* large quark polarization when the quarks carry most of the proton momentum.[4] However, the EMC experiment at CERN, using a naturally polarized muon beam and a very large polarized target, discovered that the quarks and anti-quarks in the proton carry essentially none of the proton spin on average, as reported in 1988.[5]

Very large effects have been observed in the field of spin physics, for protons both for transverse spin, and for helicity. These effects were entirely unexpected. The importance of understanding the contributions to the proton spin for longitudinal polarization has been widely recognized. The transverse spin field is both old and new. The "old" results were tantalizing with huge asymmetries seen, but without a theoretical framework for interpretation. We did not know whether these results could be interpreted within a hard scattering framework, or were effects from unknown soft physics processes. The last several years have brought considerably more information, both from theory and from experiment.

2. where we've been ....

2.1. *The Prehistory and the Awakening*

Fig. 1 shows, on the left, the Λ polarization result from 1975, and, on the right, the proton–proton elastic scattering result from 1978.

Fig. 2 shows the remarkably large asymmetries measured by Fermilab E704 for inclusive pion production with a polarized proton beam hitting an unpolarized target, from 1991.[6] Indeed, this discovery also was largely unexpected. I use the qualifier "largely" because large asymmetries in pion production had already been measured at lower energy at Argonne,[7] but most thought that these asymmetries must be a low energy phenomenon that would disappear at Fermilab. Despite the earlier observation of large Λ polarization at Fermilab.

There were many other striking results. Many hyperons were found to be produced polarized.[8] The effect was found to be most dependent on x_F, with the polarization surpassing 30% at large x_F and persisting even to $p_T = 4\,\mathrm{GeV}/c$.[9] "Persisting" refers to the expectation that this *must* be a soft physics effect that would die out at "serious" p_T. Very large Λ polarization was measured at the ISR.[10] Experiments explored polarization, analyzing power, and spin transfer, with frequent large effects. An anti-hyperon was observed to be produced polarized (anti-Ξ^-) in the fragmentation region of beam protons, which seemed to throw many possible models for the production of polarized hyperons, spin transfer, and pion

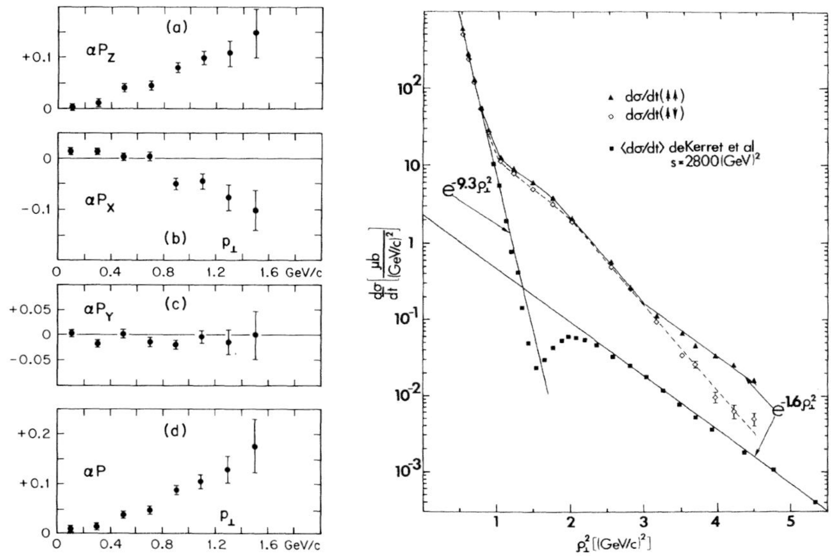

Fig. 1. Left: Λ polarization measured for inclusive production from proton–beryllium scattering at Fermilab.[1] Right: Ratio of spin-parallel to spin-antiparallel p-p elastic cross sections plotted against p_T^2, from Argonne ZGS.[2]

asymmetries, into disarray.[11] A very nice attempt to connect many or all of these phenomena is presented in Ref. 12.

The modern era *transverse* spin awakening began with Hermes results for the asymmetry of pion production in semi-inclusive DIS with a longitudinally polarized target.[13] A large π^+ azimuthal asymmetry was observed that could be from the small $(\tilde{1}5\%)$ transverse component of spin of the target due to the kinematics of the virtual photon probing the target (tied to the $e - e'$ lepton scattering plane), or from the asymmetry A_{UL} from an interference of transverse and longitudinal photon, scattering from a longitudinally polarized target, which is also tied to the lepton scattering plane.

Coming into the programs of RHIC, Hermes transversely polarized target, COMPASS, and Belle, all discussed at this workshop, we had learned that there were very large transverse spin effects for high energy, moderate p_T, scattering; and that the proton spin structure, viewed longitudinally, was also very surprising.

Cross section results from the RHIC collider were the next important contribution. The RHIC experiments measured the cross sections for for-

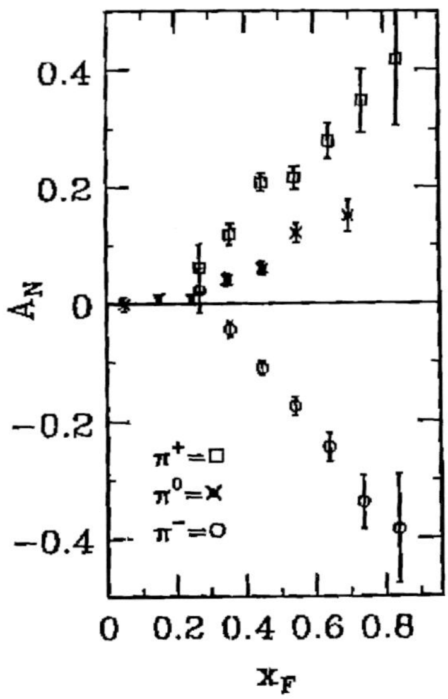

Fig. 2. Left–right asymmetry in the inclusive production of pions, using a polarized proton beam at Fermilab.[6]

ward and for mid-rapidity production of pions at $\sqrt{s} = 200\,\text{GeV}$. The perturbative predictions for the cross sections describe the data, even in the moderate p_T region where large spin effects were previously observed, $p_T = 1.5\,\text{GeV}$, and over 8 orders of magnitude out to $p_T = 14\,\text{GeV}/c$, see Fig. 3. It was noted recently [14] that the lower energy cross sections for π^0 production are *not* described by pQCD, challenging the idea of linking the lower energy spin results to a quark–gluon hard scattering framework. However, the RHIC cross sections imply that the phenomena to be measured at RHIC will be properly described under the pQCD factorization approach. Recent work [15] describes the lower energy cross sections by including multiple gluon emission in the interactions. The authors find that the contributions from this "resummation" are more important at lower energy where the measurements are closer to the transverse momentum kinematic limit, and obtain cross sections agreeing well with data for mid-rapidity ISR pion production data. This is a work in progress, with respect to rapidity dependence and to its effect on spin asymmetries. An added confirmation of our understanding of the underlying physics at RHIC energy is the excellent agreement of the NLO pQCD prediction with the direct photon unpolarized cross section, as measured by PHENIX at mid-rapidity, over 3 orders of magnitude, for $p_T = 5$ to $16\,\text{GeV}/c$, see Fig. 4.

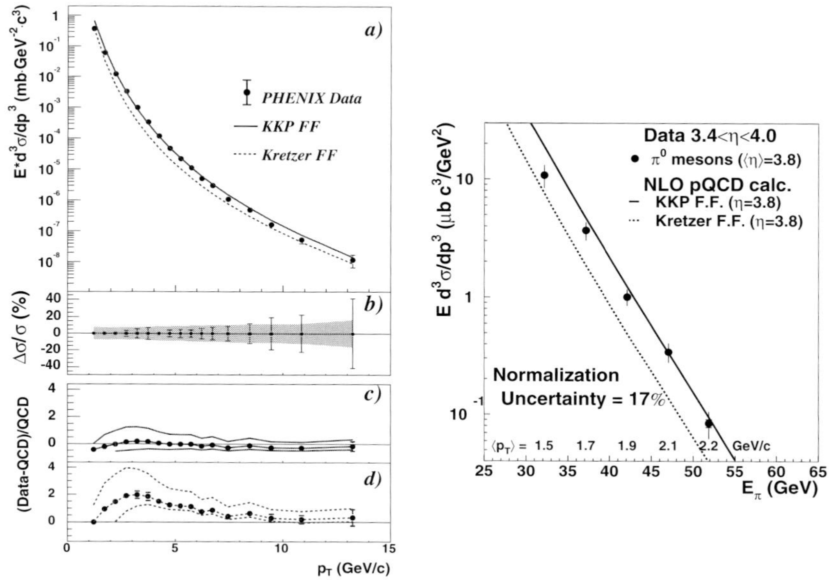

Fig. 3. Inclusive cross sections for π^0 production at RHIC, mid-rapidity from PHENIX [16] (left panel) and forward from STAR [17] (right panel). NLO pQCD predictions [18] are also shown.

2.2. *The New Transverse Spin Results*

The new results come from RHIC at $\sqrt{s} = 200\,\text{GeV}$, DIS with transversely polarized targets, and quark-antiquark fragmentation asymmetry correlations observed for unpolarized $e^+ - e^-$ annihilation at Belle.

The original RHIC result from STAR, see the left side of Fig. 5, showing a large left–right asymmetry for forward π^0 production with one proton beam transversely polarized, has been augmented with STAR measurements of backward production of π^0 with no observed asymmetry,[21] with no asymmetry observed at mid-rapidity for charged and neutral pions by PHENIX,[22] and with a large mirror asymmetry observed for forward production of π^+ and π^- by BRAHMS. The BRAHMS results are shown on the right of Fig. 5. These large asymmetries are observed in an energy and kinematic region where the cross section is well-described by pQCD, as discussed earlier. This statement has been a long time in com-

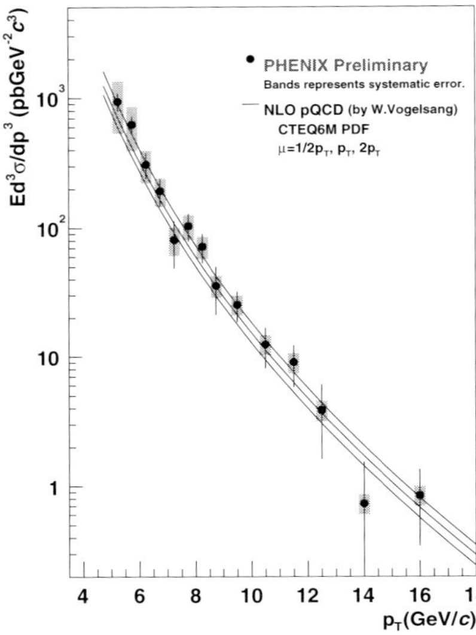

Fig. 4. Inclusive cross section for direct photon production at RHIC at mid-rapidity from PHENIX.[19] The NLO pQCD prediction[20] is also shown.

ing: **large spin asymmetries are observed in hard scattering.**

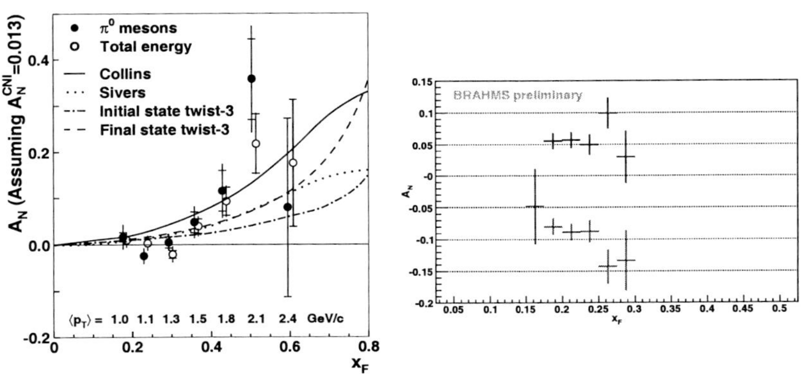

Fig. 5. Left: Analyzing power for production of forward π^0 at RHIC from STAR.[17] Right: Analyzing power for inclusive forward production of charged pions, BRAHMS at RHIC.[23]

The large RHIC asymmetries can come from a number of effects.[24] Two are (transversity×Collins) and (Sivers×analyzer). Transversity is the probability of finding polarized quarks in a transversely polarized proton,[25] and

Collins refers to the analyzing power for the polarized quarks to fragment asymmetrically.[26] Sivers is the probability of an internal transverse momentum asymmetry for the quarks or gluons in the transversely polarized proton, related to the proton spin axis;[27] for the effect to create an observed asymmetry, there would need to be an analyzing power for the scattering of the quarks or gluons, to generate a left–right asymmetry.[28] In the case of RHIC, the forward production of pions is described by pQCD as predominantly from a forward quark (from the polarized proton) scattering from a backward gluon (from the unpolarized proton), producing a forward quark which fragments into the observed pion. An important question is whether the left–right asymmetry is observed in the jet (the quark) or only in the fragment (the pion). The Collins mechanism would predict that the fragment is asymmetric, but that the jet is not. Indeed, a new detector is being built by STAR to address this question.[29] Murgia at this workshop presented a prediction that the (transversity$\times$Collins) mechanism should be small for pion production at RHIC, and suggests that the Sivers effect should be dominant.[30]

At mid-rapidity at RHIC, and moderate p_T, the asymmetries are zero, and the scattering is described in pQCD as roughly half gluon–gluon, and half gluon–quark scattering.[31] Therefore, for these kinematics ($x \sim 0.03$), the gluon spin effects are much smaller that the forward quark effects ($x \sim 0.2$–0.5). Further, the STAR observation of zero asymmetry for backward production implies that the gluon spin effect is small at low x, $x \sim 10^{-3}$. Therefore, it may be that the Sivers gluon asymmetry is small (but note the required analyzing power which could be zero).

Semi-inclusive DIS experiments can disentangle the azimuthal asymmetry effects and independently measure Sivers (with analyzer) and Collins (with transversity). The Hermes results for a transversely polarized proton target are very exciting, see Fig. 6. COMPASS reported here results for a transversely polarized deuterium target.[32] To summarize, a large Sivers effect is observed for a proton target, π^+, zero Sivers asymmetry for a proton target, π^-. Zero Sivers asymmetry is observed for the deuteron target. Therefore, there may be large quark Sivers asymmetries (k_T/orbital angular momentum) with opposite signs for u and d quarks, with cancellations that lead to zero asymmetry for proton target π^- and for deuteron target $\pi^\pm$. Large $\pi^\pm$ Collins asymmetries are observed for a proton target with opposite signs for π^+ and π^-, and no Collins asymmetry is seen for a deuteron target. A very important test will be COMPASS measurements with a transversely polarized proton target.

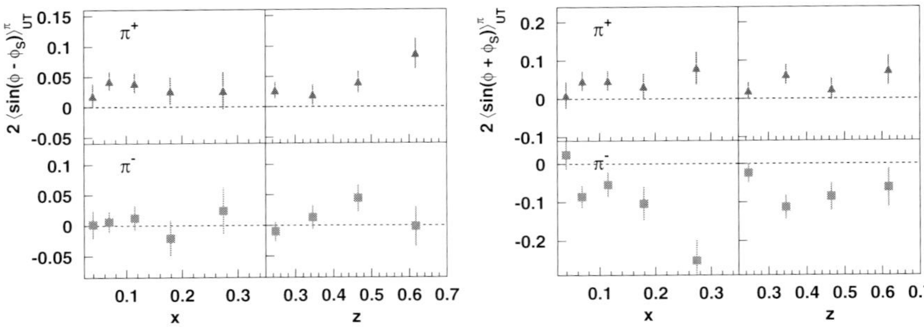

Fig. 6. Virtual photon moments for DIS semi-inclusive pion production from a transversely polarized proton target.[33] Left panel shows the Sivers moments, right panel the Collins moments; the left side of each panel gives the dependence on the quark momentum fraction x, the right side of each panel gives the dependence on the fragmentation momentum fraction z.

A RIKEN BNL Research Center team joined the Belle experiment at KEK to analyze $e^+ - e^- \to q - \bar{q}$ toward measuring the Collins effect, that polarized quarks could fragment asymmetrically. A preliminary result was presented here by Ralf Seidl.[34] The correlation, see Fig. 7, is large, of order 0.05. The correlation is quadratic in the Collins asymmetry, giving $A_{\text{Collins}} \sim 20\%$. If this result is confirmed (and to an extent it has been with the Hermes observation of non-zero Collins asymmetry), we have the direct observation that polarized quarks generate an orbital angular momentum in fragmentation.

To summarize where we are,

- the gluon Sivers internal k_T asymmetry may be small
- the quark Sivers is large
- transversity is large
- the Collins fragmentation analyzing power is large
- some *are* and all *may* be in the hard scattering domain

3. .... and where we're going

3.1. *Experiment*

Planned:

- COMPASS with a transversely polarized proton target
- SIDIS π^0, also Hermes 2×statistics
- forward jets at STAR
- spin transfer to Λ at RHIC and SIDIS
- k_T measurements with two hadrons/jets at PHENIX, STAR
- interference fragmentation at COMPASS, Hermes, STAR, PHENIX

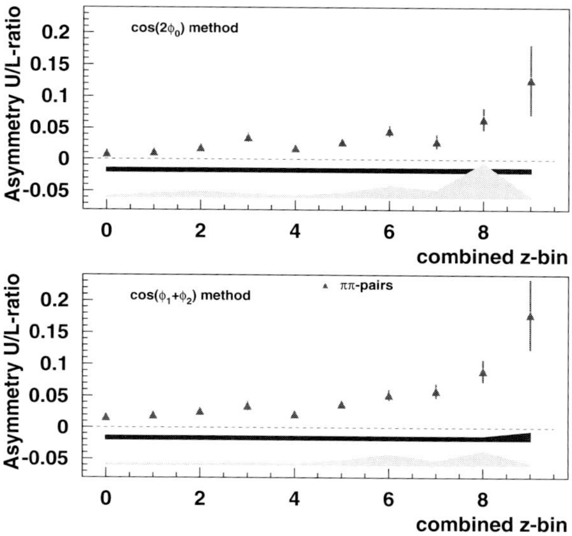

Fig. 7. Two methods are shown to extract the correlation asymmetry for $e^+ - e^- \to q - \bar{q}$, fragmenting to π^+-π^-, from Belle.[34]

Future–we now know that these spin signals are *large*:

- *study the* $e^+ - e^-$ *data!*
- exploit the particle identification and luminosity available at present and future fixed target machines
- RHIC Drell–Yan: two-spin transversity, one-spin Sivers
- $e - p$ collisions with transversely polarized protons–an eRHIC option

Here is a list of spin correlations that can be studied for $e^+ - e^-$ collisions, Belle and BaBar:

- azimuthal correlation of π and K, Λ
- azimuthal correlation of $K, \bar{\Lambda}$ and K, Λ
- correlations of π, K, $\bar{\Lambda}$ azimuth and Λ spin
- correlations of $\bar{\Lambda}$ spin and Λ spin
- correlation of Λ and Ξ
- correlations of $u\bar{u}$, $d\bar{d}$, $s\bar{s}$, $c\bar{c}$, $b\bar{b}$
- π, K, ρ, Λ, Ξ, ...
- unpolarized fragmentation functions

If the Belle spin effects are confirmed, $q^\uparrow \to h(\phi, \text{spin})$ offers to us a remarkable laboratory for studying QCD spin physics, in addition to providing needed analyzing power for probing transversity.

3.2. *Theory*

Again, a list:

- L—through generalized parton distributions; through a connection of L to Sivers asymmetries?
- How to exploit the new field of fragmentation analyzing power? $q^{\uparrow} \rightarrow h(\phi, spin)$ is bare QCD!
- Connections between Sivers, Collins, and transversity?

Huge spin effects cannot have complicated origins. This is an exciting time. We have discovered very large, unexpected, spin effects in the hard scattering domain. Large transverse spin effects appear to be a feature of hard scattering, and transverse spin physics will evidently play an important role in the future, toward understanding QCD and hadronic structure.

Acknowledgments

I would like to thank Werner Vogelsang for numerous discussions, as well as to acknowledge discussions with many colleagues in the spin field. Naomi Makins presented a very similar perspective on the field at an RBRC workshop this past June. My work was supported by U.S. DOE Medium Energy Physics and by the RIKEN BNL Research Center.

References

1. G. Bunce for the FNAL E8 Collaboration, Division of Particle Physics, Seattle (1975); G. Bunce *et al.*, *Phys. Rev. Lett.* **36** (1976) 1113.
2. D.G. Crabb *et al.*, *Phys. Rev. Lett.* **41** (1978) 1257.
3. G.L. Kane, J. Pumplin, and W. Repko, *Phys. Rev. Lett.* **41** (1978) 1689.
4. M.J. Alguard *et al.* [SLAC E80 Collaboration with V.W. Hughes], *Phys. Rev. Lett.* **37** (1976) 1261.
5. J. Ashman *et al.* [European Muon Collaboration], *Phys. Lett.* **B206** (1988) 364; *Nucl. Phys.* **B328** (1989) 1.
6. D.L. Adams *et al.* [FNAL E704 Collaboration], *Phys. Lett.* **B264** (1991) 462.
7. R.D. Klem *et al.*, *Phys. Rev. Lett.* **36** (1976) 929; CERN PS results: J. Antille *et al.*, *Phys. Lett.* **94B** (1980) 523; subsequent AGS results: B.E. Bonner *et al.*, *Phys. Rev.* **D41** (1990) 13; S. Saroff *et al.*, *Phys. Rev. Lett.* **64** (1990) 995; K. Krueger *et al.*, *Phys. Lett.* **B459** (1999) 412.
8. K. Heller, proc. *12th. International Symposium on High Energy Spin Physics* C.W. de Jager *et al.* eds., World Scientific (1996) 57.
9. B. Lundberg *et al.*, *Phys. Rev.* **D40** (1989) 3557.

10. A.M. Smith *et al.*, *Phys. Lett.* **B185** (1987) 209.
11. P.M. Ho *et al.*, *Phys. Rev. Lett.* **65** (1990) 1713.
12. K. Kubo, Y. Yamamoto and H. Toki, *Prog. Theor. Phys.* **101** (1999) 615.
13. A. Airapetian *et al.* [HERMES Collaboration], *Phys. Rev. Lett.* **84** (2000) 4047; A. Airapetian *et al.*, *Phys. Lett.* **B622** (2005) 14.
14. C. Bourrely and J. Soffer, *Eur. Phys. J.* **C36** (2004) 371.
15. D. de Florian and W. Vogelsang, *Phys. Rev.* **D71** (2005) 114004; also, considering partonic k_T, M. Anselmino *et al.*, *Phys. Rev.* **D70** (2004) 074009.
16. S.S. Adler *et al.* [PHENIX Collaboration], *Phys. Rev. Lett.* **91** (2003) 241803.
17. J. Adams *et al.* [STAR Collaboration], *Phys. Rev. Lett.* **92** (2004) 171801.
18. B. Jager *et al.*, *Phys. Rev.* **D67** (2003) 054005; D. de Florian, *Phys. Rev.* **D67** (2003) 054004; F. Aversa *et al.*, *Nucl. Phys.* **B327** (1989) 105; parton distribution functions from J. Pumplin *et al.* [CTEQ Collaboration], *JHEP* **0207** (2002) 012; fragmentation functions from S. Kretzer, *Phys. Rev.* **D62** (2000) 054001; B. Kniehl, G. Kramer and B. Potter, *Nucl. Phys.* **B582** (2000) 514; *ibid.* **B597** (2001) 337; L. Bourhis *et al.*, *Eur. Phys. J.* **C19** (2001) 89.
19. K. Okada [PHENIX Collaboration], arXiv:hep-ex/0501066.
20. L.E. Gordon and W. Vogelsang, *Phys. Rev.* **D48** (1993) 3136; *ibid.* **D50** (1994) 1901; P. Aurenche *et al.*, *Phys. Lett.* **B140** (1984) 87; *Nucl. Phys.* **B297** (1988) 661; H. Baer *et al.*, *Phys. Rev.* **D42** (1990) 61; *Phys. Lett.* **B234** (1990) 127.
21. A. Ogawa for the STAR Collaboration, proc. *Spin2004, Trieste, Italy*, eds. K. Aulenbacher *et al.*, World Scientific (2005) 337.
22. S.S. Adler *et al.* [PHENIX Collaboration], *Phys. Rev. Lett.* **95** (2005) 202001.
23. F. Videbaek for the BRAHMS Collaboration, to appear in proc. *DIS 05, Madison, USA*, nucl-ex/0508015.
24. V. Barone and P.G. Ratcliffe, *Transverse Spin Physics*, World Scientific (2003).
25. J.P. Ralston and D.E. Soper, *Nucl. Phys.* **B152** (1979) 109; X. Artru and M. Mekhfi, *Z. Phys.* **C45** (1990) 669; R.L. Jaffe and X. Ji, *Nucl. Phys.* **B375** (1992) 527.
26. J. Collins, *Nucl. Phys.* **B396** (1993) 161; J.C. Collins, S.F. Heppelmann and G.G. Ladinsky, *Nucl. Phys.* **B420** (1994) 565; A.V. Efremov, L. Mankiewicz and N.A. Tornqvist, *Phys. Lett.* **B284** (1992) 394.
27. D.W. Sivers, *Phys. Rev.* **D41** (1990) 83.
28. M. Burkardt, *Phys. Rev.* **D69** (2004) 091501.
29. L. Bland for the STAR Collaboration, *Como Transversity 2005* (2005).
30. F. Murgia *et al.*, *Como Transversity 2005*, hep-ph/0511017; M. Anselmino *et al.*, *Phys. Rev.* **D71** (2005) 014002.
31. V. Guzey, M. Strikman and W. Vogelsang, *Phys. Lett.* **B603** (2004) 173; S. Kretzer, arXiv:hep-ph/0410219.
32. V.Yu. Alexakhin *et al.*, *Phys. Rev. Lett.* **94** (2005) 202002.
33. A. Airapetian *et al.*, *Phys. Rev. Lett.* **94** (2005) 012002.
34. R. Seidl for the BELLE Collaboration, *Como Transversity 2005* (2005); hep-ex/0507063.

AUTHOR INDEX

Aidala, C., 68
Anselmino, M., 9, 188, 220, 236
Artru, X., 154
ASSIA, 112
Avakian, H., 88

Bacchetta, A., 181
Barone, V., 298
BELLE, 55
Benhizia, K., 154
Bianconi, A., 284
Bland, L., 77
Boer, D., 244
Boglione, M., 188, 220, 236
Bosted, P., 88
Bradamante, F., 3
Brodsky, S.J., 139
Bunce, G., 325
Burkardt, M., 162
Burkert, V., 88

Cafarella, A., 298
Chen, J.P., 96
CLAS, 88
Collins, J.C., 212, 236
COMPASS, 34, 49, 61
Contalbrigo, M., 104
Corianò, C., 298

D'Alesio, U., 188, 220, 236
Di Salvo, E., 306
Diehl, M., 173
Drago, A., 170

Efremov, A.V., 212, 236
Elouadrhiri, L., 88

Ferrero, A., 61

Göckeler, M., 173
Gabbert, D., 55
Gamberg, L.P., 252, 260
Goeke, K., 196, 212, 236
Goldstein, G.R., 252, 260
Grosse Perdekamp, M., 55, 212
Guzzi, M., 298

Hägler, Ph., 173
Hasuko, K., 55
HERMES, 23, 42
Horn, I., 34
Horsley, R., 173

Jiang, X., 96
JLab Hall A, 96

Kotzinian, A., 188, 228, 236

Leader, E., 220

Maggiora, M., 112
Melis, S., 220
Menzel, S., 212, 236
Meredith, B., 212
Metz, A., 196, 212, 236
Mielech, A., 49
Mukherjee, A., 292
Murgia, F., 188, 220, 236

Nikolaev, N.N., 120

Ogawa, A., 55

Pavlov, F.F., 120

Peng, J.-C., 96
PHENIX, 68
Pleiter, D., 173
Prokudin, A., 188, 236

QCDSF, 173

Radici, M., 268
Rakow, P.E.L., 173
Ratcliffe, P.G., 298

Schäfer, A., 173
Schierholz, G., 173
Schlegel, M., 196
Schmidt, I., 204
Schnell, G., 23
Schweitzer, P., 212, 236
Seidl, R., 55
Stratmann, M., 292

Teryaev, O.V., 276
Troshin, S.M., 314
Tyurin, N.E., 314

UKQCD, 173

van der Nat, P.B., 42
Vogelsang, W., 236, 292

Yuan, F., 236

Zanotti, J.M., 173
Zhu, L., 96